Augmented Lagrangian and Operator-Splitting Methods in Nonlinear Mechanics

SIAM Studies in Applied Mathematics

MARSHALL SLEMROD, Managing Editor

This series of monographs focuses on mathematics and its applications to problems of current concern to industry, government, and society. These monographs will be of interest to applied mathematicians, numerical analysts, statisticians, engineers, and scientists who have an active need to learn useful methodology.

Series List

Vol. 1. *Lie-Backlund Transformations in Applications*
Robert L. Anderson and Nail H. Abragimov

Vol. 2. *Methods and Applications of Interval Analysis*
Ramon E. Moore

Vol. 3. *Ill-Posed Problems for Integrodifferential Equations in Mechanics and Electromagnetic Theory*
Frederick Bloom

Vol. 4. *Solitons and the Inverse Scattering Transform*
Mark J. Ablowitz and Harvey Segur

Vol. 5. *Fourier Analysis of Numerical Approximations of Hyperbolic Equations*
Robert Vichnevetsky and John B. Bowles

Vol. 6. *Numerical Solution of Elliptic Problems*
Garret Birkhoff and Robert E. Lynch

Vol. 7. *Analytical and Numerical Methods for Volterra Equations*
Peter Linz

Vol. 8. *Contact Problems in Elasticity: A Study of Variational Inequalities and Finite Element Methods*
N. Kikuchi and J. T. Oden

Vol. 9. *Augmented Lagrangian and Operator-Splitting Methods in Nonlinear Mechanics*
Roland Glowinski and P. Le Tallec

Roland Glowinski and Patrick Le Tallec

Augmented Lagrangian and Operator-Splitting Methods in Nonlinear Mechanics

Society for Industrial and Applied Mathematics
siam *Philadelphia/1989*

Library of Congress Cataloging-in-Publication Data

Glowinski, R.

Augmented Lagrangian and operator-splitting methods in nonlinear mechanics / Roland Glowinski and Patrick Le Tallec.

p. cm. -- (SIAM studies in applied mathematics ; 9)

Bibliography: p.

Includes index.

ISBN 0-89871-230-0

1. Continuum mechanics. 2. Lagrangian functions. 3. Operator theory. I. Le Tallec, Patrick. II. Title. III. Title: Operator -splitting methods in nonlinear mechanics. IV. Series.

QA808.2.G56 1989

531--dc20

89-11319

CIP

Contents

Preface

When *Augmented Lagrangian Methods*, edited by M. Fortin and R. Glowinski, appeared in 1983, the authors of the present book quickly realized that a sequel was needed for a variety of reasons, including the emergence of new applications and the sophistication of existing ones; a deeper understanding of the convergence properties of augmented Lagrangian algorithms and of their relationship to operator-splitting methods such as alternating-direction methods; and the development of more efficient algorithms. The decision was made to write a volume oriented to applications in continuum mechanics that would contain these new developments. The nature and the structure of this book are derived from that choice.

This book deals with the numerical simulation of the behavior of continuous media by augmented Lagrangian and operator-splitting methods (coupled to finite-element approximations). It begins with a description of the mechanical and mathematical frameworks of the considered applications as well as a general analysis of the basic numerical methods traditionally used to study them. These ideas are then applied to specific classes of mechanical problems.

More precisely, Chapter 1 introduces the mathematical modeling of different classes of continuous materials (standard materials in small strains, incompressible viscoplastic fluids, and hyperelastic solids), and Chapter 2 introduces the variational formulations of the mechanical problems associated with these materials.

Chapter 3 gives a general presentation of augmented Lagrangian methods, discusses their relationship to operator-splitting techniques, and describes their application to linear and nonlinear eigenvalue problems.

Chapter 4 is concerned with the calculation of quasi-static flows of elastoviscoplastic materials in small strains.

Chapter 5 is devoted to limit load analysis, the purpose of which is to determine whether a given structure can sustain a certain distribution of loads without damage.

Chapter 6 studies the two-dimensional flows of incompressible viscoplastic fluids in a stream-function formulation.

Chapter 7 examines equilibrium problems in finite elasticity, in particular, the determination of the final positions of an elastic body subjected to large deformations.

Finally, Chapter 8 describes the large displacement calculations of flexible rods.

The last five chapters are independent of one another. They each treat nonlinear, possibly nondifferentiable and highly constrained problems through the systematic use of an operator-splitting strategy to confine the nonlinearities to local problems of small dimension.

We would like to thank D. Begis, J. F. Bourgat, C. H. Li, L. deLoura, S. Mani, M. Ravachol, and M. Vidrascu for their participation in the preparation of this book. We also acknowledge our particular indebtedness to J. Nohel, who initiated this project and who welcomed us to the Mathematical Research Center (M.R.C.) of the University of Wisconsin at Madison where a large part of this work was drafted. Actually, Chapter 3 was published as M.R.C. Report No. 2965 and was sponsored by the United States Army under Contract No. DAAG29-80-C-0041. We wish also to thank E. Becker, J. Cahouet, J. M. Dumay, M. Fortin, D. Gabay, M. Gurtin, G. Labadie, P. L. Lions, M. Luskin, D. Malkus, B. Mercier, J. T. Oden, J. Simo, and L. Tartar, whose collaboration or comments and suggestions were essential for many of the results presented here. We thank also Christianne Demars and L. M. Brooks for their typing of the manuscript. We acknowledge the support of the National Science Foundation via grant number INT8612680. Last, but not least, our thanks go to INRIA, CPC, and the University of Houston for their constant support in the preparation of this book.

ROLAND GLOWINSKI
PATRICK LE TALLEC
University of Houston
Houston, Texas

Chapter 1

Some Continuous Media and Their Mathematical Modeling

1. The physical problems. Most of the physical problems we consider in this book involve the study of deformations, stresses, and velocities of continuous bodies. More precisely, our goal is to compute rather accurately the effects of various given distributions of external loads on continuous media. Given such a global framework, this study should include the whole field of computational mechanics and should require a precise knowledge of the mechanical behavior of general continuous media. However, we will restrict ourselves to two types of mechanical behavior, namely, those corresponding to

(i) *standards materials* including elastic, plastic, and viscoplastic solids or fluids of different kinds, but subjected only to small deformations or to viscous flows, and

(ii) *hyperelastic materials.* Large-strain plasticity or convection-dominated flow problems will not be considered.

A complete description of the mechanical behavior of continuous media involves *kinematics*, which studies the motion of a body independently of the cause of the motion; *kinetics* and *thermodynamics*, which describe the physical conservation laws to be satisfied by any continuum; and *constitutive equations*, which characterize the mechanical properties of the material of which the continuum is composed.

Thus, the derivation of the mechanical models treated in this book begins by a brief review of the kinematics and thermodynamics of deformation (§§ 2, 3) and then considers the constitutive equations corresponding to standard materials in small strains (§ 4), to incompressible viscous fluids (§ 5), and to hyperelastic materials (§ 6).

2. Kinematics. Kinematics describes the changes of geometry undergone by continuous bodies during an evolutionary process. As soon as large displacements or large deformations become involved, then monitoring these changes of geometry becomes a delicate matter. Therefore, it is most important to work with consistent notation and definitions. All the definitions given in this chapter are standard in continuum mechanics. In particular, they can be found in Gurtin [1981] or in Marsden and Hughes [1983]. The notation chosen herein is taken mostly from Ciarlet [1988]. The ways in which it differs from those of Truesdell and Noll [1965] and Marsden and Hughes [1983] are indicated in Table 2.1.[1]

We assume once and for all that an origin **O** and an orthonormal basis $(\mathbf{e}_i)$ have been chosen in the three-dimensional Euclidian space. Within this convention, a *continuous body* can be defined as a compact, regular region B of $\mathbb{R}^3$. It is made of an infinite number of *particles*, or material points, whose positions at a time t define the *configuration* of the body. More precisely, a configuration is a smooth mapping of a body onto a region of the three-dimensional Euclidian space. Among all configurations, we will choose one, once and for all, which will not vary in time, as a *reference configuration*. We will then identify each particle of the body with its position $\mathbf{x}$ in the reference configuration. The coordinates (x_i) of $\mathbf{x}$ in the basis $(\mathbf{e}_i)$ are called the *Lagrangian coordinates* of the particle.

TABLE 2.1

Correspondence of authors' notation to Truesdell and Noll [1965] *and Marsden and Hughes* [1983].

Designation	Glowinski–Le Tallec	Truesdell–Noll	Marsden–Hughes
Material point	$\mathbf{x}$	$\mathbf{X}$	$\mathbf{X}$
Material area element	da	ds_R	dA
Unit normal vector in reference configuration	$\boldsymbol{\nu}$	$\mathbf{n}_R$	$\mathbf{N}$
Material volume element	$d\mathbf{x}$	dv_R	dV
Spatial point	$\bar{\mathbf{x}}$	$\mathbf{x}$	$\mathbf{x}$
Spatial area element	$d\bar{a}$	ds	da
Displacement field	$\mathbf{u}$	$\mathbf{u}$	$\mathbf{U}$
Material velocity	$\mathbf{v}$	$\dot{\mathbf{x}}$	$\mathbf{V}$
Material acceleration	$\boldsymbol{\gamma}$	$\ddot{\mathbf{x}}$	$\mathbf{A}$
Mass density in reference configuration	ρ	ρ_R	ρ_{ref}
Body forces density in reference configuration (per unit volume)	$\mathbf{f}$		$\rho_{\text{ref}}\,\mathbf{B}$
Piola-Kirchhoff stress vector	$\mathbf{g}$	$\mathbf{t}_R$	$\mathbf{T}$
1st Piola-Kirchhoff stress tensor	$\mathbf{t}$	$\mathbf{T}_R$	$\mathbf{P}$
2nd Piola-Kirchhoff stress tensor	$\boldsymbol{\sigma}$	$\mathbf{T}$	$\mathbf{S}$
Cauchy stress vector	$\bar{\mathbf{g}}$	$\mathbf{t}$	$\mathbf{t}$
Cauchy stress tensor	$\bar{\boldsymbol{\sigma}}$	$\mathbf{T}_\varepsilon$	$\boldsymbol{\sigma}$

[1] Marsden and Hughes [1983] use capital letters when referring to the reference configuration and lower-case letters when referring to the deformed configuration. In the same situations, we use lower-case letters and overlined lower-case letters, respectively.

The interior of the body is identified to be an open, bounded, connected subset Ω of $\mathbb{R}^3$ with boundary Γ.

A *deformation* of the body is a smooth, one-to-one, orientation-preserving vector field, defined on the reference configuration, with values in $\mathbb{R}^3$; in other words, it is a map

$$\boldsymbol{\phi}: B \to \mathbb{R}^3,$$

with

$$\det \nabla\boldsymbol{\phi} > 0 \quad \text{on } \Omega.$$

The spatial point

$$\bar{\mathbf{x}} = \boldsymbol{\phi}(\mathbf{x})$$

is the *place* occupied by $\mathbf{x}$ in the deformation $\boldsymbol{\phi}$. A mapping with domain B is a *material field*; a mapping with domain $\boldsymbol{\phi}(B)$ is a *spatial field.* The *displacement field* is the material field defined by

$$\mathbf{u}(\mathbf{x}) = \boldsymbol{\phi}(\mathbf{x}) - \mathbf{x},$$

and the field

$$\mathbf{F} = \nabla\boldsymbol{\phi},$$

or, componentwise in the basis $(\mathbf{e}_i)$,

$$F_{ij} = \frac{\partial \phi_i}{\partial x_j},$$

is called the *deformation gradient.* All these definitions are illustrated in Fig. 2.1 and permit an easy description of the changes of length, area, and volume undergone by elementary subsets of the body in a given deformation.

Indeed, let $\boldsymbol{\delta}\mathbf{x}$ be an elementary vector engraved on the body at the point $\mathbf{x}$ in the reference configuration. After deformation, this vector is transformed into the vector $\boldsymbol{\delta}\bar{\mathbf{x}}$, whose length is given by

$$|\boldsymbol{\delta}\bar{\mathbf{x}}|^2 = (\boldsymbol{\delta}\bar{\mathbf{x}})^T \boldsymbol{\delta}\bar{\mathbf{x}} = \boldsymbol{\delta}\mathbf{x}^T \mathbf{F}(\mathbf{x})^T \mathbf{F}(\mathbf{x}) \boldsymbol{\delta}\mathbf{x}.$$

The tensor $\mathbf{F}^T\mathbf{F}$, which measures the length of an elementary vector defined over the deformed configuration in terms of its definition in the reference configuration, is called the *right Cauchy-Green tensor.*

Similarly, an area element da of the reference configuration, with normal vector $\boldsymbol{\nu}$, is transformed after deformation into an area element

$$d\bar{a} = \det[\mathbf{F}(\mathbf{x})] |\mathbf{F}(\mathbf{x})^{-T}\boldsymbol{\nu}|\, da$$

with normal vector

$$\mathbf{n} = \frac{\mathbf{F}(\mathbf{x})^{-T}\boldsymbol{\nu}}{|\mathbf{F}(\mathbf{x})^{-T}\boldsymbol{\nu}|}.$$

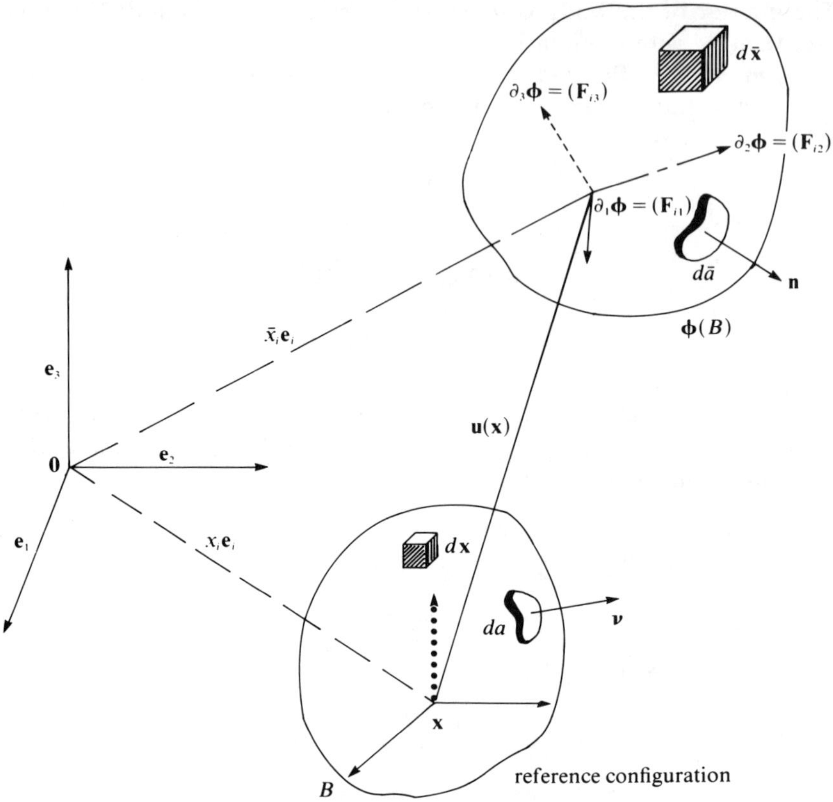

FIG. 2.1. *Illustration of definitions in* § 2.

Finally, a volume element $d\mathbf{x}$, located at the point $\mathbf{x}$ in the reference configuration, is transformed, after deformation, into the volume element

$$d\bar{\mathbf{x}} = \det [\mathbf{F}(\mathbf{x})]\, d\mathbf{x}.$$

A *motion* of the body is now defined by a function

$$\boldsymbol{\phi} : B \times \mathbb{R} \to \mathbb{R}^3,$$

where $\boldsymbol{\phi}(\cdot, t)$ is the deformation undergone by the body at time t. For a given time, the changes of geometry of the body are computed as indicated above. Moreover, velocities and accelerations are introduced as follows.

The *velocity* of a particle $\mathbf{x}$ at time t is defined as its time rate of change of position, which, in Lagrangian coordinates, is given by

$$\mathbf{v}(\mathbf{x}, t) = \dot{\boldsymbol{\phi}}(\mathbf{x}, t),$$

where the overdot $(\dot{\ })$ denotes partial differentiation with respect to time, holding the material coordinates constant (*material derivative*).

The *acceleration* of a particle is the time rate of change of its velocity and is given by

$$\boldsymbol{\gamma}(\mathbf{x}, t) = \ddot{\boldsymbol{\phi}}(\mathbf{x}, t).$$

3. Kinetics and thermodynamics of deformation. The kinematic variables that describe the motion of any continuum are related through fundamental mechanical principles to the external loads acting upon it.

3.1. Conservation of mass. The mass of a body is a measure of its inertia and is supposed to be invariant during deformation. In other words, denoting by ρ the mass density of a continuous body in its reference configuration, we have

$$\boxed{\rho = \rho(\mathbf{x}).} \tag{3.1}$$

3.2. Conservation of momentum. The *forces* acting on a body measure the external actions that lead to its deformation. A *system of forces* acting at a given time on a given regular subbody $\mathcal{V}_R$ of a body B ($\mathcal{V}_R$ being defined by its occupied volume in the reference configuration) consists of

(i) *body forces*, represented by the material vector field $\mathbf{f}$,

$$\mathbf{f}: \mathcal{V}_R \to \mathbb{R}^3,$$

which are the forces exerted throughout the volume of the deformed body, measured per unit volume of the reference configuration; and

(ii) *surface forces*, represented by the vector field $\mathbf{g}$,

$$\mathbf{g}: \partial\mathcal{V}_R \to \mathbb{R}^3,$$

which are the contact forces exerted on the boundary $\partial\mathcal{V}_R$ of the subregion $\mathcal{V}_R$, by the exterior of $\mathcal{V}_R$, measured per unit surface of $\partial\mathcal{V}_R$ in the reference configuration.

These fields, sometimes referred to as the *nominal system of forces*, satisfy Cauchy's theorem, whose derivation can be found in most books on continuum mechanics.

THEOREM 3.1. (*Cauchy's theorem, Lagrangian coordinates.*) *There exists a material, second-order tensor field* $\mathbf{t}$ *called the first Piola–Kirchhoff stress tensor, such that the surface force* $\mathbf{g}$, *acting on the boundary of any subbody* $\mathcal{V}_R$, *with normal unit vector* $\boldsymbol{\nu}$ (*pointed outward from* $\mathcal{V}_R$), *is equal to*

$$\mathbf{g} = \mathbf{t}\boldsymbol{\nu}.$$

Furthermore, we have:

$$\boxed{\begin{aligned} \mathbf{t}\mathbf{F}^T = \mathbf{F}\mathbf{t}^T \quad & \text{in } \Omega, \\ \rho\ddot{\boldsymbol{\phi}} = \mathbf{f} + \operatorname{div}[\mathbf{t}] \quad & \text{in } \Omega, \\ \mathbf{t}\boldsymbol{\nu} = \mathbf{g} \quad & \text{on } \Gamma. \end{aligned}} \tag{3.2}$$

□

Also of interest to the following discussion is the *second Piola–Kirchhoff stress tensor* defined by

$$\boldsymbol{\sigma} = \mathbf{F}^{-1}\mathbf{t}.$$

Note that the notation div [**t**] denotes the vector whose components in the basis $(\mathbf{e}_i)$ are

$$[\operatorname{div}(\mathbf{t})]_i = \sum_{j=1}^{3} \frac{\partial t_{ij}}{\partial x_j}.$$

A consequence of Cauchy's theorem is the virtual work theorem, which is used extensively in many practical situations.

THEOREM 3.2. (*Virtual work theorem.*) *If the first Piola–Kirchhoff stress tensor field* **t** *is piecewise continuously differentiable and if the applied forces* **f** *and* **g** *are absolutely continuous (on* Ω *and* Γ*, resp.) Cauchy's theorem implies that*

$$\int_\Omega [\rho\ddot{\boldsymbol{\phi}} \cdot \mathbf{w} + \mathbf{t} \cdot \nabla\mathbf{w}]\, d\mathbf{x} = \int_\Omega \mathbf{f} \cdot \mathbf{w}\, d\mathbf{x} + \int_{\Gamma_2} \mathbf{g} \cdot \mathbf{w}\, da$$

for any continuously differentiable displacement field **w** *that is defined on B and has zero trace on* $\Gamma_1 = \Gamma - \Gamma_2$. □

Above, the dot product $\mathbf{t} \cdot \nabla\mathbf{w}$ of two tensors having matrix representations t_{ij} and $(\nabla\mathbf{w})_{ij}$ in the orthonormal basis $(\mathbf{e}_i)$ denotes the quantity

$$\mathbf{t} \cdot \nabla\mathbf{w} = \sum_{i,j=1}^{3} t_{ij}(\nabla\mathbf{w})_{ij}.$$

The virtual work theorem states that for any virtual displacement, the work developed by the internal stresses and the inertia forces is equal to the work developed by the external loads. This theorem is sometimes taken as an alternative formulation of the axiom of force and moment balance. The proof is very simple and consists of multiplying the equilibrium equations (3.2) by **w** and integrating by parts on each subdomain where **t** is continuously differentiable. Boundary integrals will cancel each other, except on Γ_2, where the term $\mathbf{t}\boldsymbol{\nu}$ can be replaced by **g**.

As indicated in Marsden and Hughes [1983], the importance of the virtual work theorem rests on three facts. First, it is believed to remain valid even when the regularity assumptions on the forces and on the stresses do not hold, or when the classical (strong) formulation of the differential equations (3.2) does not make sense; second, the resulting equality coincides with the weak form of the equilibrium equations (3.2), to which many existing mathematical tools can be applied; and, third, as illustrated in this book, it is the basis of many numerical methods.

4. Standard materials in small strains.

4.1. The assumption of small strains. This assumption, which is commonly made in the study of the deformations of solid bodies, supposes that the displacement fields $\mathbf{u}$ and their spatial gradient $\nabla\mathbf{u} = \partial\mathbf{u}/\partial\mathbf{x}$ remain small during deformation. This leads to the following approximations.

(i) The configuration of the body remains fixed in time. In other words, for any particle, the positions before and after deformation, respectively, $\mathbf{x}$ and $\bar{\mathbf{x}}$, can be identified, i.e.,

$$\bar{\mathbf{x}} \approx \mathbf{x} \qquad \forall \mathbf{x} \in \Omega;$$

(ii) the right Cauchy–Green tensor reduces to

$$\mathbf{F}^T\mathbf{F} \approx \mathbf{I} + \nabla\mathbf{u} + (\nabla\mathbf{u})^T,$$

which leads to the definition of the *linearized strain tensor*

$$\mathbf{E} = \tfrac{1}{2}[\nabla\mathbf{u} + (\nabla\mathbf{u})^T].$$

Then, the first and the second Piola–Kirchhoff stress tensor can be identified as

$$\mathbf{t} = \mathbf{F}\boldsymbol{\sigma} \approx \boldsymbol{\sigma},$$

and the same holds for Cauchy stress tensor

$$\bar{\boldsymbol{\sigma}} = \frac{\mathbf{F}\boldsymbol{\sigma}\mathbf{F}^T}{\det(\mathbf{F})} \approx \boldsymbol{\sigma}.$$

In that context, the conservation of mass and Cauchy's theorem reduce to

$$(4.1) \qquad \boxed{\begin{array}{ll} \rho = \rho(\mathbf{x}) & \text{(mass)}, \\ \left.\begin{array}{l} \rho\dot{\mathbf{v}} = \mathbf{f} + \operatorname{div}[\boldsymbol{\sigma}] \\ \boldsymbol{\sigma}^T = \boldsymbol{\sigma} \end{array}\right\} & \text{(momentum)}, \end{array}}$$

together with the boundary conditions

$$(4.2) \qquad \boxed{\boldsymbol{\sigma}\boldsymbol{\nu} = \mathbf{g} \quad \text{on } \Gamma_2,}$$

where Γ_2 denotes the part of the boundary Γ of Ω on which a given surface force is imposed.

4.2. Definition of standard materials. We will follow the terminology of Germain [1973] or of Suquet [1982]. The functions introduced below may implicitly depend on the material coordinates of the body. In addition, for a real convex function $\mathcal{W}$, defined over $\mathbb{R}^m$, we will denote by $\partial\mathcal{W}/\partial\mathbf{T}(\mathbf{T}_0)$ or by $\partial\mathcal{W}(\mathbf{T}_0)$ its subgradient at $\mathbf{T}_0$. This subgradient is given as

$$\frac{\partial\mathcal{W}}{\partial\mathbf{T}}(\mathbf{T}_0)=\{\mathbf{S}\in\mathbb{R}^m, \mathbf{S}\cdot(\mathbf{T}-\mathbf{T}_0)\leq\mathcal{W}(\mathbf{T})-\mathcal{W}(\mathbf{T}_0), \forall\mathbf{T}\in\mathbb{R}^m\}$$

and reduces to the classical gradient of $\mathcal{W}$ if $\mathcal{W}$ is differentiable.

DEFINITION 4.1. A standard material is characterized by the existence of two convex material functions, the specific free energy $\mathcal{W}$, and the internal dissipation potential $\mathcal{D}_1$, depending only on the local values of the linearized strain tensor $\mathbf{E}$, the temperature T, and additional state variables $\boldsymbol{\alpha}$, such that we have

$$(4.3)\qquad \begin{cases}\boldsymbol{\sigma}=\rho\dfrac{\partial\mathcal{W}}{\partial\mathbf{E}}+\dfrac{\partial\mathcal{D}_1}{\partial\dot{\mathbf{E}}},\\[2ex] \mathbf{0}=\rho\dfrac{\partial\mathcal{W}}{\partial\boldsymbol{\alpha}}+\dfrac{\partial\mathcal{D}_1}{\partial\dot{\boldsymbol{\alpha}}}.\end{cases}\qquad \text{(constitutive laws)}$$

In the case of kinematic constraints imposed on $\boldsymbol{\alpha}$ or $\mathbf{E}$, the above laws have to be modified in order to include the Lagrange multipliers associated with these constraints.

It is important to note that this definition is by no means universal but is only a convenient framework in which to model experimental data obtained on certain classes of materials. While it can include the classical models of plasticity or viscoplasticity in small strains, it is far too restrictive when large strains or high velocities are involved. In fact, the convexity assumption on the free energy $\mathcal{W}$ is in general incompatible with large strains.

4.3. Examples. When the temperature can be eliminated, standard materials lead to simplified mechanical models. This occurs either when the internal dissipation potential $\mathcal{D}_1$ and the partial derivatives of the free-energy potential with respect to $\mathbf{E}$ and $\boldsymbol{\alpha}$ do not depend on the temperature (temperature-insensitive materials), or when an isothermal (resp., isentropic) evolution process where temperature (resp., entropy) does not vary in time is considered.

We then obtain simplified, purely mechanical models where, by writing the conservation of mass, the virtual work theorem, and the constitutive laws, we can solve the problems described in § 1. Among these, we find different models of elasticity, elastoviscoplasticity, and viscoplasticity.

4.3.1. Compressible linear elasticity (solid in small strains). While this is one of the simplest and oldest models, in many practical situations it is both powerful and accurate. It is characterized by

$$\begin{cases} \text{state variable: } \mathbf{E} \text{ (linearized strain tensor)}, \\ \text{free-energy potential: } \mathcal{W} = \dfrac{1}{2\rho} \mathbf{E} \cdot \mathbf{AE} \text{ (quadratic form in } \mathbf{E}), \\ \text{internal dissipation potential: } \mathcal{D}_1 = 0. \end{cases}$$

From the definition of $\mathcal{W}$ as a quadratic form of the symmetric, linearized strain tensor $\mathbf{E}$, one can also choose the infinitesimal elasticity tensor $\mathbf{A}$ with the following symmetry properties:

$$A_{ijkl} = A_{klij} = A_{jikl}, \qquad 1 \leq i, j, k, l \leq 3.$$

In this case, there is only one constitutive law, given by

$$\boxed{\boldsymbol{\sigma} = \mathbf{A}\, \mathbf{E}.} \tag{4.4}$$

For isotropic bodies, due to the symmetry requirement, this constitutive law has to take the form

$$\sigma_{ii} = \lambda_e \left(\sum_{k=1}^{3} E_{kk} \right) + 2\mu_e E_{ii}, \qquad \sigma_{ij} = 2\mu_e E_{ij} \quad \text{for } i \neq j,$$

where λ_e and μ_e are the *Lamé constants.*

4.3.2. Norton viscoplastic material. In small strains, a viscoplastic material of Norton's type corresponds to

$$\begin{cases} \text{state variable: } \mathbf{E} \text{ (linearized strain tensor)}, \\ \text{constraint tr: } \mathbf{E} = 0 \text{ (incompressibility)}, \\ \text{free-energy potential: } \mathcal{W} = 0, \\ \text{internal dissipation potential: } \mathcal{D}_1 = \dfrac{\lambda^{1-s}}{s} (k\sqrt{2})^s |\dot{\mathbf{E}}|^s. \end{cases}$$

Here, λ, k, and s are strictly positive material constants with k characterizing the mechanical strength of the material and the exponent s $(s > 1)$ being a measure of its viscosity. The constitutive law associated with this model is

$$\boxed{\boldsymbol{\sigma} = \lambda^{1-s} (k\sqrt{2})^s |\dot{\mathbf{E}}|^{s-2} \dot{\mathbf{E}} - p\mathbf{Id},} \tag{4.5}$$

with p, the hydrostatic pressure, being a Lagrange multiplier of the incompressibility constraint.

In Fig. 4.1 we show the stress–strain relationship for a Norton viscoplastic bar under shear.

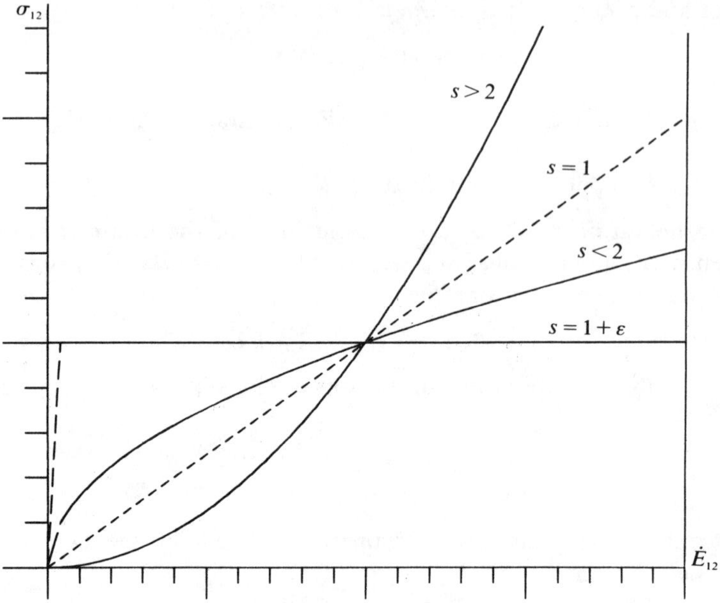

FIG. 4.1. *Stress–strain relationship for a bar under shear obeying Norton's viscoplastic material law.*

In (4.5), stresses are related only to deformation rates, which means that such materials behave like fluids.

4.3.3. Bingham viscoplastic material. In small strains, a Bingham viscoplastic material corresponds to

$$\begin{cases} \text{state variable: } \mathbf{E} \text{ (linearized strain tensor)}, \\ \text{constraint: tr } \mathbf{E}=0 \text{ (incompressibility)}, \\ \text{internal dissipation potential: } \mathcal{D}_1=\sqrt{2}g|\dot{\mathbf{E}}|+\mu|\dot{\mathbf{E}}|^2, \\ \text{free-energy potential: } \mathcal{W}=0, \end{cases}$$

where g and μ are real positive constants representing the rigidity and the viscosity of the material. The associated constitutive law is

$$\boxed{\boldsymbol{\sigma}=(\sqrt{2}g|\dot{\mathbf{E}}|^{-1}+2\mu)\dot{\mathbf{E}}-p\mathbf{Id}.} \tag{4.6}$$

In Fig. 4.2 we show the stress–strain relationship of a Bingham viscoplastic bar under shear.

4.3.4. Elastoviscoplasticity (solids in small strains.) A general model for elastoviscoplastic solids in small strains can be obtained by combining the elastic and the viscoplastic models introduced in §§ 4.3.1. and 4.3.2.

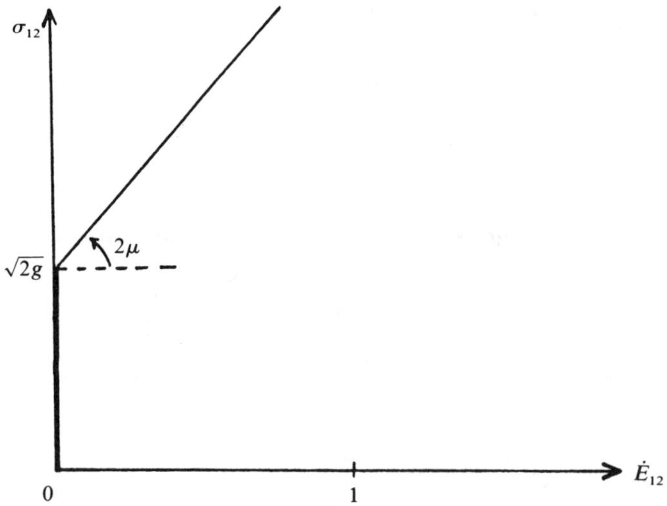

FIG. 4.2. *Stress-strain-rate relationship for a bar under shear* ($\dot{E}_{ij} = \dot{E}_{12}\delta_{1i}\delta_{2j}$) *obeying Bingham's viscoplastic material law.*

For that purpose, let us first introduce a closed and convex set $\mathbf{C}$ in the space $\mathbb{R}^9_{\text{sym}}$ of the second-order symmetric tensors operating on $\mathbb{R}^3$. The interior of this set will correspond to the stress tensors for which the material behavior is mostly elastic. To $\mathbf{C}$, we associate the real function ψ_c defined on $\mathbb{R}^9_{\text{sym}}$ by

$$\psi_c(\mathbf{E}) = \sup_{\boldsymbol{\sigma} \in \mathbf{C}} \boldsymbol{\sigma} \cdot \mathbf{E}. \tag{4.7}$$

By definition, this is the polar function to the indicator I_c of $\mathbf{C}$. We will suppose that $\mathbf{C}$ contains the null tensor so that ψ_c is a positive lower semicontinuous convex function.

A generalized Maxwell-Norton elastoviscoplastic solid in small strains is defined by

$$\begin{cases} \text{state variables:} & \begin{cases} \mathbf{E} \text{ (linearized strain tensor)}, \\ \mathbf{E}^{\text{an}} \text{ (anelastic part of } \mathbf{E}), \end{cases} \\ \text{constraint: tr } \mathbf{E}^{\text{an}} = 0 \text{ (for plastically incompressible bodies only)}, \\ \text{free-energy potential: } \mathcal{W}(\mathbf{E} - \mathbf{E}^{\text{an}}) = \dfrac{1}{2\rho} (\mathbf{E} - \mathbf{E}^{\text{an}}) \cdot \mathbf{A}(\mathbf{E} - \mathbf{E}^{\text{an}}), \\ \text{internal dissipation potential: } \mathcal{D}_1 = \dfrac{\lambda^{1-s}}{s} |\psi_c(\dot{\mathbf{E}}^{\text{an}})|^s. \end{cases}$$

For plastically incompressible bodies, the convex set $\mathbf{C}$ must be invariant under translation along diagonal tensors; otherwise, the constitutive law would be incompatible with the constraint tr $(\mathbf{E}^{an}) = 0$. In any case, we obtain as constitutive laws

(4.8) $$\begin{cases} \boldsymbol{\sigma} = \mathbf{A}(\mathbf{E} - \mathbf{E}^{an}), \\ \boldsymbol{\sigma} \in \partial\left[\dfrac{\lambda^{1-s}}{s} |\psi_c(\dot{\mathbf{E}}^{an})|^s\right]. \end{cases}$$

By inverting these equations and differentiating the first equation with respect to time, we can eliminate $\dot{\mathbf{E}}^{an}$ and obtain a single constitutive law, involving only $\boldsymbol{\sigma}$ and $\mathbf{E}$, of the form

(4.9) $$\boxed{(\dot{\mathbf{E}} - \mathbf{A}^{-1}\dot{\boldsymbol{\sigma}}) \in \partial\left[\frac{\lambda}{q} |j_c(\boldsymbol{\sigma})|^q\right].}$$

Here, q is the conjugate exponent of s ($sq = s + q$) and $j_c(\boldsymbol{\sigma})$ is the gauge function of the set $\mathbf{C}$, that is

$$j_c(\boldsymbol{\sigma}) = \inf\left\{\mu \in \mathbb{R}, \mu > 0, \frac{\sigma}{\mu} \in \mathbf{C}\right\}.$$

In Fig. 4.3 we show the relaxation and creep responses of a generalized Maxwell–Norton elastoviscoplastic solid under tension.

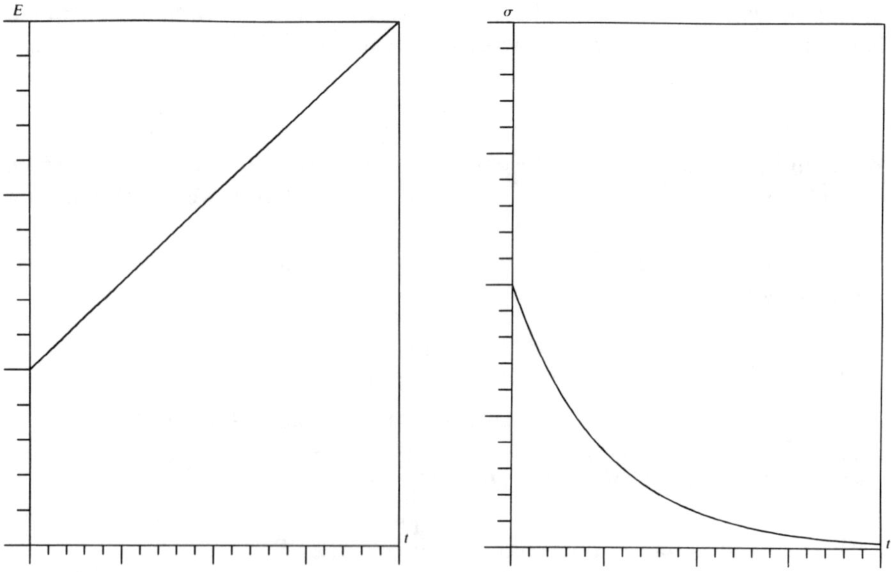

FIG. 4.3. *Relaxation and creep test for bar under tension obeying Maxwell viscoelasticity* ($\dot{\mathbf{E}} = \dot{\sigma}/E + \sigma/\mu$).

Remark 4.2. For $s=1$, the above constitutive law reduces to the *classical Prandtl–Reuss flow rule* for perfectly elastoplastic materials, so that we have

$$(4.10) \qquad (\dot{\mathbf{E}}-\mathbf{A}^{-1}\dot{\boldsymbol{\sigma}}) \in \partial[I_c(\boldsymbol{\sigma})],$$

where I_c is the indicator of $\mathbf{C}$, i.e., the function with value 0 if $\boldsymbol{\sigma}$ belongs to $\mathbf{C}$ and with value $+\infty$ if not.

Remark 4.3. Maxwell viscoelasticity corresponds to

$$\begin{cases} \rho \mathscr{W}(\mathbf{E}) = \dfrac{\nu E}{2(1-2\nu)(1+\nu)}(\operatorname{tr}\mathbf{E})^2 + \dfrac{E}{2(1+\nu)}\operatorname{tr}\mathbf{E}^2, \\ \mathbf{C} = \{\boldsymbol{\sigma} \in \mathbb{R}^9_{\text{sym}},\ |\boldsymbol{\sigma} - \tfrac{1}{3}(\operatorname{tr}\boldsymbol{\sigma})\mathbf{Id}| < k\sqrt{2}\}, \\ s=2, \end{cases}$$

which leads to the constitutive equation

$$(4.11) \qquad \boxed{\dot{E}_{ij} = \frac{(1+\nu)}{E}\dot{\sigma}_{ij} - \frac{\nu}{E}(\operatorname{tr}\dot{\boldsymbol{\sigma}})\delta_{ij} + \lambda(k\sqrt{2})^{-2}\left[\sigma_{ij} - \frac{1}{3}(\operatorname{tr}\boldsymbol{\sigma})\delta_{ij}\right].}$$

E and ν denote the Young modulus and the Poisson ratio of the material, respectively.

Remark 4.4. As in § 4.3.1, the fourth-order elasticity tensor $\mathbf{A}$ is supposed to satisfy the symmetry properties

$$A_{ijkl} = A_{klij} = A_{jikl}, \qquad 1 \le i, j, k, l \le 3.$$

Moreover, for physical reasons (see the discussion of the Baker–Eriksen inequality in Truesdell and Noll [1965, Eq. (51.22)]), this tensor must be positive-definite and hence invertible.

5. Incompressible viscous fluids.

5.1. Virtual work theorem in present configuration. Fluids do not remember any past configuration. Therefore, in the study of their flows, it is natural to express the conservation laws in terms of the present configuration of the fluid. Denoting by $\bar{\rho}$ the mass per unit volume of the fluid in this present configuration, and using the kinematical notation and the results of § 2, the *principle of conservation of mass* becomes

$$(5.1) \qquad \bar{\rho}(\bar{\mathbf{x}}, t) = \frac{\rho(\mathbf{x})}{\det(\mathbf{F}(\mathbf{x}, t))}.$$

Differentiating (5.1) with respect to time, we get the classical Eulerian form of this principle

$$\boxed{\frac{\partial \bar{\rho}}{\partial t}+\sum_{i=1}^{N} \frac{\partial}{\partial \bar{x}_i}(\bar{\rho}\bar{\mathbf{v}})=0.} \tag{5.2}$$

Similarly, through the change of variables $\mathbf{x}\rightarrow\bar{\mathbf{x}}$ and introducing the Cauchy stress tensor

$$\bar{\boldsymbol{\sigma}}=\frac{\mathbf{t}\mathbf{F}^T}{\det(\mathbf{F})},$$

the virtual work theorem becomes

$$\boxed{\begin{aligned}&\int_{\boldsymbol{\phi}(\Omega)}\left\{\bar{\rho}\dot{\bar{\mathbf{v}}}\cdot\mathbf{w}+\sum_{i,j=1}^{N}\bar{\sigma}_{ij}\frac{\partial w_i}{\partial \bar{x}_j}\right\}d\bar{\mathbf{x}}=\int_{\boldsymbol{\phi}(\Omega)}\bar{\mathbf{f}}\cdot\mathbf{w}\,d\bar{\mathbf{x}}+\int_{\boldsymbol{\phi}(\Gamma_2)}\bar{\mathbf{g}}\cdot\mathbf{w}\,d\bar{a}\\ &\forall\mathbf{w}\in\mathbf{C}^1(\boldsymbol{\phi}(\Omega))\quad\text{with }\mathbf{w}=\mathbf{0}\text{ on }\boldsymbol{\phi}(\Gamma-\Gamma_2).\end{aligned}} \tag{5.3}$$

The body forces $\bar{\mathbf{f}}$, surface tractions $\bar{\mathbf{g}}$, and velocity $\bar{\mathbf{v}}$ that are now expressed with respect to the present configuration are related to the equivalent quantities $\mathbf{f}$, $\mathbf{g}$, and $\mathbf{v}$ introduced in §§ 2 and 3 for the reference configuration by the relations

$$\bar{\mathbf{f}}(\bar{\mathbf{x}},t)=\mathbf{f}(\mathbf{x},t)\,d\mathbf{x}/d\bar{\mathbf{x}},$$

$$\bar{\mathbf{g}}(\bar{\mathbf{x}},t)=\mathbf{g}(\mathbf{x},t)\,da/d\bar{a},$$

$$\bar{\mathbf{v}}(\bar{\mathbf{x}},t)=\mathbf{v}(\mathbf{x},t).$$

Moreover, $\dot{\bar{\mathbf{v}}}$ represents the acceleration of the fluid, that is, the material time derivative of its velocity (see § 2). In Eulerian coordinates, we have

$$\dot{\bar{\mathbf{v}}}=\frac{\partial\bar{\mathbf{v}}}{\partial t}+\sum_{i=1}^{N}\bar{v}_i\frac{\partial\bar{\mathbf{v}}}{\partial x_i}.$$

5.2. Definition of a viscoplastic incompressible fluid. Viscoplastic fluids are materials whose memory reduces to the knowledge of their present local velocity and which resist any local variation of this velocity. More precisely, denoting by $\mathbf{D}(\bar{\mathbf{v}})$ the deformation-rate tensor of the fluid, which can be defined by

$$\mathbf{D}(\bar{\mathbf{v}})=\frac{1}{2}\left\{\frac{\partial\bar{\mathbf{v}}}{\partial\bar{\mathbf{x}}}+\left(\frac{\partial\bar{\mathbf{v}}}{\partial\bar{\mathbf{x}}}\right)^T\right\},$$

we have the following definition.

DEFINITION 5.1. A viscoplastic incompressible fluid is a material characterized by the mechanical constitutive law

$$(5.4)\qquad \boxed{(\bar{\boldsymbol{\sigma}} - \bar{p}\mathbf{Id}) \in \partial\mathcal{D}_1(\mathbf{D}(\bar{\mathbf{v}})), \qquad \operatorname{tr}\mathbf{D}(\bar{\mathbf{v}}) = 0,}$$

where $\mathcal{D}_1$ is a real positive convex isotropic function of $\mathbf{D}$ vanishing at the origin.

We recall that the hydrostatic pressure $\bar{p}$ is a multiplier of the incompressibility constraint $\operatorname{tr}(\mathbf{D}(\bar{\mathbf{v}})) = 0$, and that an isotropic function of a symmetric matrix $\mathbf{D}$ is a function that depends only on the invariants of $\mathbf{D}$.

As was the case for Definition 4.1, Definition 5.1 is only a convenient, consistent framework in which to model certain classes of materials. It does include the most classical models of viscous fluids; in particular, *Newtonian* and *Bingham fluids*, given by

$$(5.5)\qquad \mathcal{D}_1(\mathbf{D}) = \mu|\mathbf{D}|^2,$$

$$(5.6)\qquad \mathcal{D}_1(\mathbf{D}) = \mu|\mathbf{D}|^2 + g\sqrt{2}|\mathbf{D}|,$$

respectively, fall within this category.

5.3. Navier–Stokes equations. Let us consider the flow of a Newtonian fluid inside a given domain $\boldsymbol{\phi}(\Omega)$. Suppose that the velocity $\bar{\mathbf{v}}_1$ of the fluid is known on the boundary $\boldsymbol{\phi}(\Gamma)$ of $\boldsymbol{\phi}(\Omega)$; in other words, suppose that Γ_2 is empty. If we employ the virtual work theorem (5.3) and then integrate by parts, apply the constitutive laws (5.4) and (5.5), and divide by $\bar{\rho}$, we observe that the velocity $\bar{\mathbf{v}}$ of the fluid inside the domain $\boldsymbol{\phi}(\Omega)$ satisfies the following set of equations:

$$(5.7)\qquad \boxed{\begin{aligned} &\frac{\partial\bar{\mathbf{v}}}{\partial t} + \sum_{i=1}^{N} \bar{v}_i \frac{\partial\bar{\mathbf{v}}}{\partial\bar{x}_i} - \frac{\mu}{\rho}\sum_{j=1}^{N}\frac{\partial^2\bar{\mathbf{v}}}{\partial\bar{x}_j^2} + \frac{1}{\rho}\frac{\partial\bar{p}}{\partial\bar{\mathbf{x}}} = \frac{\mathbf{f}}{\bar{\rho}} \quad \text{on } \boldsymbol{\phi}(\Omega),\\ &\sum_{i=1}^{N}\frac{\partial\bar{v}_i}{\partial\bar{x}_i} = 0 \quad \text{on } \boldsymbol{\phi}(\Omega),\\ &\bar{\mathbf{v}} = \bar{\mathbf{v}}_1 \quad \text{on } \boldsymbol{\phi}(\Gamma).\end{aligned}}$$

These equations governing the unsteady flows of Newtonian incompressible fluids are called the *Navier–Stokes equations*.

6. Finite elasticity.

6.1. Weakness of the small-strain assumption. In the preceding sections, we have derived mechanical models that can handle a wide variety of elastic or plastic behaviors but are mainly restricted to small strains. Therefore, they are no longer valid for highly deformable materials such as rubber or frozen soils; highly flexible structures such as slender, three-dimensional bodies; buckling situations; or singular problems arising from small-strain theory and leading to solutions with unbounded deformations such as cracks or discontinuous loads.

In such situations, new models are needed. In the following discussion, we will consider only the case of hyperelasticity. While this restriction simplifies a number of difficult, purely mechanical problems such as the choice of adequate state variables and the derivation of consistent and realistic constitutive equations, it nonetheless also allows the description of many nonlinear phenomena encountered in practice such as buckling, loss of ellipticity, surface discontinuities, and cavitation. Furthermore, we will suppose that temperature can be eliminated from our models, as was done in § 4.3.

Finite elasticity models are then simply obtained by adding, to the general conservation laws of § 3, constitutive laws that relate the first Piola–Kirchhoff stress tensor $\mathbf{t}$ and the deformation gradient matrix $\mathbf{F}$.

6.2. Definition of hyperelasticity. Hyperelastic materials are characterized by the constitutive equations

$$\boxed{\mathbf{t}(\mathbf{x}, t) = \rho \frac{\partial \mathcal{W}}{\partial \mathbf{F}}(\mathbf{x}, \mathbf{F}(\mathbf{x}, t)) \qquad \text{(compressible materials)},} \tag{6.1}$$

$$\boxed{\mathbf{t}(\mathbf{x}, t) = \rho \frac{\partial \mathcal{W}}{\partial \mathbf{F}}(\mathbf{x}, \mathbf{F}(\mathbf{x}, t)) - p \frac{\partial(\det \mathbf{F}(\mathbf{x}, t))}{\partial \mathbf{F}} \qquad \text{(incompressible materials)}.} \tag{6.2}$$

The function $\mathcal{W}$ corresponds to the *specific internal elastic energy* of the body, also called the *stored energy function.* In (6.2), the hydrostatic pressure p is a Lagrange multiplier of the incompressibility constraint.

6.3. Admissibility conditions. Any function $\mathcal{W}$ will not be suitable for defining the internal elastic energy. The constitutive laws that are associated with $\mathcal{W}$ must satisfy two requirements (Gurtin [1981, p. 11]), first,

$$\boxed{\mathcal{W}(\mathbf{x}, \mathbf{QF}) = \mathcal{W}(\mathbf{x}, \mathbf{F}) \qquad \text{(frame indifference)}} \tag{6.3}$$

for any rotation $\mathbf{Q}$ and deformation gradient $\mathbf{F}$, and second,

$$\boxed{\mathcal{W}(\mathbf{x}, \mathbf{FQ}) = \mathcal{W}(\mathbf{x}, \mathbf{F}) \qquad \text{(symmetry)}} \tag{6.4}$$

for any $\mathbf{Q}$ in the symmetry group $\mathcal{G}$ of the body at point $\mathbf{x}$ and for any deformation gradient $\mathbf{F}$.

These two admissibility requirements assume a much simpler form for those isotropic bodies in which the symmetry group $\mathcal{G}$ contains all the rotations of $\mathbb{R}^3$. Then frame indifference and symmetry are equivalent to the following.

Admissibility conditions for isotropic hyperelasticity. The specific internal elastic energy $\mathcal{W}(\mathbf{x}, \mathbf{F})$ must be an isotropic function of the right Cauchy-Green tensor $\mathbf{F}^T\mathbf{F}$

$$\boxed{\begin{aligned} &\mathcal{W}(\mathbf{x}, \mathbf{F}) = \mathcal{W}(\mathbf{x}, I_1, I_2, J), \\ &I_1 = \text{tr}\,(\mathbf{F}^T\mathbf{F}), \\ &I_2 = \tfrac{1}{2}\{(\text{tr}\,\mathbf{F}^T\mathbf{F})^2 - \text{tr}\,(\mathbf{F}^T\mathbf{F})^2\}, \\ &J = \det \mathbf{F}. \end{aligned}} \tag{6.5}$$

The proof of this equivalence is very elementary and can be found in most books on continuum mechanics (Marsden and Hughes [1983]).

6.4. Examples of isotropic hyperelastic materials. These materials will be characterized by the associated specific internal elastic energy $\mathcal{W}(\mathbf{F})$, which will be given as an isotropic function of the deformation gradient $\mathbf{F}$. The real numbers C_{ij}, μ_i, and α_j that will appear are material constants.

6.4.1. Incompressible isotropic examples.

Neo-Hookian (see Knowles and Sternberg [1980])

$$\rho\mathcal{W}(\mathbf{F}) = C_{10}(I_1 - 3).$$

Mooney-Rivlin (see Chadwick and Haddon [1972])

$$\rho\mathcal{W}(\mathbf{F}) = C_{10}(I_1 - 3) + C_{01}(I_2 - 3);$$

typical numerical values are

$$\begin{cases} C_{10} = .183 \text{ MPa}, \\ C_{01} = .0034 \text{ MPa}. \end{cases}$$

Haines-Wilson [1979]

$$\begin{aligned} \rho\mathcal{W}(\mathbf{F}) = C_{10}(I_1 - 3) + C_{01}(I_2 - 3) + C_{20}(I_1 - 3)^2 + C_{02}(I_2 - 3)^2 \\ + C_{11}(I_1 - 3)(I_2 - 3) + C_{30}(I_1 - 3)^3. \end{aligned}$$

Ogden [1972]

$$\rho \mathscr{W}(\mathbf{F}) = \sum_{i=1}^{n} \mu_i \alpha_i^{-1}(\operatorname{tr}((\mathbf{F}^T\mathbf{F})^{\alpha_i/2}) - 3);$$

typical numerical values are

$$\begin{cases} n = 3, \quad \alpha_1 = 1.3, \quad \alpha_2 = 5, \quad \alpha_3 = -2, \\ \mu_1 = .63 \text{ MPa}, \quad \mu_2 = .0012 \text{ MPa}, \\ \mu_3 = -.01 \text{ MPa}. \end{cases}$$

Knowles–Sternberg [1980]

$$\rho \mathscr{W}(\mathbf{F}) = \frac{\mu}{2\alpha}\left\{\left(1 + \frac{\alpha}{n}(I_1 - 3)\right)^n - 1\right\}.$$

All these materials are constrained by the incompressibility relation

$$J = \det \mathbf{F} = 1.$$

6.4.2. Compressible isotropic examples.

Ogden [1976]

$$\rho \mathscr{W}(\mathbf{F}) = \sum_{i=1}^{3} \mu_i \operatorname{tr}(\mathbf{F}^T\mathbf{F})^{\alpha_i/2} + \mu_4 \alpha_4^{-2} J^{-\alpha_4}$$

$$+ \left(\mu_4 \alpha_4^{-1} \sum_{i=1}^{3} \alpha_i \mu_i\right) \ln J + \mu_5 J^{2/3}(J-1) \operatorname{tr}(\mathbf{F}^T\mathbf{F})^{-2}.$$

Saint-Venant Kirchhoff (see Ciarlet [1987])

$$\rho \mathscr{W}(\mathbf{F}) = \frac{\lambda}{2}\left(\operatorname{tr}\frac{1}{2}(\mathbf{F}^T\mathbf{F} - \mathbf{I}d)\right)^2 + \mu \operatorname{tr}\left(\frac{1}{2}(\mathbf{F}^T\mathbf{F} - \mathbf{I}d)\right)^2.$$

Ciarlet-Geymonat [1982]

$$\rho \mathscr{W}(\mathbf{F}) = C_1(I_1 - 3) + C_2(I_2 - 3) + a(J^2 - 1) - (2C_1 + 4C_2 + 2a) \ln J.$$

Chapter 2

Variational Formulations of the Mechanical Problems

1. Preliminaries.

1.1. Generalities and synopsis. The mechanical equations derived in the first chapter are formal. The minimal degree of regularity of the unknown functions (displacements, stresses, velocities, etc.) is not specified. Moreover, one must decide whether the derivatives involved in the conservation laws and in the constitutive laws should be considered in the distribution (weak) sense or in the classical Frechet (strong) sense. Therefore, a specific topological structure must be given to these mechanical equations if we want to ensure that they will lead to well-posed mathematical problems.

The choice of this structure might be delicate and is based both on mechanical and on mathematical arguments. However, in all cases,

(i) the derivatives appearing in the conservation laws should be taken as distributional derivatives, because these laws are written initially under an integral form;

(ii) the chosen topology should be strong enough to give a mathematical meaning to the physical variables; and

(iii) the chosen topology should be weak enough to ensure that all physical solutions are solutions of the corresponding mathematical problem.

Bearing these principles in mind, in this chapter we will derive precise mathematical formulations of the mechanical problems that were introduced in Chapter 1. We will concentrate on *temperature-independent, static* (or more generally *quasi-static*) *problems.* In such problems, temperature can be eliminated (see Chap. 1, § 4.3) and the inertia terms

$$\int_{\mathcal{V}_R} \rho \ddot{\boldsymbol{\phi}}\, d\mathbf{x} \quad \text{and} \quad \int_{\mathcal{V}_R} \rho \boldsymbol{\phi} \times \ddot{\boldsymbol{\phi}}\, d\mathbf{x}$$

can be neglected in the laws of force and moment balance.

Remark 1.1. The inertia terms can be neglected in the equilibrium equations when the considered body is at equilibrium (no motion), or when these terms are very small compared with the other terms that appear. For example, this is usually the case for physical situations in which the time scale is very long compared with the relaxation time of the considered materials, such as, for example, soils freezing.

Remark 1.2. Considering quasi-static problems only is consistent with the numerical augmented Lagrangian techniques that will be described later in this book. Indeed, from the numerical point of view, a dynamic, evolutionary process can be reduced to a sequence of static model problems through an implicit or a semiimplicit time discretization. The mathematical and numerical analysis presented in the following paragraphs and chapters will apply to these discretized problems.

1.2 Basic concepts of Sobolev spaces. We assume the reader is familiar with the standard concepts of functional analysis such as reflexive Banach spaces, separability, Hilbert spaces, continuous imbedding, compactness, weak topologies, and $L^p(\Omega)$ spaces. All of these concepts are defined in references such as Adams [1975, Chap. 1, pp. 4–9, 22]. In what follows, then, we will simply recall the definitions and the results that are relevant to our analysis. In the following parts of this book, the set Ω, defined in general as the interior of the continuous body under consideration in its *reference configuration*, is supposed to be *open, bounded, and connected* in $\mathbb{R}^3$ and to have a *Lipschitz-continuous boundary* Γ. Occasionally, Ω will correspond only to a plane section of the considered three-dimensional body; it will then be an open, bounded, connected subset of $\mathbb{R}^2$ with a Lipschitz-continuous boundary. Such subsets Ω of $\mathbb{R}^N$ ($N=2$ or 3) are equipped with the classical Lebesgue measure of $\mathbb{R}^N$, denoted by $d\mathbf{x}$, and their closure is denoted by B. The space $\mathcal{D}(\Omega)$ is the linear space of real functions infinitely differentiable and with compact support in Ω. The C^k norms defined for any k and any compact subset of Ω induce a topology on $\mathcal{D}(\Omega)$. A topological dual space $\mathcal{D}'(\Omega)$ can then be defined; it is called the *space of distributions* on Ω. Of course, any locally Lebesgue integrable real function f defined on Ω can be identified with a distribution by

$$\langle f, \phi\rangle = \int_\Omega f(\mathbf{x})\phi(\mathbf{x})\, d\mathbf{x} \qquad \forall \phi \in \mathcal{D}(\Omega),$$

with $\langle\cdot,\cdot\rangle$ denoting the duality pairing between $\mathcal{D}'(\Omega)$ and $\mathcal{D}(\Omega)$. For $\alpha = \{\alpha_1, \alpha_2, \cdots, \alpha_N\}$ in $\mathbb{N}^N$, and $|\alpha| = \alpha_1 + \cdots + \alpha_N$, the *$\alpha$ derivative of a distribution* f is defined by

$$\langle \partial^\alpha f, \phi\rangle = (-1)^{|\alpha|}\left\langle f, \frac{\partial^{|\alpha|}\phi}{\partial x_1^{\alpha_1}\cdots \partial x_N^{\alpha_N}}\right\rangle \qquad \forall \phi \in \mathcal{D}(\Omega).$$

For a function f of class $C^{|\alpha|}$, this concept corresponds to the usual definition of the partial derivative $\partial^\alpha f$.

For $m \in \mathbb{N}$ and $p \in \mathbb{R}$, $1 \leq p \leq +\infty$, we define the Sobolev space

$$W^{m,p}(\Omega) = \{v \in L^p(\Omega), \partial^\alpha v \in L^p(\Omega)\ \forall |\alpha| \leq m\},$$

which is a Banach space for the norm

$$\|v\|_{m,p} = \left\{ \sum_{\alpha \leq m} \int_\Omega |\partial^\alpha v|^p \, d\mathbf{x} \right\}^{1/p} \quad \text{if } p < +\infty,$$

$$\|v\|_{m,\infty} = \sup_{|\alpha| \leq m} \left\{ \sup_{\mathbf{x} \in \Omega} \text{ess} \, |\partial^\alpha v(\mathbf{x})| \right\} \quad \text{if } p = +\infty.$$

We also provide $W^{m,p}(\Omega)$ with the seminorm (if $p < +\infty$)

$$|v|_{m,p} = \left\{ \sum_{|\alpha| = m} \int_\Omega |\partial^\alpha v|^p \, d\mathbf{x} \right\}^{1/p}$$

For $p = 2$, $W^{m,2}(\Omega)$ is denoted by $H^m(\Omega)$, which is a Hilbert space for the scalar product

$$(u, v)_m = \sum_{|\alpha| \leq m} \int_\Omega \partial^\alpha u \, \partial^\alpha v \, d\mathbf{x}.$$

Finally, we introduce

$$\mathbf{W}^{m,p}(\Omega) = [W^{m,p}(\Omega)]^N,$$

which is also a Banach space, equipped with the norm

$$\|\mathbf{v}\|_{m,p} = \left\{ \sum_{i=1}^N \|v_i\|_{m,p}^p \right\}^{1/p};$$

the corresponding seminorm is denoted by $|\mathbf{v}|_{m,p}$.

An example of a very clear and concise introduction to these concepts can be found in Girault and Raviart [1986]. A more detailed treatment is given in Adams [1975]. Below, we only state, without proof, the basic properties of such Sobolev spaces.

THEOREM 1.1. *For $1 < p < +\infty$, $W^{m,p}(\Omega)$ is a separable reflexive Banach space.*

THEOREM 1.2. (*Sobolev imbedding theorem.*) *Let p, $1 \leq p \leq +\infty$, be given:*

(i) *for $0 \leq m < \dfrac{N}{p}$, $W^{m,p}(\Omega)$ is continuously imbedded in $L^q(\Omega)\left(\dfrac{1}{q} = \dfrac{1}{p} - \dfrac{m}{N}\right)$;*

(ii) *for $m = \dfrac{N}{p}$, $W^{m,p}(\Omega)$ is continuously imbedded in any $L^q(\Omega)$, $1 \leq q < +\infty$;*

(iii) *for $m > \dfrac{N}{p}$, $W^{m,p}(\Omega)$ is continuously imbedded in $C^0(B)$.*

THEOREM 1.3. (*Kondrasov theorem.*) *Let p, $1 \le p \le +\infty$, be given:*

(i) *for $0 \le m \le N/p$, the injection from $W^{m,p}(\Omega)$ in $L^q(\Omega)$ is compact for any q with $q < Np/(N-mp)$;*

(ii) *for $m > N/p$, the injection from $W^{m,p}(\Omega)$ in $C^0(B)$ is compact.* □

THEOREM 1.4. (*Trace theorem.*) *We call the trace operator the extension to $W^{m,p}(\Omega)$ of the map defined on $C^\infty(B)$ that associates to each function of $C^\infty(B)$ its restriction on Γ. The trace operator is continuous from $W^{m,p}(\Omega)$ into $L^q(\Gamma)$ if m is greater or equal to 1 and if q satisfies*

(i) *for* $1 \le m < \dfrac{N}{p}$, $1 \le q \le \dfrac{(N-1)p}{N-mp}$;

(ii) *for* $m \ge \dfrac{N}{p}$, $1 \le q < +\infty$.

Moreover, its range is dense in $L^q(\Gamma)$. □

A direct consequence of the classical Green's identity and of the trace theorem is the generalized *Green's formula*:

$$\begin{cases} \text{For } v \in W^{1,p}(\Omega), \text{ for } u \in W^{1,p^*}(\Omega), \dfrac{1}{p^*} + \dfrac{1}{p} = 1, \text{ we have} \\ \displaystyle\int_\Omega u \frac{\partial v}{\partial x_i}\, d\mathbf{x} + \int_\Omega v \frac{\partial u}{\partial x_i}\, d\mathbf{x} = \int_\Gamma uv\, \boldsymbol{\nu} \cdot \mathbf{e}_i\, da, \end{cases}$$

where $\boldsymbol{\nu}$ is the unit outward normal vector.

On these Sobolev spaces, the divergence operator has interesting properties, which were derived by Temam [1977] for $p = 2$ and extended to arbitrary p by Geymonat and Suquet [1986].

THEOREM 1.5. *Let p, $1 < p < +\infty$, and $g \in L^p(\Omega)$ be given such that $\int_\Omega g\, d\mathbf{x} = 0$. Then, there exists $\mathbf{u}$ in $\mathbf{W}^{1,p}(\Omega)$ such that*

$$\begin{cases} \mathbf{u} = \mathbf{0} \quad \text{on } \Gamma, \\ \operatorname{div} \mathbf{u} = g \quad \text{on } \Omega, \\ \|\mathbf{u}\|_{1,p} < C(\Omega, p) \|g\|_{0,p}, \end{cases}$$

$C(\Omega, p)$ being a strictly positive real number depending only on Ω and p. □

COROLLARY 1.6. *Let Γ_1 be a measurable subset of Γ that is not dense in Γ. Let p, $1 < p < +\infty$, and $g \in L^p(\Omega)$ be given. Then, there exists $\mathbf{u}$ in $\mathbf{W}^{1,p}(\Omega)$ such that*

$$\begin{cases} \mathbf{u} = \mathbf{0} \quad \text{on } \Gamma_1, \\ \operatorname{div} \mathbf{u} = g \quad \text{on } \Omega, \\ \|\mathbf{u}\|_{1,p} < C(\Omega, \Gamma_1, p) \|g\|_{0,p}. \end{cases}$$ □

Finally, we close this section by recalling the classical inequality of Korn (Duvaut and Lions [1972], Fichera [1972]), which was extended by Geymonat and Suquet [1986] to the case p arbitrary and Γ Lipschitz-continuous.

THEOREM 1.7. (*Inequality of Korn.*) *Let* p, $1<p<+\infty$, *be given. Let* Γ_1 *be a measurable subset of* Γ *whose interior in* Γ *is not empty. Then, there exists a strictly positive number* $C(\Omega, p, \Gamma_1)$, *depending only on* Ω, p, *and* Γ_1 *such that*

$$\begin{cases} \left[\sum_{i,j=1}^{N} \int_\Omega \left|\frac{1}{2}\left(\frac{\partial u_j}{\partial x_i}+\frac{\partial u_j}{\partial x_i}\right)\right|^p d\mathbf{x}\right]^{1/p} \geq C(\Omega, p, \Gamma_1)\|\mathbf{u}\|_{1,p} \\ \forall \mathbf{u} \in \mathbf{W}^{1,p}(\Omega) \quad \text{with } \mathbf{u}=\mathbf{0} \text{ on } \Gamma_1. \end{cases}$$

□

1.3. Basic tools for existence theory. Existence theory is important from the numerical point of view because it helps in the choice of consistent approximation techniques and numerical methods for the solution of the underlying problem. In this book, existence results will be based mainly on three fundamental theorems.

THEOREM 1.8. (*Lax–Milgram lemma.*) *Let* V *be a Hilbert space,* $a(\cdot,\cdot)$ *a continuous bilinear form on* $V\times V$, *and* $L(\cdot)$ *a continuous linear form on* V. *If* $a(\cdot,\cdot)$ *is* V*-elliptic, that is, if there exists a strictly positive real number* α *such that*

$$a(\mathbf{v},\mathbf{v}) \geq \alpha\|\mathbf{v}\|_V^2 \quad \forall \mathbf{v}\in V,$$

then there exists a unique $\mathbf{u}$ *in* V *depending linearly and continuously on* L *such that*

$$a(\mathbf{u},\mathbf{v}) = L(\mathbf{v}) \qquad \forall \mathbf{v}\in V.$$

□

THEOREM 1.9. (*Generalized Weierstrass theorem.*) *Let* V *be a reflexive Banach space. Let* K *be a nonempty, weakly sequentially closed subset of* V. *Let* $\Pi: K\to \mathbb{R}\cup\{+\infty\}$ *be weakly lower semicontinuous and attain a finite value on* K.[1] *If* K *is bounded or if* Π *is coercive on* K (*i.e.*, $\lim_{\|v\|\to+\infty}\Pi(\mathbf{v})=+\infty$), *then there exists* $\mathbf{u}$ *in* K *that minimizes* Π *on* K. □

Before stating the third theorem, it is necessary to introduce the following concepts of *convex analysis*. Let X be a topological vector space, X^* denote the dual space of X, and ϕ be a real convex function defined on X. Then, the *subgradient* $\partial\phi(x)$ of ϕ at x is defined by

$$\partial\phi(x) = \{x^*\in X^*, \langle x^*, y-x\rangle \leq \phi(y)-\phi(x)\ \forall y\in X\},$$

where the brackets $\langle\cdot,\cdot\rangle$ denote the duality pairing between X and X^*. The *Legendre–Fenchel transform* of ϕ is the real convex function ϕ^* defined on X^* by

$$\phi^*(x^*) = \sup_{y\in X}\{\langle x^*, y\rangle - \phi(y)\}.$$

[1] Functions from K to $\mathbb{R}\cup\{+\infty\}$ that attain finite values on K are said to be proper over K.

These concepts are detailed in Ekeland and Temam [1976, Chap. 3] and lead to the following theorem.

THEOREM 1.10. (*Convex analysis.*) *Let X and Y be two locally convex Hausdorff topological vector spaces in duality, respectively, with X^* and Y^*. Let Φ from $X \times Y$ into $\mathbb{R} \cup \{+\infty\}$ be convex lower semicontinuous and denote by Φ^* its Legendre-Fenchel transform. The primal problem $\mathcal{P}$ and the dual problem $\mathcal{P}^*$ are defined, respectively, by*

$$\mathcal{P}: \text{minimize } \Phi(x, 0) \quad \text{on } X;$$

$$\mathcal{P}^*: \text{maximize } \{-\Phi^*(0, y^*)\} \quad \text{on } Y^*.$$

In that framework the following properties are equivalent:

(i) $(0, \bar{y}) \in \partial\Phi(\bar{x}, 0)$;

(ii) $\Phi(\bar{x}, 0) = -\Phi^*(0, \bar{y})$;

(iii) $\bar{x}$ *is a solution of* $\mathcal{P}$, $\bar{y}$ *is a solution of* $\mathcal{P}^*$, *and* $\min \mathcal{P} = \max \mathcal{P}^*$.

Moreover, if there exists x_0 in X such that $\Phi(x_0, \cdot)$ is continuous at $y = 0$, and if $\Phi(x, 0)$ is bounded below on X, the problem $\mathcal{P}^*$ has a solution, and we have

$$\inf \mathcal{P} = \max \mathcal{P}^*. \qquad \square$$

We finally recall that a convex and lower semicontinuous function ϕ defined on a Banach space X is weakly lower semicontinuous, that is, satisfies

$$\phi(\bar{x}) \leq \liminf \phi(x_n) \quad \text{if } x_n \to \bar{x} \text{ weakly in } X.$$

2. Quasi-static viscoplasticity.

2.1. Mechanical background. In the notation of Chapter 1, quasi-static flows of incompressible viscoplastic solids in small strains were characterized by

(i) small strains: displacement fields and their spatial gradient remain small during deformation;

(ii) quasi-static flows: the inertia terms in the laws of force and moment balance are neglected;

(iii) incompressible viscoplastic materials: the constitutive laws reduce to the purely mechanical relations

$$\boldsymbol{\sigma} = \lambda^{1-s}(k\sqrt{2})^s|\dot{\mathbf{E}}|^{s-2}\dot{\mathbf{E}} - p\mathbf{Id} \quad \text{(Norton)}, \qquad 1 < s < +\infty, \tag{2.1}$$

$$\boldsymbol{\sigma} = (\sqrt{2}g|\dot{\mathbf{E}}|^{-1} + 2\mu)\dot{\mathbf{E}} - p\mathbf{Id} \quad \text{(Bingham)}, \tag{2.2}$$

together with the incompressibility constraint $\operatorname{tr}(\mathbf{E}) = 0$.

Recall that $\dot{\mathbf{E}}$ denotes the time-derivative of the linearized strain tensor $\mathbf{E}$, and $\boldsymbol{\sigma}$ denotes the second Piola–Kirchhoff stress tensor; λ, k, g, and μ are material constants that may depend on the material coordinates $\mathbf{x}$ of the body.

The compressible case corresponds to the simpler model in which the incompressibility constraint and the hydrostatic pressure p have been suppressed. In all these models, temperature has been implicitly eliminated (Chap. 1, § 4.3).

Both constitutive laws (2.1) and (2.2) are of the type

$$(\boldsymbol{\sigma}+p\mathbf{Id}) \in \partial\mathcal{D}_1(\mathbf{x}, \dot{\mathbf{E}}), \tag{2.3}$$

where $\partial\mathcal{D}_1(\mathbf{x}, \mathbf{E})$ represents the subgradient of the internal dissipation potential $\mathcal{D}_1(\mathbf{x}, \cdot)$ at deformation rate $\dot{\mathbf{E}}$, $\mathcal{D}_1$ being convex with respect to $\dot{\mathbf{E}}$. For Norton and Bingham materials, we have, respectively,

$$\mathcal{D}_1(\mathbf{x}, \dot{\mathbf{E}}) = \frac{\lambda^{1-s}}{s}(k(\mathbf{x})\sqrt{2})^s|\dot{\mathbf{E}}|^s,$$

$$\mathcal{D}_1(\mathbf{x}, \dot{\mathbf{E}}) = \sqrt{2}g(\mathbf{x})|\dot{\mathbf{E}}| + \mu(\mathbf{x})|\dot{\mathbf{E}}|^2.$$

Under the form (2.3), the viscoplastic constitutive laws still make sense for $\dot{\mathbf{E}} = 0$, whereas the original formulations do not.

The physical problem we want to solve consists of computing the velocity field inside a viscoplastic body when this body flows in a quasi-static way under the action of given body forces $\mathbf{f}$, given surface tractions $\mathbf{g}$ applied on Γ_2, and imposed velocity $\mathbf{v}_1$ given on $\Gamma_1 = \Gamma - \Gamma_2$ (Fig. 2.1). To solve this

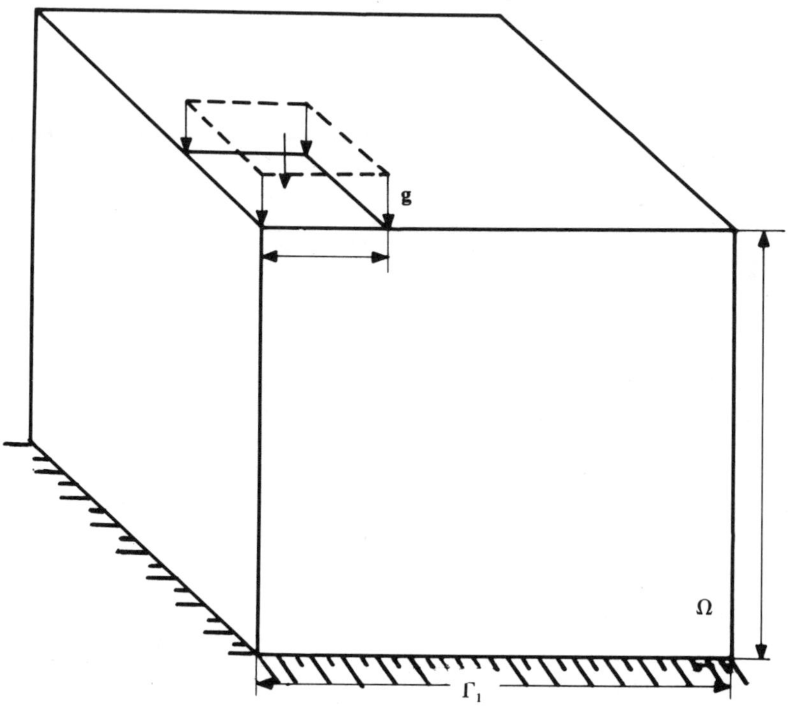

FIG. 2.1. *The physical problem of the computation of the velocity field inside a viscoplastic body given surface tractions* **g**.

problem, we will use only the constitutive law (2.3) and the virtual work theorem. In the quasi-static case in small strains, if we neglect the inertia terms, identify the two Piola–Kirchhoff stress tensors **t** and $\boldsymbol{\sigma}$, and use the symmetry of $\boldsymbol{\sigma}$, this theorem reduces to

$$\int_\Omega \boldsymbol{\sigma} \cdot \mathbf{D}(\mathbf{w})\, d\mathbf{x} = \int_\Omega \mathbf{f} \cdot \mathbf{w}\, d\mathbf{x} + \int_{\Gamma_2} \mathbf{g} \cdot \mathbf{w}\, da \qquad \forall \mathbf{w} \in V,$$

where $\mathbf{D}(\mathbf{w})$ is given by $\mathbf{D}(\mathbf{w}) = \frac{1}{2}(\nabla \mathbf{w} + \nabla \mathbf{w}^T)$ and V is a space of test functions with zero trace on Γ_1.

2.2. Formulation as a minimization problem. If we restrict the set V of test functions to the set X of divergence-free vector fields and if we take the constitutive laws into account, the above equation reduces formally to

$$\int_\Omega \partial \mathcal{D}_1(\mathbf{x}, \mathbf{D}(\mathbf{v})) \cdot \mathbf{D}(\mathbf{w})\, d\mathbf{x} = \int_\Omega \mathbf{f} \cdot \mathbf{w}\, d\mathbf{x} + \int_{\Gamma_2} \mathbf{g} \cdot \mathbf{w}\, da \qquad \forall \mathbf{w} \in X,$$

$\mathbf{v}$ being the unknown velocity. Because $\mathcal{D}_1$ is convex in **D**, this is formally equivalent to

$$\text{minimize } J(\mathbf{w}) \quad \text{on } X_1,$$

where X_1 is a set deduced from X by translation and where the functional $J(\cdot)$ is defined by

$$J(\mathbf{w}) = \int_\Omega \mathcal{D}_1(\mathbf{x}, \mathbf{D}(\mathbf{w}))\, d\mathbf{x} - \int_\Omega \mathbf{f} \cdot \mathbf{w}\, d\mathbf{x} - \int_{\Gamma_2} \mathbf{g} \cdot \mathbf{w}\, da. \tag{2.4}$$

This leads to the following mathematical formulation of quasi-static viscous flow problems for incompressible viscoplastic materials

(2.5) Minimize the rate of energy dissipation $J(\mathbf{w})$ over the set X_1 of kinematically admissible velocity fields,

with X_1 given by

$$X_1 = \{\mathbf{w} \in \mathbf{W}^{1,s}(\Omega),\ \operatorname{div} \mathbf{w} = 0,\ \mathbf{w} = \mathbf{v}_1 \text{ on } \Gamma_1\}. \tag{2.6}$$

For compressible materials, the problem is the same within the suppression of the divergence-free constraint in the definition of X_1.

2.3. Existence and characterization of solutions.

THEOREM 2.1. *Let Ω be open, bounded, and connected in $\mathbb{R}^3$ with Lipschitz-continuous boundary Γ. Suppose that the interior of Γ_1 is not empty and that $\mathbf{v}_1$ is the trace of a function of $\mathbf{W}^{1,s}(\Omega)$, $1<s<+\infty$, and satisfies*

$$\int_\Gamma \mathbf{v}_1 \cdot \boldsymbol{\nu}\, da = 0,$$

whenever $\Gamma_1=\Gamma$. Assume moreover that the external body forces $\mathbf{f}$ and surface tractions $\mathbf{g}$ are in $\mathbf{L}^{s^*}(\Omega)$ and $\mathbf{L}^{s^*}(\Gamma_2)$, respectively, with $ss^*=s+s^*$, and that the internal dissipation potential is measurable in $\mathbf{x}$, convex in $\mathbf{D}$, and satisfies

$$C_1|\mathbf{D}|^s \le \mathscr{D}_1(\mathbf{x}, \mathbf{D}) \le C_2 + C_3|\mathbf{D}|^s \tag{2.7}$$

almost everywhere in Ω and for any symmetric 3×3 matrix $\mathbf{D}$ with zero trace (note that the C_i are strictly positive real constants). Then, there exists a velocity field $\mathbf{v}$ that minimizes J over X_1. This solution is unique if $\mathscr{D}_1$ is strictly convex in $\mathbf{D}$.

Proof. To apply the generalized Weierstrass theorem, we first check that X_1 is a nonempty, weakly sequentially closed subset of $\mathbf{W}^{1,s}(\Omega)$.

For that purpose, consider the application h from $\mathbf{W}^{1,s}(\Omega)$ into $\mathbf{L}^s(\Gamma_1)\times L^s(\Omega)$ defined by

$$h(\mathbf{w}) = \{\mathrm{tr}\,(\mathbf{w}-\mathbf{v}_1), \mathrm{div}\,(\mathbf{w})\}.$$

From the assumed regularity of $\mathbf{v}_1$, the trace theorem, and the definition of the divergence operator, it can be seen that h is linear and continuous on $\mathbf{W}^{1,s}(\Omega)$. It is therefore weakly sequentially continuous on $\mathbf{W}^{1,s}(\Omega)$, and its kernel, X_1, is convex and weakly sequentially closed in $\mathbf{W}^{1,s}(\Omega)$. Moreover, by assumption, we can always find $\mathbf{w}_1$ in $\mathbf{W}^{1,s}(\Omega)$ with

$$\begin{cases} \mathbf{w}_1 = \mathbf{v}_1 & \text{on } \Gamma_1, \\ \displaystyle\int_\Gamma \mathbf{w}_1 \cdot \boldsymbol{\nu}\, d\mathbf{x} = 0. \end{cases}$$

Then, by applying Theorem 1.5 of § 1.2, there exists $\mathbf{w}$ in $\mathbf{W}^{1,s}(\Omega)$ with

$$\begin{cases} \mathbf{w} = 0 & \text{on } \Gamma, \\ \mathrm{div}\, \mathbf{w} = -\mathrm{div}\, \mathbf{w}_1 & \text{on } \Omega. \end{cases}$$

The sum $\mathbf{w}+\mathbf{w}_1$ belongs to X_1, which is therefore nonempty.

We now check the continuity of J on X_1. Since $\mathscr{D}_1(\mathbf{x}, \cdot)$ is convex and uniformly bounded on the space of symmetric real 3×3 matrices with zero trace, J is convex and bounded on the space of divergence-free vector fields of $\mathbf{W}^{1,s}(\Omega)$. Therefore, $J(\cdot)$ is continuous and thus weakly lower semicontinuous on this space, which contains X_1.

Finally, from the assumed coerciveness of $\mathscr{D}_1(\mathbf{x}, \cdot)$, we have

$$J(\mathbf{w}) \ge C_1\|\mathbf{D}(\mathbf{w})\|^s_{0,s,\Omega} - \|\mathbf{f}\|_{0,s^*,\Omega}\|\mathbf{w}\|_{0,s,\Omega} - \|\mathbf{g}\|_{0,s^*,\Gamma_2}\|\mathbf{w}\|_{0,s,\Gamma_2},$$

which, from the Korn's inequality, yields the coerciveness of J on X_1 for the $\mathbf{W}^{1,s}(\Omega)$ topology.

The existence of solutions follows, then, from a direct application of the generalized Weierstrass theorem on $\mathbf{W}^{1,s}(\Omega)$.

To check the uniqueness of the solution when the internal dissipation potential is strictly convex, consider a solution $\mathbf{v}$ of the above problem, an arbitrary element $\mathbf{w}$ of X_1 different from $\mathbf{v}$, and compute $J((\mathbf{w}+\mathbf{v})/2)$. From the strict convexity of $\mathcal{D}_1(\mathbf{x},\cdot)$ in $\mathbf{D}$, we have

$$J(\mathbf{w})+J(\mathbf{v})>2J\left(\frac{\mathbf{w}+\mathbf{v}}{2}\right).$$

Because X_1 is convex, and because $\mathbf{v}$ is a solution of (2.5), this implies

$$J(\mathbf{w})+J(\mathbf{v})>2J\left(\frac{\mathbf{w}+\mathbf{v}}{2}\right)\geq 2J(\mathbf{v}),$$

which gives

$$J(\mathbf{w})>J(\mathbf{v}) \qquad \forall \mathbf{w}\in X_1, \quad \mathbf{w}\neq\mathbf{v}. \qquad \square$$

Remark 2.1. For Norton materials, the internal dissipation potential is strictly convex and satisfies (2.7) with:

$$C_2=0, \quad C_1=\frac{\lambda^{1-s}}{s}(k_{\min}\sqrt{2})^s, \quad C_3=\frac{\lambda^{1-s}}{s}(k_{\max}\sqrt{2})^s.$$

The same holds for Bingham materials, with

$$C_1=\mu_{\min}, \quad C_2=\sqrt{2}g_{\min}, \quad C_3=(\sqrt{2}g_{\max}+\mu_{\max}), \quad s=2.$$

However, (2.7) is still valid, and, therefore, Theorem 2.1 is still applicable to materials associated with potentials that are nondifferentiable and not strictly convex such as

$$\mathcal{D}_1(\mathbf{x},\mathbf{D})=\frac{\lambda^{1-s}}{s}(\sigma_0(\mathbf{x}))^s \sup_{1\leq i\leq 3}\{|D_i|^s\},$$

where D_i denotes the eigenvalues of $\mathbf{D}$. This corresponds to Tresca-type viscoplasticity (see Chap. 4, § 6).

In its derivation, the variational formulation of viscoplastic flow problems was taken only as a formal equivalent of the constitutive laws and of the virtual work theorem. But, in fact, solutions of this minimization problem do rigorously satisfy these equations, as indicated by the next theorem.

THEOREM 2.2. *Under the assumptions of Theorem* 2.1, *for any minimizer* $\mathbf{v}$ *of the rate of energy dissipation* J *over the set* X_1 *of kinematically admissible velocity fields, there exists a deviatoric stress tensor field* $\boldsymbol{\sigma}_D$ *in* $[L^{s^*}(\Omega)]^{3\times 3}$ *and a hydrostatic pressure field* p *in* $L^{s^*}(\Omega)$ *with*:

$$
(2.8)\quad \begin{cases} \displaystyle\int_\Omega (\boldsymbol{\sigma}_D - p\mathbf{Id}) \cdot \mathbf{D}(\mathbf{w})\, d\mathbf{x} = \int_\Omega \mathbf{f} \cdot \mathbf{w}\, d\mathbf{x} + \int_{\Gamma_2} \mathbf{g} \cdot \mathbf{w}\, da \qquad \forall \mathbf{w} \in V \\ \qquad\qquad\qquad\qquad\qquad\qquad\qquad\qquad \textit{(virtual work theorem)}, \\ \boldsymbol{\sigma}_D \in \partial \mathcal{D}_1(\mathbf{x}, \mathbf{D}(\mathbf{v})) \qquad \textit{a.e. on } \Omega \qquad \textit{(constitutive law)}, \\ V = \{\mathbf{w} \in \mathbf{W}^{1,s}(\Omega),\ \mathbf{w} = 0 \textit{ on } \Gamma_1\}. \end{cases}
$$

Proof. The proof is based on the standard results of convex analysis recalled at the beginning of this chapter. In view of their application, we introduce the spaces

$$
X = \{\mathbf{w} \in \mathbf{W}^{1,s}(\Omega),\ \mathbf{w} = \mathbf{0} \text{ on } \Gamma_1,\ \operatorname{div} \mathbf{w} = 0\}, \qquad X^* \text{ the dual space of } X;
$$

$$
Y = \{\mathbf{D} \in [L^s(\Omega)]^{3\times 3},\ \mathbf{D}^T = \mathbf{D},\ \operatorname{tr} \mathbf{D} = 0\};
$$

$$
Y^* = \{\boldsymbol{\sigma} \in [L^{s^*}(\Omega)]^{3\times 3},\ \boldsymbol{\sigma}^T = \boldsymbol{\sigma},\ \operatorname{tr} \boldsymbol{\sigma} = 0\} \qquad \left(\frac{1}{s} + \frac{1}{s^*} = 1\right);
$$

the duality pairing

$$
\langle \mathbf{D}, \boldsymbol{\sigma} \rangle = \int_\Omega \sum_{ij} D_{ij} \sigma_{ij}\, d\mathbf{x};
$$

the functional

$$
\Phi(\mathbf{w}, \mathbf{D}) = \int_\Omega \mathcal{D}_1(\mathbf{x}, \mathbf{D}(\mathbf{v}) + \mathbf{D}(\mathbf{w}) - \mathbf{D})\, d\mathbf{x} - \int_\Omega \mathbf{f} \cdot (\mathbf{v} + \mathbf{w})\, d\mathbf{x} - \int_{\Gamma_2} \mathbf{g} \cdot (\mathbf{v} + \mathbf{w})\, da;
$$

and the orthogonal matrix

$$
\mathbf{Q} \in \mathbb{R}^{3\times 3}, \quad \mathbf{Q}\mathbf{Q}^T = \mathbf{Id}, \quad Q_{3j} = \frac{1}{\sqrt{3}}.
$$

We first verify that the bilinear form $\langle \mathbf{D}, \boldsymbol{\sigma} \rangle$ defines a *duality pairing* between Y and Y^*. To do this we observe that, for any $\mathbf{D}$ in Y, we can bijectively associate a tensor $\boldsymbol{\sigma}(\mathbf{D})$ in Y^* by setting

$$
\begin{cases} \sigma_{ij}(\mathbf{D}) = |D_{ij}|^{s-2} D_{ij} \quad \text{if } i \neq j, \\ \sigma_{ii}(\mathbf{D}) = \sum_j Q_{ji} \left| \sum_k Q_{jk} D_{kk} \right|^{s-2} \left(\sum_\ell Q_{j\ell} D_{\ell\ell} \right). \end{cases}
$$

Since $\mathbf{Q}^T\mathbf{Q} = \mathbf{Id}$ and $Q_{3j} = 1/\sqrt{3}$, we have

$$\begin{aligned}\sum_i \sigma_{ii}(\mathbf{D}) &= \sqrt{3}\sum_i Q_{3i}\sigma_{ii}\\ &= \sqrt{3}\sum_{i,j}(Q_{3i}Q_{ji})\left|\sum_k Q_{jk}D_{kk}\right|^{s-2}\left(\sum_\ell Q_{j\ell}D_{\ell\ell}\right)\\ &= \sqrt{3}\sum_j \delta_{3j}\left|\sum_k Q_{jk}D_{kk}\right|^{s-2}\left(\sum_\ell Q_{j\ell}D_{\ell\ell}\right)\\ &= \left(\frac{1}{\sqrt{3}}\right)^{s-2}|\mathrm{tr}\,\mathbf{D}|^{s-1} = 0,\end{aligned}$$

together with

$$\langle \mathbf{D}, \boldsymbol{\sigma}(\mathbf{D})\rangle = \int_\Omega \sum_{ij}\sigma_{ij}D_{ij}\,d\mathbf{x} = \sum_{i\neq j}\left\|D_{ij}\right\|_{0,s}^{s} + \sum_j\left\|\sum_\ell Q_{j\ell}D_{\ell\ell}\right\|_{0,s}^{s}.$$

This product can only vanish for $\mathbf{D} = \mathbf{0}$. Conversely, by inverting the above mapping, we are able to construct, for any $\boldsymbol{\sigma}$ in Y^*, a tensor $\mathbf{D}(\boldsymbol{\sigma})$ in Y such that the product $\langle \mathbf{D}(\boldsymbol{\sigma}), \boldsymbol{\sigma}\rangle$ only vanishes for $\boldsymbol{\sigma} = 0$. Since $\langle \mathbf{D}, \boldsymbol{\sigma}\rangle$ is continuous on $Y \times Y^*$, the fact that products only vanish for zero tensors implies that this bilinear form indeed defines a duality pairing between Y and Y^*.

In addition, it follows from (2.7) that $\Phi(\mathbf{0}, \mathbf{D})$ is convex and finite on Y and thus continuous on Y; moreover, $\Phi(\mathbf{w}, \mathbf{0})$, which is equal to $J(\mathbf{v}+\mathbf{w})$ by construction, is bounded below on X by $J(\mathbf{v})$ because $\mathbf{v}$ is assumed to be a minimizer of J over $X_1 = X + \mathbf{v}_1$.

Therefore, we can apply Theorem 1.10, stated in § 1.3 of this chapter, and prove the existence of a solution $(-\boldsymbol{\sigma}_D)$ in Y^* to the dual problem $\mathcal{P}^*$, which satisfies

$$\inf \mathcal{P} = \max \mathcal{P}^* = -\Phi^*(\mathbf{0}, -\boldsymbol{\sigma}_D).$$

The primal problem, which consists of the minimization of $\Phi(\mathbf{w}, \mathbf{0}) = J(\mathbf{v}+\mathbf{w})$ over $X = X_1 - \mathbf{v}_1$, also has a solution $\mathbf{w} = \mathbf{0}$, since $\mathbf{v}$ minimizes J over K. Applying Theorem 1.10 again, the fact that both the primal and the dual problem have a solution with $\inf \mathcal{P} = \max \mathcal{P}^*$ is equivalent to

$$(\mathbf{0}, -\boldsymbol{\sigma}_D) \in \partial\Phi(\mathbf{0}, \mathbf{0}).$$

By definition of the subgradient, this means that

$$(2.9)\quad \begin{cases} -\displaystyle\int_\Omega \boldsymbol{\sigma}_D\cdot\mathbf{D}\,d\mathbf{x} \le \int_\Omega \mathcal{D}_1(\mathbf{x}, \mathbf{D}(\mathbf{v}+\mathbf{w}) - \mathbf{D})\,d\mathbf{x} - \int_\Omega \mathbf{f}\cdot(\mathbf{v}+\mathbf{w})\,d\mathbf{x}\\ \qquad -\displaystyle\int_{\Gamma_2}\mathbf{g}\cdot(\mathbf{v}+\mathbf{w})\,da - J(\mathbf{v}) \qquad \forall(\mathbf{w}, \mathbf{D}) \in X \times Y.\end{cases}$$

Employing (2.9) for $\mathbf{w}$ arbitrary and $\mathbf{D} = \mathbf{D}(\mathbf{w})$, we get

$$-\int_\Omega \boldsymbol{\sigma}_D\cdot\mathbf{D}(\mathbf{w})\,d\mathbf{x} \le -\int_\Omega \mathbf{f}\cdot\mathbf{w}\,d\mathbf{x} - \int_{\Gamma_2}\mathbf{g}\cdot\mathbf{w}\,da \qquad \forall \mathbf{w} \in X.$$

If we replace $\mathbf{w}$ by $-\mathbf{w}$ and use the linearity of the above expression with respect to $\mathbf{w}$, this yields

$$\text{(2.10)} \qquad \int_\Omega \boldsymbol{\sigma}_D \cdot \mathbf{D}(\mathbf{w})\, d\mathbf{x} = \int_\Omega \mathbf{f} \cdot \mathbf{w}\, d\mathbf{x} + \int_{\Gamma_2} \mathbf{g} \cdot \mathbf{w}\, da \qquad \forall \mathbf{w} \in X.$$

Now, employing (2.9) with $\mathbf{w} = 0$ and $\mathbf{D} = -\mathbf{H}$ gives

$$\text{(2.11)} \int_\Omega \boldsymbol{\sigma}_D \cdot \mathbf{H}\, d\mathbf{x} \leq \int_\Omega \{\mathscr{D}_1(\mathbf{x}, \mathbf{D}(\mathbf{v}) + \mathbf{H}) - \mathscr{D}_1(\mathbf{x}, \mathbf{D}(\mathbf{v}))\}\, d\mathbf{x} \qquad \forall \mathbf{H} \in Y,$$

which can only hold if (Ekeland and Temam [1976, pp. 21, 271])

$$\boldsymbol{\sigma}_D \in \partial \mathscr{D}_1(\mathbf{x}, \mathbf{D}(\mathbf{v})) \qquad \text{a.e. in } \Omega.$$

The theorem is thus proved if, in (2.10), $\boldsymbol{\sigma}_D$ and X can be replaced by $(\boldsymbol{\sigma}_D - p\mathbf{Id})$ and V. For this purpose, recall that Corollary 1.6 (resp., Thm. 1.5 if $\Gamma_1 = \Gamma$) proves that the divergence operator is a closed linear surjection from V onto $L^s(\Omega)$ (resp., onto $L_0^s(\Omega)$, with

$$L_0^s(\Omega) = \left\{ g \in L^s(\Omega), \int_\Omega g\, d\mathbf{x} = 0 \right\} \Big).$$

By construction, the kernel of the divergence operator is X. From the closed-range theorem (Yoshida [1968, p. 205]), the transpose of the divergence operator is thus an isomorphism from $(L^s(\Omega))^* = L^{s^*}(\Omega)$ (resp., $(L_0^s(\Omega))^* = L^{s^*}(\Omega)/\mathbb{R}$) onto the orthogonal of X in V^*. But, from (2.10), it can be seen that the linear functional defined on V by

$$L(\mathbf{w}) = \int_\Omega \boldsymbol{\sigma}_D \cdot \mathbf{D}(\mathbf{w})\, d\mathbf{x} - \int_\Omega \mathbf{f} \cdot \mathbf{w}\, d\mathbf{x} - \int_{\Gamma_2} \mathbf{g} \cdot \mathbf{w}\, da$$

belongs precisely to this orthogonal. Therefore, there exists a unique element p in $L^{s^*}(\Omega)$ (resp., $L^{s^*}(\Omega)/\mathbb{R}$) depending linearly and continuously on L with

$$L(\mathbf{w}) = \langle \operatorname{div}^T(p), \mathbf{w} \rangle = \int_\Omega p \operatorname{div} \mathbf{w}\, d\mathbf{x} \qquad \forall \mathbf{w} \in V,$$

which are exactly the equilibrium equations we are looking for. □

Remark 2.2. In this theory we did not suppose $\mathscr{D}_1$ to the Frechet-differentiable on $\mathbb{R}^{3\times 3}$; rather, what was needed was the generalized subgradient of $\mathscr{D}_1$ which is defined, although possibly empty, for every convex function. The numerical techniques will have to respect this possible lack of differentiability.

Remark 2.3. Under the assumptions of Theorem 2.1, for an arbitrary potential $\mathscr{D}_1$, nothing can be said about the continuous dependence of the solution $(\mathbf{v}, \boldsymbol{\sigma}_D, p)$ as a function of the data $(\mathbf{v}_1, \mathbf{f}, \mathbf{g})$. Nevertheless, from the coerciveness of $\mathscr{D}_1$, and Korn's inequality, if $\mathbf{v}_1 = \mathbf{0}$, $\mathbf{v}$ satisfies

$$\|\mathbf{v}\|_{0,s} \leq C(\|\mathbf{f}\|_{0,s^*} + \|\mathbf{g}\|_{0,s^*,\Gamma_2})^{1/(s-1)}.$$

Moreover, if $\mathcal{D}_1$ is uniformly convex in $\mathbf{D}$, i.e., if $\mathcal{D}_1$ satisfies

$$\begin{cases}(\boldsymbol{\sigma}_1-\boldsymbol{\sigma}_2)\cdot(\mathbf{D}_1-\mathbf{D}_2)\geq\psi(|\mathbf{D}_1-\mathbf{D}_2|) & \forall\boldsymbol{\sigma}_i\in\partial\mathcal{D}_1(\mathbf{x},\mathbf{D}_i),\quad \mathbf{D}_i\in\mathbb{R}^{3\times3},\\ \mathbf{D}_i^T=\mathbf{D}_i,\quad \operatorname{tr}(\mathbf{D}_i)=0,\quad i=1,2,\end{cases}$$

with ψ being a strictly monotone function from $\mathbb{R}_+\to\mathbb{R}_+$ satisfying $\psi(0)=0$, then the solution $\mathbf{v}$ of (2.5) depends continuously on the data $(\mathbf{v}_1,\mathbf{f},\mathbf{g})$.

Remark 2.4. The demonstration techniques of Theorems 2.1 and 2.2 can also be used to prove the existence and uniqueness of the solution $\{\mathbf{u},p\}$ of the problem

$$\text{(2.12)}\quad\begin{cases}\text{Find } \{\mathbf{u},p\} \text{ in } \mathbf{H}^1(\Omega)\times L_0^2(\Omega) \text{ such that}\\ \displaystyle\int_\Omega\{\alpha\mathbf{u}\cdot\mathbf{v}+\nu\nabla\mathbf{u}\cdot\nabla\mathbf{v}-p\operatorname{div}\mathbf{v}\}\,d\mathbf{x}=\langle\mathbf{f},\mathbf{v}\rangle \qquad \forall\mathbf{v}\in\mathbf{H}_0^1(\Omega);\\ \operatorname{div}\mathbf{u}=0,\qquad \mathbf{u}|_\Gamma=\mathbf{u}_1|_\Gamma,\end{cases}$$

with $\mathbf{f}$ an arbitrary element of $\mathbf{H}^{-1}(\Omega)=(\mathbf{H}_0^1(\Omega))^*$.

Indeed, considering $\mathbf{f}$ in $\mathbf{H}^{-1}(\Omega)$ and adding the term $\frac{1}{2}\alpha|v|^2$ to the functional $J(\cdot)$ does not affect the validity of the two proofs above if we set

$$X=\{\mathbf{w}\in\mathbf{H}_0^1(\Omega),\ \operatorname{div}\mathbf{w}=0\},$$

$$Y=\{\mathbf{D}\in(\mathbf{L}^2(\Omega))^{N\times N},\ \operatorname{tr}(\mathbf{D})=0\},$$

$$\phi(\mathbf{w},\mathbf{D})=\frac{1}{2}\int_\Omega\{\alpha|\mathbf{v}+\mathbf{w}|^2+\nu|\nabla(\mathbf{v}+\mathbf{w})-\mathbf{D}|^2\}\,d\mathbf{x}-\langle\mathbf{f},\mathbf{v}+\mathbf{w}\rangle.$$

Problems such as (2.12) are important to our discussion for two reasons. First, they naturally appear in the time-discretization of the Navier–Stokes equations (the equations that model the evolution of a Newtonian viscous fluid flowing inside a given domain). Second, we will use such models to introduce, in Chapter 3, the numerical decomposition techniques presented in this monograph. A further discussion of problems similar to (2.12) can be found in Tartar [1978], Temam [1977], or Girault and Raviart [1986].

3. Time-dependent flows of viscoplastic fluids. Let us consider an incompressible viscoplastic fluid flowing inside a given domain $\boldsymbol{\phi}(\Omega)$ (its present configuration). To simplify the notation used in this section, we will continue to denote its present configuration $\boldsymbol{\phi}(\Omega)$ by Ω and its boundary by Γ. The problem consists of finding, at any positive time, the velocity $\bar{\mathbf{v}}$ of the fluid when its initial value $\mathbf{v}_0$, its trace $\mathbf{v}_1$ on Γ_1, the applied body forces $\bar{\mathbf{f}}$, and the applied surface tractions $\bar{\mathbf{g}}$ on $\Gamma_2=\Gamma-\Gamma_1$ are known (see Fig. 2.1).

By employing the virtual work theorem in the configuration $\boldsymbol{\phi}(\Omega)$ (Chap. 1, (5.3)) and taking into account the definition of a viscoplastic fluid (Chap. 1,

(5.4)), we obtain the formal equation satisfied by the unknown velocity $\bar{\mathbf{v}}$,

$$\begin{aligned}\int_\Omega \bar{\rho}\left(\frac{\partial \bar{\mathbf{v}}}{\partial t}+\sum_{i=1}^N \bar{v}_i \frac{\partial \bar{\mathbf{v}}}{\partial \bar{x}_i}\right)\cdot \mathbf{w}\, d\bar{\mathbf{x}}+\int_\Omega \partial \mathscr{D}_1(\mathbf{D}(\bar{\mathbf{v}}))\cdot \mathbf{D}(\mathbf{w})\, d\bar{\mathbf{x}} \\ =\int_\Omega \bar{\mathbf{f}}\cdot \mathbf{w}\, d\bar{\mathbf{x}}+\int_{\Gamma_2} \bar{\mathbf{g}}\cdot \mathbf{w}\, d\bar{a}.\end{aligned} \tag{3.1}$$

This equation holds for almost all positive times t and for any divergence-free function $\mathbf{w}$ with zero trace on Γ_1. Usually, in (3.1), the convection terms $\bar{\rho}\bar{v}_i\, \partial\bar{\mathbf{v}}/\partial\bar{x}_i$ are small compared with the other terms and can be neglected. On the other hand, (3.1) is only formal because $\partial\mathscr{D}_1$ is a subgradient and not a function.

If we neglect the convection terms and if we use the correct mathematical definition of $\partial\mathscr{D}_1$, (3.1) leads to the following variational inequality, which corresponds to the mathematical formulation of the time-dependent flow problems of incompressible viscoplastic fluids.

$$\left\{\begin{aligned}&\text{Find } \bar{\mathbf{v}}:\mathbb{R}^+\to \mathbf{W}^{1,s}(\Omega) \text{ such that, almost everywhere in } t \text{ we have:}\\ &\operatorname{div}\bar{\mathbf{v}}(t)=0 \quad \text{on } \Omega;\\ &\bar{\mathbf{v}}(t)=\mathbf{v}_1(t) \quad \text{on } \Gamma_1;\\ &\int_\Omega \bar{\rho}\frac{\partial \bar{\mathbf{v}}}{\partial t}(t)\cdot(\mathbf{w}-\bar{\mathbf{v}}(t))\, d\bar{\mathbf{x}}+\int_\Omega [\mathscr{D}_1(\mathbf{D}(\mathbf{w}))-\mathscr{D}_1(\mathbf{D}(\bar{\mathbf{v}}(t)))]\, d\bar{\mathbf{x}}\\ &\qquad \ge \int_\Omega \bar{\mathbf{f}}(t)\cdot(\mathbf{w}-\bar{\mathbf{v}}(t))\, d\bar{\mathbf{x}}+\int_{\Gamma_2} \bar{\mathbf{g}}(t)\cdot(\mathbf{w}-\bar{\mathbf{v}}(t))\, d\bar{a} \qquad \forall \mathbf{w}\in X;\\ &\bar{\mathbf{v}}(0)=\mathbf{v}_0, \qquad \operatorname{div}\mathbf{v}_0=0.\end{aligned}\right. \tag{3.2}$$

Here, $\mathbf{v}_1(t)$, $\bar{\mathbf{f}}(t)$, and $\bar{\mathbf{g}}(t)$ are known elements of $\mathbf{W}^{1,s}(\Omega)$, $\mathbf{L}^{s^*}(\Omega)$, and $\mathbf{L}^{s^*}(\Gamma_2)$, respectively, and we have

$$X=\{\mathbf{w}\in \mathbf{W}^{1,s}(\Omega), \operatorname{div}\mathbf{w}=0, \mathbf{w}=0 \text{ on } \Gamma_1\}.$$

The formulation (3.2) is still incomplete because it does not specify the regularity of $\bar{\mathbf{v}}$ as a function of time. This regularity will depend on the dissipation potential $\mathscr{D}_1$ and on the data $\mathbf{v}_0$, $\mathbf{v}_1$, $\bar{\mathbf{f}}$, and $\bar{\mathbf{g}}$.

For example, for a Bingham fluid, if $\mathbf{v}_0$, $\mathbf{v}_1$, $\bar{\mathbf{f}}$, and $\bar{\mathbf{g}}$ belong, respectively, to $\mathbf{L}^2(\Omega)$, $L^2(0, t_1; \mathbf{H}^1(\Omega))$, $L^2(0, t_1; \mathbf{L}^2(\Omega))$, $L^2(0, t_1; \mathbf{L}^2(\Gamma_2))$, then from Duvaut and Lions [1972], the solution $\bar{\mathbf{v}}$ between times 0 and t_1 satisfies

$$\begin{aligned}(\bar{\mathbf{v}}-\mathbf{v}_1)\in L^2(0, t_1; X), \qquad \frac{\partial \bar{\mathbf{v}}}{\partial t}\in L^2(0, t_1; X^*),\\ X=\{\mathbf{w}\in \mathbf{H}^1(\Omega), \mathbf{w}=0 \text{ on } \Gamma_1, \operatorname{div}\mathbf{w}=0\}.\end{aligned} \tag{3.3}$$

In that framework, existence and uniqueness of solutions can be proved (see Duvaut and Lions [1972]).

The regularity of $\bar{\mathbf{v}}$ as a function of t has little influence on the numerical methods used for the solution of (3.2).[2] Indeed, most numerical methods restrict (3.2) to times $n\Delta t$, $n \in \mathbb{N}$, and replace the time-derivative $\partial\bar{\mathbf{v}}/\partial t$ by the finite difference

$$\frac{\bar{\mathbf{v}}^{n+1}-\bar{\mathbf{v}}^{n}}{\Delta t}$$

where, for all n, $\bar{\mathbf{v}}^n$ is an approximation of $\bar{\mathbf{v}}(n\Delta t)$. Thus, these methods reduce (3.2) to a finite sequence of quasi-static problems with unknown $\bar{\mathbf{v}}^n$ in a similar manner to the problems treated in § 2.

4. Elastoviscoplasticity in small strains.

4.1. Mechanical equations. Let us consider a continuous body made of an elastoviscoplastic material that occupies a domain $\Omega \subset \mathbb{R}^3$ in its reference configuration, that is fixed on the part Γ_1 of the boundary Γ of Ω, and that is subjected to given body forces $\mathbf{f}$ and surface tractions $\mathbf{g}$ applied on the part $\Gamma_2 = \Gamma - \Gamma_1$ of its boundary. This body can, for example, represent very well a concrete dam anchored in bedrock and subjected to its own weight $\mathbf{f} = \rho\boldsymbol{\gamma}$ and to hydrostatic pressure forces $\mathbf{g} = -p\boldsymbol{\nu}$ (Fig. 4.1).

The problem consists, then, of determining in the time interval $[0, t_1]$ and for given initial values $\mathbf{u}_0$ and $\boldsymbol{\sigma}_0$ the history $\mathbf{u}(\cdot, t)$ and $\boldsymbol{\sigma}(\cdot, t)$ of the displacement and of the second Piola–Kirchhoff stress tensor inside this body. If, as is reasonable, we neglect the inertial effects, we assume that the strains and displacements remain small, and we suppose that the temperature fields are given, then, as seen in Chapter 1, this problem is governed by the virtual work theorem (Chap. 1, Thm. 3.2) and by the elastoviscoplastic constitutive law (Chap. 1, (4.9)). Employing the virtual work theorem as we did in § 2.1, these governing equations become

$$(4.1) \qquad \int_\Omega \boldsymbol{\sigma} \cdot \mathbf{D}(\mathbf{w})\, d\mathbf{x} = \int_\Omega \mathbf{f} \cdot \mathbf{w}\, d\mathbf{x} + \int_{\Gamma_2} \mathbf{g} \cdot \mathbf{w}\, da \quad \forall \mathbf{w} \in V \quad \text{a.e. in } (0, t_1),$$

$$(4.2) \qquad (\dot{\mathbf{E}} - \mathbf{A}^{-1}\dot{\boldsymbol{\sigma}}) \in \frac{\partial}{\partial \boldsymbol{\sigma}}\left[\frac{\lambda}{q}|j_c(\mathbf{x}, \boldsymbol{\sigma})|^q\right] \qquad \text{a.e. in } \Omega \times (0, t_1),$$

where $\mathbf{D}(\mathbf{w})$ is given by $\mathbf{D}(\mathbf{w}) = \frac{1}{2}(\nabla\mathbf{w} + \nabla\mathbf{w}^T)$ and V is a space of test functions with zero trace on Γ_1. Recall, in addition, that an overdot denotes a partial differentiation with respect to time; that $\mathbf{E}$ and $\mathbf{A}$ represent, respectively, the linearized strain tensor $\mathbf{E} = \mathbf{D}(\mathbf{u})$ and the fourth-order elasticity tensor; that λ

[2] It has, however, direct consequences on the truncation error associated with this time-discretization scheme.

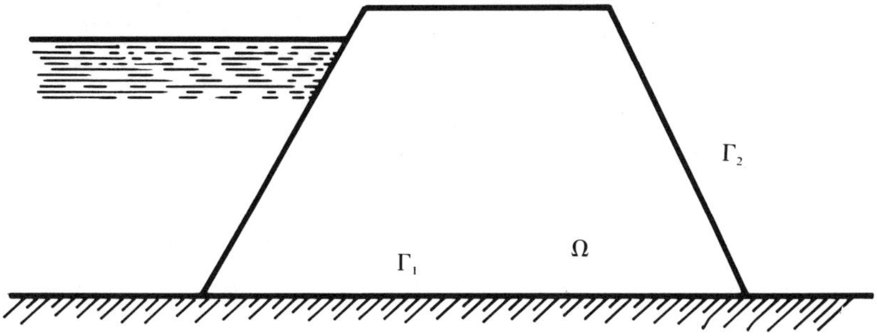

FIG. 4.1. *The physical problem of computing displacement and stress field inside an elastoviscoplastic dam given external forces.*

and q are material constants measuring its viscosity; and that $j_c(\mathbf{x}, \boldsymbol{\sigma})$ is the gauge function of the set $\mathbf{C}(\mathbf{x})$ of locally admissible elastic stresses, that is

$$j_c(\mathbf{x}, \boldsymbol{\sigma}) = \inf\left\{\mu > 0, \frac{\boldsymbol{\sigma}}{\mu} \in \mathbf{C}(\mathbf{x}) \subset \mathbb{R}^9_{\text{sym}}\right\},$$

with $\mathbb{R}^9_{\text{sym}}$ the set of second-order symmetric tensors operating on $\mathbb{R}^3$.

In (4.2), plastically incompressible materials were characterized by sets $\mathbf{C}(\mathbf{x})$ (or, equivalently, by functions j_c) which depend only on the deviatoric parts $\boldsymbol{\sigma}_D = (\boldsymbol{\sigma} - \text{tr}\,(\boldsymbol{\sigma})\mathbf{Id}/3)$ of the stress tensors. Among these materials, the best known are the *linearly viscoelastic Maxwell models,* for which the constitutive law (4.2) reduces to

$$\dot{E}_{ij} - \frac{(1+\nu)}{E}\dot{\sigma}_{ij} + \frac{\nu}{E}\text{tr}\,(\dot{\boldsymbol{\sigma}})\delta_{ij} = \lambda(k\sqrt{2})^{-2}(\boldsymbol{\sigma}_D)_{ij}. \tag{4.3}$$

4.2. Variational formulation. For $1 < s < +\infty$, let us introduce the spaces

$$\begin{cases} \Sigma^s = \{\boldsymbol{\tau}\colon \Omega \to R^9_{\text{sym}}, \text{tr}\,(\boldsymbol{\tau}) \in L^2(\Omega), \boldsymbol{\tau}_D \in (L^s(\Omega))^9, \|\boldsymbol{\tau}\|_{\Sigma^s} = \|\text{tr}\,(\boldsymbol{\tau})\|_{0,2} \\ \qquad + \|\boldsymbol{\tau}_D\|_{0,s}\} \qquad \text{(plastically incompressible case)}, \\ \Sigma^s = \{\boldsymbol{\tau}\colon \Omega \to R^9_{\text{sym}}, \boldsymbol{\tau} \in (L^s(\Omega))^9\} \qquad \text{(plastically compressible case)}, \\ V^s = \{\mathbf{w}, \mathbf{w} = 0 \text{ on } \Gamma_1, \mathbf{D}(\mathbf{w}) \in \Sigma^s, \|\mathbf{w}\|_{V^s} = \|\mathbf{D}(\mathbf{w})\|_{\Sigma^s}\}, \end{cases} \tag{4.4}$$

$\|\mathbf{w}\|_{V^s}$ being a norm on V^s from Theorems 1.5 and 1.7.

If we now employ the virtual work theorem with $V = V^{m^*}$, $m^* = m/(m-1)$, and $m = \max\,(2, q)$, and if we integrate the constitutive law (4.2) over Ω, assuming $\boldsymbol{\sigma}(\cdot, t)$ in Σ^m and taking into account the definition of the subgradient, we then obtain the following variational formulation for the quasistatic evolution problem in elastoviscoplasticity.

$$
(4.5)\quad \begin{cases}
\text{Find } \{\mathbf{u}, \boldsymbol{\sigma}\} : [0, t_1] \to V^{m^*} \times \Sigma^m \text{ such that} \\
\displaystyle\int_\Omega \boldsymbol{\sigma} \cdot \mathbf{D}(\mathbf{w})\, d\mathbf{x} = \int_\Omega \mathbf{f} \cdot \mathbf{w}\, d\mathbf{x} + \int_{\Gamma_2} \mathbf{g} \cdot \mathbf{w}\, da \quad \forall \mathbf{w} \in V^{m^*} \quad \text{a.e. in } (0, t_1); \\
\displaystyle\int_\Omega (\dot{\mathbf{E}}(\mathbf{u}) - \mathbf{A}^{-1}\dot{\boldsymbol{\sigma}}) \cdot (\boldsymbol{\tau} - \boldsymbol{\sigma})\, d\mathbf{x} \\
\qquad \displaystyle\le \int_\Omega \frac{\lambda}{q} \{|j_c(\mathbf{x}, \boldsymbol{\tau})|^q - |j_c(\mathbf{x}, \boldsymbol{\sigma})|^q\}\, d\mathbf{x} \quad \forall \boldsymbol{\tau} \in \Sigma^m \quad \text{a.e. in } (0, t_1); \\
\mathbf{u}(\mathbf{x}, 0) = \mathbf{u}_0(\mathbf{x}), \qquad \boldsymbol{\sigma}(\mathbf{x}, 0) = \boldsymbol{\sigma}_0(\mathbf{x}).
\end{cases}
$$

4.3. Existence result (incompressible case). The derivation of the variational formulation (4.5) is rather natural. Moreover, to further justify (4.5), we state without proof an existence result proved in Blanchard and Le Tallec [1986]. Introducing

$$
\begin{cases}
\displaystyle\psi_t(\boldsymbol{\tau}) = \int_\Omega \frac{1}{q} \lambda(\mathbf{x}, t) |j_c(\mathbf{x}, \boldsymbol{\tau})|^q\, d\mathbf{x}, \\
\displaystyle S^m(t) = \Big\{ \boldsymbol{\tau} \in \Sigma^m, \int_\Omega \boldsymbol{\tau} \cdot \mathbf{D}(\mathbf{w})\, d\mathbf{x} \\
\qquad\qquad \displaystyle = \int_\Omega \mathbf{f}(\mathbf{x}, t) \cdot \mathbf{w}\, d\mathbf{x} + \int_{\Gamma_2} \mathbf{g}(\mathbf{x}, t) \cdot \mathbf{w}\, da \ \forall \mathbf{w} \in V^{m^*} \Big\}, \\
I_t^m : \Sigma^m \to \mathbb{R} \cup \{+\infty\} = \text{indicator function of } S^m(t) \\
\qquad (\text{i.e., } I_t^m(\boldsymbol{\tau}) = 0 \text{ if } \boldsymbol{\tau} \in S^m(t),\ I_t^m(\boldsymbol{\tau}) = +\infty \text{ if not}),
\end{cases}
$$

we have the following theorem.

THEOREM 4.1. *Suppose the following hold.*

(i) Ω *is open, bounded, and connected in* $\mathbb{R}^3$ *with a Lipschitz-continuous boundary* Γ, *and that* $\Gamma_1 \subset \Gamma$ *and is smooth and not empty*;

(ii) $\lambda = \lambda(\mathbf{x}, t) \in W^{1,1}(0, t_1; L^\infty(\Omega))$, $\lambda(\mathbf{x}, t) \ge \bar{\lambda} > 0$ *a.e. in* $\Omega \times (0, t_1)$;

(iii) $j_c = j_c(\mathbf{x}, \boldsymbol{\sigma}_D)$ *measurable in* $\mathbf{x}$;

(iv) *there exist two strictly positive numbers* C_1 *and* C_2 *such that*

$$
\begin{cases}
|\boldsymbol{\sigma}_D| \le C_1 \Rightarrow \boldsymbol{\sigma}_D \in \mathbf{C}(\mathbf{x}), \\
\boldsymbol{\sigma}_D \in \mathbf{C}(\mathbf{x}) \Rightarrow |\boldsymbol{\sigma}_D| \le C_2;
\end{cases}
$$

(v) $\mathbf{A} = \mathbf{A}(\mathbf{x}, t) \in W^{1,\infty}(0, t_1; (L^\infty(\Omega))^{81})$, *and there exists* $C_3 > 0$ *such that*

$$
\mathbf{E} \cdot \mathbf{A}(\mathbf{x}, t)\mathbf{E} > C_3 |\mathbf{E}|^2 \quad \forall \mathbf{E} \in \mathbb{R}^{3\times 3} \quad \textit{a.e. in } \Omega \times (0, t_1);
$$

(vi) $\{\mathbf{f}, \mathbf{g}\} \in W^{1,m}(0, t_1; \mathbf{L}^m(\Omega) \times \mathbf{L}^m(\Gamma_2))$;

(vii) $\{\mathbf{u}_0, \boldsymbol{\sigma}_0\} \in V^{m^*} \times S^m(0)$.

Then, the variational problem (4.5) *has a solution* $\{\mathbf{u}, \boldsymbol{\sigma}\}$ *in* $W^{1,m^*}(0, t_1; V^{m^*}) \times W^{1,2}(0, t_1; \Sigma^2)$ *with* $\boldsymbol{\sigma}_D$ *in* $L^m(0, t_1; \Sigma^m)$. *Moreover,* $\boldsymbol{\sigma}$ *is the unique solution of the stress evolution problem*

$$(4.6)\qquad \begin{cases} -\mathbf{A}^{-1}\dot{\boldsymbol{\sigma}} \in \{\partial\psi_t(\boldsymbol{\sigma}) + \partial I_t^m(\boldsymbol{\sigma})\} & \text{in } \Sigma^{m^*} \quad \text{a.e. in } (0, t_1), \\ \boldsymbol{\sigma}(\mathbf{x}, 0) = \boldsymbol{\sigma}_0(\mathbf{x}). \end{cases}$$

□

Remark 4.1. The decomposition (4.6) is the theoretical background upon which the application of augmented Lagrangian techniques to elastoviscoplasticity is based (see Chap. 4). In (4.6), the duality pairing $\langle \mathbf{E}, \boldsymbol{\tau} \rangle$ corresponds to the product $\int_\Omega \mathbf{E} \cdot \boldsymbol{\tau}\, d\mathbf{x}$.

Remark 4.2. Theorem 4.1 applies to linearly viscoelastic Maxwell models. More generally, it applies to the so-called *Maxwell–Norton materials* whose constitutive laws are given by

$$\dot{E}_{ij} - \frac{1+\nu}{E}\dot{\sigma}_{ij} + \frac{\nu}{E}\operatorname{tr}(\dot{\boldsymbol{\sigma}})\delta_{ij} = \frac{\lambda}{q}(k\sqrt{2})^{-q}|\boldsymbol{\sigma}_D|^{q-1}(\boldsymbol{\sigma}_D)_{ij}.$$

It can also be extended to plastically compressible materials simply by replacing $\boldsymbol{\sigma}_D$ with $\boldsymbol{\sigma}$ in assumptions (iii) and (iv).

5. Static finite elasticity.

5.1. The physical problem. The physical problem consists of the determination of the final equilibrium position $\boldsymbol{\phi}(\mathbf{x}) = \mathbf{x} + \mathbf{u}(\mathbf{x})$ of any particle $\mathbf{x}$ of a hyperelastic body that occupies a set Ω in its reference configuration and that is subjected to a given distribution of external loads and imposed displacements. The body forces are of intensity $\mathbf{f}$ per unit volume in the reference configuration, and surface tractions $\mathbf{g}$, measured per unit area in the reference configuration, are prescribed on a portion Γ_2 of the boundary Γ of Ω. The displacement takes on the prescribed values $\mathbf{u}_1$ on the complementary part Γ_1 of Γ_2 in Γ (Fig. 5.1).

The equations that model this physical problem can be obtained simply by taking into account the specific form of the constitutive laws in finite hyperelasticity (Chap. 1, § 6) and by using the virtual work theorem to obtain the laws of force and moment balance. They are

$$(5.1)\qquad \int_\Omega \mathbf{t} \cdot \nabla\mathbf{v}\, d\mathbf{x} = \int_\Omega \mathbf{f} \cdot \mathbf{v}\, d\mathbf{x} + \int_{\Gamma_2} \mathbf{g} \cdot \mathbf{v}\, da \qquad \forall \mathbf{v} \in V$$

virtual work theorem
(static case);

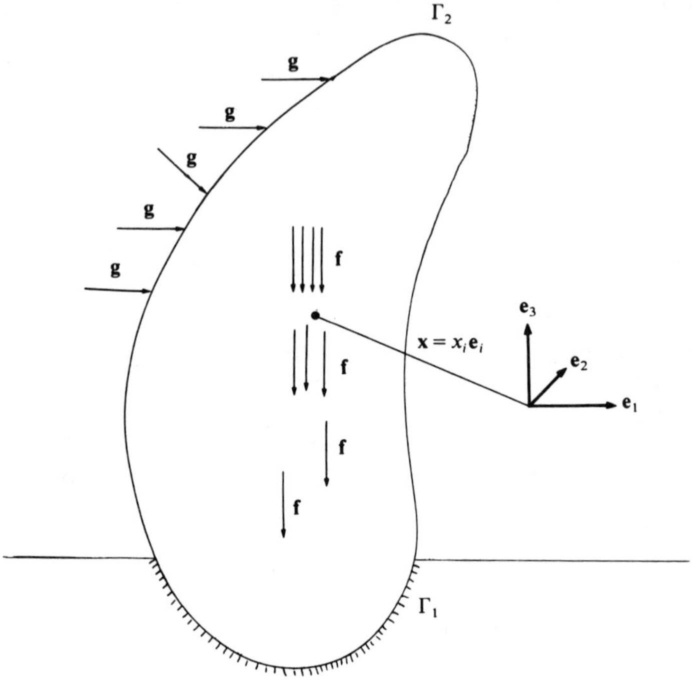

FIG. 5.1. *Body in its reference configuration.*

(5.2)	$\mathbf{t}=\rho\dfrac{\partial \mathscr{W}}{\partial \mathbf{F}}(\mathbf{x},\mathbf{F}),$	constitutive laws (compressible materials),
(5.3)	$\begin{cases}\mathbf{t}=\rho\dfrac{\partial \mathscr{W}}{\partial \mathbf{F}}(\mathbf{x},\mathbf{F})-p\dfrac{\partial(\det \mathbf{F})}{\partial \mathbf{F}}\\ \det \mathbf{F}=1\end{cases}$	constitutive laws (incompressible materials).

Moreover, the displacement $\mathbf{u}$ must be kinematically admissible, that is, the deformation $\boldsymbol{\phi}$ must be one to one, and we must have

$$(5.4)\qquad \begin{cases}\det \mathbf{F}>0 & (\boldsymbol{\phi} \text{ is orientation-preserving}),\\ \mathbf{u}=\mathbf{u}_1 \text{ on } \Gamma_1 & (\text{boundary conditions on } \mathbf{u}).\end{cases}$$

Recall that $\mathbf{t}$ represents the first Piola–Kirchhoff stress tensor, which characterizes the contact forces (in the present configuration) applied through a surface which is defined in the reference configuration. Moreover, ρ denotes the mass density in the reference configuration, $\mathbf{F}$ the deformation gradient ($\mathbf{F} = \nabla\boldsymbol{\phi} = \mathbf{Id} + \nabla\mathbf{u}$), and $\mathcal{W}$ the free-energy potential which, in hyperelasticity, is a function of $\mathbf{x}$ and $\mathbf{F}(\mathbf{x})$ alone. For incompressible materials, the effect of incompressibility manifests itself in the presence of the term

$$p\frac{\partial \det \mathbf{F}}{\partial \mathbf{F}}$$

in the constitutive equation. Here, p denotes the hydrostatic pressure.

As is customary in finite elasticity, we will drop the global invertibility requirement on $\boldsymbol{\phi}$ and require only local invertibility as imposed by (5.4). This simplifies the problem but can be justified only by checking a posteriori the global invertibility of solutions of the resulting simplified problem. In certain cases, we may get inadmissible solutions (Fig. 5.2).

5.2. Weak formulations. A natural way to get a variational formulation of the equilibrium equations (5.1), (5.2), and (5.4) associated with compressible hyperelastic materials is simply to eliminate the stress tensor $\mathbf{t}$ between the virtual work theorem (5.1) and the constitutive law (5.2). This leads to a

Reference configuration Inadmissible deformation (locally invertible)

FIG. 5.2. *An example of a physically inadmissible solution of* (5.1)–(5.4).

problem of the form

$$(5.5)\qquad \left\{ \begin{array}{l} \text{Find } \mathbf{u} \in K \text{ such that, for any } \mathbf{v} \text{ in } V, \text{ we have} \\ \displaystyle\int_\Omega \rho \frac{\partial \mathcal{W}(\mathbf{x}, \mathbf{Id}+\nabla\mathbf{u})}{\partial \mathbf{F}} \cdot \nabla\mathbf{v}\, d\mathbf{x} = \int_\Omega \mathbf{f}\cdot\mathbf{v}\, d\mathbf{x} + \int_{\Gamma_2} \mathbf{g}\cdot\mathbf{v}\, da, \end{array} \right.$$

which is completely specified as soon as the set K of kinematically admissible displacement fields and the space V of test functions (virtual displacements) are defined. Adding the constraint (5.4) and proceeding by analogy with the linear case, we may choose

$$(5.6)\qquad V = \{\mathbf{v} \in \mathbf{W}^{1,s}(\Omega), \mathbf{v} = \mathbf{0} \text{ on } \Gamma_1\},$$

$$(5.7)\qquad K = \{\mathbf{v} \in \mathbf{W}^{1,s}(\Omega), \mathbf{v} = \mathbf{u}_1 \text{ on } \Gamma_1, \det(\mathbf{Id}+\nabla\mathbf{v}) > 0 \text{ a.e. on } \Omega\}.$$

The superscript s is such that the integral

$$\int_\Omega \rho \frac{\partial \mathcal{W}}{\partial \mathbf{F}}(\mathbf{Id}+\nabla\mathbf{u}) \cdot \nabla\mathbf{v}\, d\mathbf{x}$$

makes sense for any $\mathbf{u}$ in K and any $\mathbf{v}$ in V.

The variational equation (5.5), together with definitions (5.6) and (5.7) for K and V, corresponds to the classical weak formulation of equilibrium problems in compressible finite elasticity.

The same technique, applied to incompressible materials, leads to problems of the form

$$\left\{ \begin{array}{l} \text{Find } \mathbf{u} \text{ in } K, p \text{ in } P \text{ such that, for any } \mathbf{v} \text{ in } V, \text{ we have} \\ \displaystyle\int_\Omega \left\{ \rho \frac{\partial \mathcal{W}}{\partial \mathbf{F}}(\mathbf{x}, \mathbf{Id}+\nabla\mathbf{u}) - p \frac{\partial \det}{\partial \mathbf{F}}(\mathbf{Id}+\nabla\mathbf{u}) \right\} \cdot \nabla\mathbf{v}\, d\mathbf{x} = \int_\Omega \mathbf{f}\cdot\mathbf{v}\, d\mathbf{x} + \int_{\Gamma_2} \mathbf{g}\cdot\mathbf{v}\, da. \end{array} \right.$$

But, in such a formulation, the specification of the set P of hydrostatic pressures is an unnecessarily difficult problem. To avoid this difficulty, it suffices to realize that, formally, the expression $p\partial(\det \mathbf{F})/\partial\mathbf{F}$ is the general form of an operator $\mathbf{A}$ acting on $(L^s(\Omega))^{3\times 3}$, an operator whose kernel contains all elements $\mathbf{G}$ such that

$$\frac{\partial \det}{\partial \mathbf{F}}(\mathbf{Id}+\nabla\mathbf{u}) \cdot \mathbf{G} = [\operatorname{adj}(\mathbf{Id}+\nabla\mathbf{u})]^T \cdot \mathbf{G} = 0 \qquad \text{a.e. on } \Omega.$$

Here, adj $\mathbf{F}$ represents the adjugate of matrix $\mathbf{F}$ (the transpose of the matrix of cofactors). The weak formulation of equilibrium problems in incompressible hyperelasticity then becomes

(5.8)
$$\boxed{\begin{aligned}&\text{Find } \mathbf{u}\in K,\ \mathbf{A}\in (L^{s^*}(\Omega))^{3\times 3}, \text{ such that}\\&\text{(i)}\ \int_\Omega \left\{\rho\frac{\partial \mathscr{W}}{\partial \mathbf{F}}(\mathbf{x}, \mathbf{Id}+\nabla\mathbf{u})+\mathbf{A}\right\}\cdot\nabla\mathbf{v}\,d\mathbf{x}\\&\qquad\qquad=\int_\Omega \mathbf{f}\cdot\mathbf{v}\,d\mathbf{x}+\int_{\Gamma_2}\mathbf{g}\cdot\mathbf{v}\,da \qquad \forall\mathbf{v}\in V,\\&\text{(ii)}\ \int_\Omega \mathbf{A}\cdot\mathbf{G}\,d\mathbf{x}=0 \qquad \forall\mathbf{G}\in dY(\mathbf{Id}+\nabla\mathbf{u}),\end{aligned}}$$

with

(5.9)
$$\boxed{V=\{\mathbf{v}\in\mathbf{W}^{1,s}(\Omega),\ \mathbf{v}=\mathbf{0} \text{ on } \Gamma_1\},}$$

(5.10)
$$\boxed{K=\{\mathbf{v}\in\mathbf{W}^{1,s}(\Omega),\ \mathbf{v}=\mathbf{u}_1 \text{ on } \Gamma_1,\ \det(\mathbf{Id}+\nabla\mathbf{v})=1 \text{ a.e. on } \Omega\},}$$

(5.11)
$$\boxed{dY(\mathbf{F})=\{\mathbf{G}\in(L^s(\Omega))^{3\times 3},\ \operatorname{adj}(\mathbf{F})^T\cdot\mathbf{G}=0 \text{ a.e. on } \Omega\}.}$$

Observe that since the external loads **f** and **g** are measured in the reference configuration, they might be known functions of the displacement field **u**. For example, for pressure-type surface forces, **g** remains normal to the deformed surface $\boldsymbol{\phi}(\Gamma_2)$, which means that it is given in the reference configuration by

$$\mathbf{g}=-\mathbf{g}(\mathbf{x})(\det\mathbf{F})\mathbf{F}^{-T}\boldsymbol{\nu}.$$

The particular case in which both **f** and **g** are independent of **u** is called *dead loading.*

Unfortunately, there is no general existence theory for the weak formulations (5.5) or (5.8) of equilibrium problems in finite elasticity.

5.3. Energetic formulations. For dead loads, the *total potential energy* of a hyperelastic body subjected to a displacement field **u** is defined as

(5.12)
$$\boxed{J(\mathbf{u})=\int_\Omega \rho\mathscr{W}(\mathbf{x}, \mathbf{Id}+\nabla\mathbf{u})\,d\mathbf{x}-\int_\Omega \mathbf{f}\cdot\mathbf{u}\,d\mathbf{x}-\int_{\Gamma_2}\mathbf{g}\cdot\mathbf{u}\,da.}$$

More generally, for conservative loads, where **f** and **g** represent the gradient in $\mathbf{L}^{s^*}(\Omega)\times\mathbf{L}^{s^*}(\Gamma_2)$ of a given potential energy $[-V(\mathbf{u})]$, the total potential energy of a hyperelastic body is

$$J(\mathbf{u}) = \int_\Omega \rho \mathscr{W}(\mathbf{x}, \mathbf{Id} + \nabla \mathbf{u})\, d\mathbf{x} + V(\mathbf{u}).$$

Now, from the virtual work theorem and the specific form of the constitutive laws in hyperelasticity, stable equilibrium positions of hyperelastic bodies under conservative loading formally correspond to those displacement fields that minimize the total potential energy $J(\cdot)$ over the set K of kinematically admissible displacement fields. Indeed, equations (5.5) or (5.8) formally express that any equilibrium position $\mathbf{u}$ corresponds to a stationary point of $J(\cdot)$ where its gradient

$$\frac{dJ}{d\mathbf{u}} \cdot \mathbf{v} = \int_\Omega \rho \frac{\partial \mathscr{W}}{\partial \mathbf{F}} (\mathbf{x}, \mathbf{Id} + \nabla \mathbf{u}) \cdot \nabla \mathbf{v}\, d\mathbf{x} - \int_\Omega \mathbf{f} \cdot \mathbf{v}\, d\mathbf{x} - \int_{\Gamma_2} \mathbf{g} \cdot \mathbf{v}\, da$$

is normal at $\mathbf{u}$ to the set K. And, if $\mathbf{u}$ is not a minimizer, stability can be violated for an adequate smooth perturbation of this position.

This formal remark is the basis of the energetic formulation of equilibrium problems in finite elasticity which, for conservative loading, is

(5.13) | Minimize the total potential energy $J(\cdot)$ over the set K of kinematically admissible displacement fields. |

The existence theory that follows defines K as

$$\begin{cases} K = \{\mathbf{v} \in \mathbf{W}^{1,s}(\Omega), \mathbf{v} = \mathbf{u}_1 \text{ on } \Gamma_1, \operatorname{adj}(\mathbf{Id} + \nabla \mathbf{v}) \in (L^q(\Omega))^{3\times 3}, & \text{(compressible case)} \\ \det(\mathbf{Id} + \nabla \mathbf{v}) \in L^r(\Omega), \det(\mathbf{Id} + \nabla \mathbf{v}) > 0 \text{ a.e. on } \Omega\}; \end{cases} \tag{5.14}$$

$$\begin{cases} K = \{\mathbf{v} \in \mathbf{W}^{1,s}(\Omega), \mathbf{v} = \mathbf{u}_1 \text{ on } \Gamma_1, \operatorname{adj}(\mathbf{Id} + \nabla \mathbf{v}) \in (L^q(\Omega))^{3\times 3}, & \text{(incompressible case)} \\ \det(\mathbf{Id} + \nabla \mathbf{v}) = 1 \text{ a.e. on } \Omega\}. \end{cases} \tag{5.15}$$

The exponents s, q, and r are determined from coerciveness assumptions on the free-energy potential $\mathscr{W}$. In any case, we must have

$$s > \frac{3}{2}, \quad \frac{1}{q} + \frac{1}{s} < \frac{4}{3}, \quad r > 1.$$

Remark 5.1. Equation (5.14) (resp., (5.15) for incompressible materials) can also be used in the weak formulation (5.5) (resp. (5.8)) of the equilibrium problems instead of equation (5.7) (resp. (5.10)). We then get new weak formulations for these problems.

5.4. Existence results. Under adequate regularity assumptions on the free-energy potential $\mathcal{W}$, Ball [1977] was able to prove existence results for the energetic formulations. For compressible hyperelastic materials under dead loading, he proved the following.

THEOREM 5.1. *Let Ω be open, bounded, connected in $\mathbb{R}^3$, and satisfy a strong Lipschitz condition.*[3] *Let its boundary Γ be split into two measurable subsets Γ_1 and Γ_2, with empty intersection, Γ_1 having positive measure in Γ. Let $\mathbf{u}_1$ be measurable, $\mathbf{f}$ be given in $\mathbf{L}^{s^*}(\Omega)$, and $\mathbf{g}$ be given in $\mathbf{L}^{s^*}(\Gamma_2)$. Let K be defined by (5.14), let $J(\cdot)$ be given by (5.12), and suppose that there exists an element $\mathbf{w}$ in K with $J(\mathbf{w})<+\infty$.*

Suppose also that the free-energy potential $\mathcal{W}(\mathbf{x}, \mathbf{F})$ is given as a function $\mathcal{G}$ defined on $\Omega\times\mathbb{R}^{3\times3}\times\mathbb{R}^{3\times3}\times(\mathbb{R}_+-\{0\})$ with values in $\mathbb{R}\cup\{+\infty\}$ and such that

$$\mathcal{W}(\mathbf{x}, \mathbf{F}) = \mathcal{G}(\mathbf{x}, \mathbf{F}, \operatorname{adj}\mathbf{F}, \det\mathbf{F}),$$

$$\mathcal{G}(\mathbf{x}, \cdot, \cdot, \cdot) \text{ is continuous and convex on } \mathbb{R}^{3\times3}\times\mathbb{R}^{3\times3}\times(\mathbb{R}_+-\{0\}),$$

$$\mathcal{G}(\cdot, \mathbf{F}, \mathbf{G}, \delta) \text{ is measurable on } \Omega,$$

$$\mathcal{G}(\mathbf{x}, \mathbf{F}, \mathbf{G}, \delta) \ge a(\mathbf{x}) + C_1|\mathbf{F}|^s + C_2|\mathbf{G}|^q + C_3|\delta|^r; \qquad (\textit{coercivity}),$$

$$C_i > 0, \quad i = 1, 3, \quad a \in L^1(\Omega),$$

$$\lim \mathcal{G}(\mathbf{x}, \mathbf{F}^n, \mathbf{G}^n, \delta^n) = +\infty \quad \text{if } \mathbf{F}^n \to \mathbf{F}, \mathbf{G}^n \to \mathbf{G}, \delta^n \to 0.$$

Then, there exists a minimizer of the total potential energy J over the set K of kinematically admissible displacement fields.

Proof. Let us briefly outline the proof, which is based on the generalized Weierstrass theorem. First introduce

$$H = \mathbf{W}^{1,s}(\Omega)\times(L^q(\Omega))^{3\times3}\times L^r(\Omega),$$

$$K_a = \{\{\mathbf{v}, \mathbf{G}, \delta\}\in H, \mathbf{v} = \mathbf{u}_1 \text{ on } \Gamma_1, \mathbf{G} = \operatorname{adj}(\mathbf{Id}+\nabla\mathbf{v}), \delta = \det(\mathbf{Id}+\nabla\mathbf{v})\},$$

$$\mathcal{G}_a(\mathbf{x}, \mathbf{F}, \mathbf{G}, \delta) = \begin{cases} \mathcal{G}(\mathbf{x}, \mathbf{F}, \mathbf{G}, \delta) & \text{if } \delta > 0, \\ +\infty & \text{if not,} \end{cases}$$

$$J_a(\mathbf{v}, \mathbf{G}, \delta) = \int_\Omega \rho\mathcal{G}(\mathbf{x}, \mathbf{Id}+\nabla\mathbf{v}, \mathbf{G}, \delta)\, d\mathbf{x} - \int_\Omega \mathbf{f}\cdot\mathbf{v}\, d\mathbf{x} - \int_{\Gamma_2} \mathbf{g}\cdot\mathbf{v}\, da.$$

From the sequentially weak continuity of the antisymmetric mappings adj and det on $(L^s(\Omega))^{3\times3}$ and $(L^s(\Omega))^{3\times3}\times(L^q(\Omega))^{3\times3}$, respectively (Ball [1977, p. 371]), K_a is weakly sequentially closed in H.

From the continuity of $\mathcal{G}(\mathbf{x}, \cdot)$ and its behavior as δ goes to zero, $\mathcal{G}_a(\mathbf{x}, \cdot)$ is lower semicontinuous on $\mathbb{R}^{3\times3}\times\mathbb{R}^{3\times3}\times\mathbb{R}$. From the coercivity of $\mathcal{G}$ and Fatou's lemma, this implies that J_a is lower semicontinuous on H. Moreover,

[3] In the sense of Necas [1967].

since $\mathcal{G}(\mathbf{x},\cdot)$ is convex, J_a is also convex. Thus, being convex and lower semicontinuous, it is also weakly sequentially lower semicontinuous on H.

Finally, the coercivity of $\mathcal{G}_a$ on $\mathbb{R}^{3\times 3}\times\mathbb{R}^{3\times 3}\times\mathbb{R}$ implies the coercivity of J_a on H; moreover, K_a is nonempty because it contains $\{\mathbf{w}, \operatorname{adj}(\mathbf{Id}+\nabla\mathbf{w}), \det(\mathbf{Id}+\nabla\mathbf{w})\}$.

The generalized Weierstrass theorem can be applied, and there exists $\{\mathbf{u}, \operatorname{adj}(\mathbf{Id}+\nabla\mathbf{u}), \det(\mathbf{Id}+\nabla\mathbf{u})\}$ in K_a which minimizes J_a on K_a. Since $J_a(\mathbf{w}, \operatorname{adj}(\nabla\mathbf{w}+\mathbf{Id}), \det(\mathbf{Id}+\nabla\mathbf{w}))=J(\mathbf{w})$ is finite, $J_a(\mathbf{u}, \operatorname{adj}(\mathbf{Id}+\nabla\mathbf{u}), \det(\mathbf{Id}+\nabla\mathbf{u}))=J(\mathbf{u})$ is also finite, which implies, by construction of J_a, that $\mathbf{u}$ belongs to K and therefore minimizes J over K. □

The same technique yields a similar existence result for the incompressible case.

THEOREM 5.2. *Let Ω be open, bounded, connected in $\mathbb{R}^3$, and satisfy a strong Lipschitz condition. Let its boundary Γ be split into two measurable subsets Γ_1 and Γ_2, with empty intersection, Γ_1 having positive measure in Γ. Let $\mathbf{u}_1$ be measurable, $\mathbf{f}$ be given in $\mathbf{L}^{s^*}(\Omega)$, and $\mathbf{g}$ be given in $\mathbf{L}^{s^*}(\Gamma_2)$. Let K be defined by (5.15) (incompressible case), let $J(\cdot)$ be given by (5.12), and suppose that there exists an element $\mathbf{w}$ in K with $J(\mathbf{w})<+\infty$.*

Suppose also that the free-energy potential $\mathcal{W}(\mathbf{x},\mathbf{F})$ is given as a real function $\mathcal{G}$ defined on $\Omega\times\mathbb{R}^{3\times 3}\times\mathbb{R}^{3\times 3}$ such that

$$\mathcal{W}(\mathbf{x},\mathbf{F})=\mathcal{G}(\mathbf{x},\mathbf{F},\operatorname{adj}(\mathbf{F})),$$

$$\mathcal{G}(\mathbf{x},\cdot,\cdot) \text{ is continuous and convex on } \mathbb{R}^{3\times 3}\times\mathbb{R}^{3\times 3} \qquad (\textit{polyconvexity}),$$

$$\mathcal{G}(\cdot,\mathbf{F},\mathbf{G}) \text{ is measurable on } \Omega,$$

$$\mathcal{G}(\mathbf{x},\mathbf{F},\mathbf{G})\ge a(\mathbf{x})+C_1|\mathbf{F}|^s+C_2|\mathbf{G}|^q,\quad C_1>0,\quad C_2>0,$$

$$a(\mathbf{x})\in L^1(\Omega) \qquad (\textit{coercivity}).$$

Then there exists a minimizer of the total potential energy J over the set K of kinematically admissible displacement fields. □

Remark 5.2. Some of the assumptions of these existence theorems can be slightly relaxed (Ball [1977]); however, in Ball's theory, the free-energy potential must always be polyconvex (i.e., $\mathcal{G}$ convex) and coercive.

Remark 5.3. The main limitation of the above existence theorems lies in the fact that it cannot be proved that the minimizers of J on K do satisfy some sort of weak equation of equilibrium. The topology taken for K is too weak to prove such a result (Le Tallec and Oden [1981], Le Dret [1985]).

Chapter 3

Augmented Lagrangian Methods for the Solution of Variational Problems

1. Introduction and synopsis.

1.1. Introduction. Duality principles play an important role in mechanics, physics, mathematical economy, and other branches of science and engineering; their role in mechanics for proving existence results for some particular classes of nonlinear problems has already been illustrated in Chapter 2, § 2. Another field in which such principles play a pivotal role is mathematical programming, i.e., the science (and sometimes the art) of minimizing or maximizing functions over sets of various kinds. Actually, some optimization techniques are founded on the application of these duality principles, most of them using those vectors called multipliers (Lagrange multipliers, John-Kuhn-Tucker multipliers, etc.). Such multiplier methods are discussed in Arrow, Hurwicz, and Uzawa [1958] (motivated by economical equilibria) and in Glowinski, Lions, and Tremolières [1976], [1981] (motivated by nonlinear mechanics). Unfortunately, these multiplier methods (at least the original one) converge linearly at best. Therefore, the resulting algorithms may be slow and, to improve the speed of convergence, one may think of conjugate-gradient variants of the original algorithms.

An alternative might be, as suggested in Hestenes [1969] and Powell [1969], to improve the conditioning of the original problem by using an appropriate augmented Lagrangian formulation. This leads to optimization techniques usually referred to as augmented Lagrangian methods, which converge faster. There is a large literature devoted to augmented Lagrangian methods; two books that contain a large number of references pertinent to augmented Lagrangian methods and also information of a historical character are Bertsekas [1982] and Gill, Murray, and Wright [1981, Chap. 6]. During the 1970s it was realized by some authors (see, for example, Glowinski and Marrocco [1974], [1975], Polyak [1979]) that these methods were also ideally suited to take advantage of the decomposition principles that often appear in physics, mechanics, economics, etc. In fact, some well-known methods of numerical analysis such as the Douglas–Rachford and Peaceman–Rachford alternating-direction methods, and also some power methods for computing eigenvalues and eigenvectors of symmetric matrices can be derived from the coupling of some decomposition principles and augmented Lagrangian methods. Such relations are discussed in, for example, Gabay [1979], [1982], [1983] and Fortin and Glowinski [1982], [1983] as well as in §§ 5 and 6 of this chapter.

1.2. Synopsis. The main goal of this chapter is to introduce, without any pretentions of being exhaustive, these augmented Lagrangian methods and decomposition principles.

First, § 2 introduces augmented Lagrangian methods for the solution of a simple, finite-dimensional problem having its origin in quadratic programming. These methods mainly consist of introducing an augmented Lagrangian formulation and an associated dual formulation of the model problem, and proposing four different algorithms for its numerical solution, each of which can be interpreted as a descent method operating on this dual formulation.

Then, § 3 generalizes the ideas of § 2 to an infinite-dimensional framework in the particular case of the Stokes problem.

In § 3, the problem under consideration already has a built-in primal–dual structure. Section 4 indicates how decomposition principles can be used to transform a very large class of nonlinear minimization problems into problems that have this primal–dual structure, and which therefore can be solved using several variants of the augmented Lagrangian methods presented in § 2.

Section 5 proves that such a combination of decomposition principles and augmented Lagrangian methods, which is fundamental to this book, can also be interpreted as an alternating-direction method acting on the dual formulation of the original minimization problem. Such an interpretation can be very useful in the derivation, the understanding, and the analysis of the numerical methods proposed in this monograph.

In § 6, we apply, as an illustration, the techniques discussed in §§ 4 and 5 to the solution of a nonlinear eigenvalue problem from quantum physics. Finally, an application to the solution of liquid crystal problems is described in § 7, together with various comments and further references.

More applications of augmented Lagrangian methods will be discussed in the remaining chapters of this monograph. In particular, we will present some applications in finite elasticity where, in the opinion of the authors, this combination of decomposition principles and augmented Lagrangian formulations has provided very efficient methods of solution.

2. Augmented Lagrangian methods in quadratic programming.

2.1. A finite-dimensional model problem. In an attempt to simplify the presentation, in this section we shall limit ourselves to a particularly simple finite-dimensional problem, as follows.

Let $\mathbf{A}$ be an $N \times N$ symmetric, positive-definite matrix, and suppose that $\mathbf{b} \in \mathbb{R}^N$; with $\mathbf{A}$ and $\mathbf{b}$ is associated the quadratic functional $J: \mathbb{R}^N \to \mathbb{R}$ defined by

$$J(\mathbf{v}) = \tfrac{1}{2}(\mathbf{A}\mathbf{v}, \mathbf{v}) - (\mathbf{b}, \mathbf{v}), \tag{2.1}$$

where $(\cdot, \cdot)$ denotes the canonical Euclidian inner product in $\mathbb{R}^N$. Let $\mathbf{B}$ be a linear mapping from $\mathbb{R}^N$ into $\mathbb{R}^M$ and thus be identifiable with an $M \times N$ matrix, and consider $\mathbf{c} \in \mathbb{R}^M$ such that

$$\mathbf{c} \in R(\mathbf{B}), \tag{2.2}$$

where $R(\mathbf{B}) = \{\mathbf{q} \mid \mathbf{q} \in \mathbb{R}^M, \exists \mathbf{v} \in \mathbb{R}^N \text{ such that } \mathbf{q} = \mathbf{B}\mathbf{v}\}$.

We consider the minimization problem

$$\boxed{\begin{cases} \text{Find } \mathbf{u} \in H \text{ such that} \\ J(\mathbf{u}) \le J(\mathbf{v}) \qquad \forall \mathbf{v} \in H, \end{cases}} \tag{2.3}$$

where

$$H = \{\mathbf{v} \mid \mathbf{v} \in \mathbb{R}^N, \mathbf{B}\mathbf{v} = \mathbf{c}\}. \tag{2.4}$$

Since (2.2) implies $H \neq \emptyset$, it is a classical result that (2.3) admits a unique solution.

2.2. Augmented Lagrangian formulation. Following a well-known technique, we introduce a Lagrange multiplier $\mathbf{p} \in \mathbb{R}^M$ that transforms (2.3) into an unconstrained problem, namely,

$$\underset{\mathbf{v} \in \mathbb{R}^N}{\text{minimize}} \{J(\mathbf{v}) + (\mathbf{p}, \mathbf{B}\mathbf{v} - \mathbf{c})\}. \tag{2.5}$$

We also use $(\cdot, \cdot)$ for the inner product in $\mathbb{R}^M$, there being no danger of ambiguity.

The Lagrange multiplier $\mathbf{p}$ appears as an extra unknown which may, for example, be obtained through the solution of a saddle-point problem. More precisely, we define $\mathscr{L}: \mathbb{R}^N \times \mathbb{R}^M \to \mathbb{R}$ by

$$\mathscr{L}(\mathbf{v}, \mathbf{q}) = J(\mathbf{v}) + (\mathbf{q}, \mathbf{B}\mathbf{v} - \mathbf{c}), \tag{2.6}$$

and we recall that $\{\mathbf{u}, \mathbf{p}\}$ will be a saddle-point of $\mathscr{L}$ on $\mathbb{R}^N \times \mathbb{R}^M$ if

$$\mathscr{L}(\mathbf{u}, \mathbf{q}) \le \mathscr{L}(\mathbf{u}, \mathbf{p}) \le \mathscr{L}(\mathbf{v}, \mathbf{p}) \quad \forall \mathbf{v} \in \mathbb{R}^N, \quad \mathbf{q} \in \mathbb{R}^M, \tag{2.7}$$

and also that (2.7) implies

$$\min_{\mathbf{v} \in \mathbb{R}^N} \max_{\mathbf{q} \in \mathbb{R}^M} \mathscr{L}(\mathbf{v}, \mathbf{q}) = \max_{\mathbf{q} \in \mathbb{R}^M} \min_{\mathbf{v} \in \mathbb{R}^N} \mathscr{L}(\mathbf{v}, \mathbf{q}) = \mathscr{L}(\mathbf{u}, \mathbf{p}). \tag{2.8}$$

The following classical result (see, e.g., Glowinski [1984, Chap. 1]) is essential to the subsequent discussion.

THEOREM 2.1. *The solution* $\mathbf{u}$ *of* (2.3) *is characterized by the existence of* $\mathbf{p} \in \mathbb{R}^M$ *such that*

$$\begin{cases} \mathbf{Au}+\mathbf{B}^t\mathbf{p}=\mathbf{b}, \\ \mathbf{Bu}=\mathbf{c}. \end{cases} \tag{2.9}$$

The relations (2.9) *also characterize all the saddle-points of* $\mathscr{L}$ *on* $\mathbb{R}^N \times \mathbb{R}^M$. *The multiplier* $\mathbf{p}$ *is unique if and only if* $\mathbf{B}$ *is surjective* (*i.e.*, rank $\mathbf{B}=M$). □

Following Hestenes [1969] and Powell [1969], we introduce the augmented Lagrangian $\mathscr{L}_r$ defined, for $r>0$, by

$$\boxed{\begin{cases} \mathscr{L}_r(\mathbf{v},\mathbf{q})=J(\mathbf{v})+(\mathbf{q},\mathbf{Bv}-\mathbf{c})+\dfrac{r}{2}|\mathbf{Bv}-\mathbf{c}|^2 \\ \qquad =\mathscr{L}(\mathbf{v},\mathbf{q})+\dfrac{r}{2}|\mathbf{Bv}-\mathbf{c}|^2, \end{cases}} \tag{2.10}$$

where, in (2.10), $|\cdot|$ denotes the canonical Euclidian norm on $\mathbb{R}^M$. It can easily be proved that any saddle-point of $\mathscr{L}_r$ is a saddle-point of $\mathscr{L}$ and that the converse also holds. This is due to the fact that $r|\mathbf{Bv}-\mathbf{c}|^2$ vanishes when the constraint $\mathbf{Bv}=\mathbf{c}$ is satisfied.

Remark 2.1. It should be observed that, for $\mathbf{q}=\mathbf{0}$, we have

$$\mathscr{L}_r(\mathbf{v},\mathbf{0})=J(\mathbf{v})+\frac{r}{2}|\mathbf{Bv}-\mathbf{c}|^2, \tag{2.11}$$

this being the classical penalized function relative to the constraint $\mathbf{Bv}=\mathbf{c}$. The advantage of the augmented Lagrangian is that, because of the term $(\mathbf{q},\mathbf{Bv}-\mathbf{c})$, the exact solution of problem (2.3) can be determined without making r tend to infinity, which, using ordinary penalization methods, would have the effect of causing a deterioration in the conditioning of the system to be solved. Furthermore, the addition of the quadratic term $\frac{1}{2}r|\mathbf{Bv}-\mathbf{c}|^2$ to the Lagrangian $\mathscr{L}$ will improve the convergence properties of the duality algorithms described later.

2.3. A first algorithm for saddle-point calculations.

2.3.1. Description of the algorithm. It follows from § 2.2 that there is an equivalence between solving (2.3) and finding a saddle-point of $\mathscr{L}_r$ on $\mathbb{R}^N \times \mathbb{R}^M$. Following Arrow, Hurwicz, and Uzawa [1958], Glowinski, Lions, and Tremolières [1976], [1981], Ekeland and Temam [1974], [1976], Ciarlet [1982], Fortin and Glowinski [1982], [1983], etc., such a saddle-point can be calculated using the following algorithm, the variants of which we shall denote under the general name of Uzawa's algorithm.

ALGORITHM (2.12)-(2.14).

(2.12) $$\mathbf{p}^0 \in \mathbb{R}^M, \text{ specified arbitrarily;}$$

with $\mathbf{p}^n$ known, calculate $\mathbf{u}^n$, then $\mathbf{p}^{n+1}$, by

(2.13) $$\begin{cases} \mathcal{L}_r(\mathbf{u}^n, \mathbf{p}^n) \le \mathcal{L}_r(\mathbf{v}, \mathbf{p}^n) & \forall \mathbf{v} \in \mathbb{R}^N, \\ \mathbf{u}^n \in \mathbb{R}^N, \end{cases}$$

(2.14) $$\mathbf{p}^{n+1} = \mathbf{p}^n + \rho_n(\mathbf{B}\mathbf{u}^n - \mathbf{c}), \qquad \rho_n > 0.$$

We observe that (2.13) is equivalent to

(2.15) $$(\mathbf{A} + r\mathbf{B}^t\mathbf{B})\mathbf{u}^n + \mathbf{B}^t\mathbf{p}^n = \mathbf{b} + r\mathbf{B}^t\mathbf{c}.$$

2.3.2. Convergence results for $\{\mathbf{u}^n\}_{n\ge 0}$. We now prove the following theorem regarding the convergence of Algorithm (2.12)-(2.14).

THEOREM 2.2. *For $0 < \alpha_0 \le \rho_n \le 2r$ and for all $\mathbf{p}^0 \in \mathbb{R}^M$, the sequence $\mathbf{u}^n$ defined by Algorithm* (2.12)-(2.14) *converges to the solution $\mathbf{u}$ of* (2.3).

Proof. Following Glowinski, Lions, and Tremolières [1976], [1981], suppose that $\{\mathbf{u}, \mathbf{p}\}$ is a saddle-point of $\mathcal{L}_r$. From Theorem 2.1, this saddle-point is characterized by

(2.16) $$(\mathbf{A} + r\mathbf{B}^t\mathbf{B})\mathbf{u} + \mathbf{B}^t\mathbf{p} = \mathbf{b} + r\mathbf{B}^t\mathbf{c},$$

(2.17) $$\mathbf{B}\mathbf{u} = \mathbf{c} \Leftrightarrow \mathbf{p} = \mathbf{p} + \rho_n(\mathbf{B}\mathbf{u} - \mathbf{c}) \quad \forall n.$$

We introduce $\bar{\mathbf{u}}^n = \mathbf{u}^n - \mathbf{u}$, $\bar{\mathbf{p}}^n = \mathbf{p}^n - \mathbf{p}$. Then, by subtracting (2.16) from (2.15) and (2.17) from (2.14), we obtain

(2.18) $$(\mathbf{A} + r\mathbf{B}^t\mathbf{B})\bar{\mathbf{u}}^n + \mathbf{B}^t\bar{\mathbf{p}}^n = \mathbf{0},$$

(2.19) $$\bar{\mathbf{p}}^{n+1} = \bar{\mathbf{p}}^n + \rho_n\mathbf{B}\bar{\mathbf{u}}^n.$$

We deduce from (2.19) that

$$|\bar{\mathbf{p}}^{n+1}|^2 = |\bar{\mathbf{p}}^n|^2 + 2\rho_n(\bar{\mathbf{p}}^n, \mathbf{B}\bar{\mathbf{u}}^n) + \rho_n^2|\mathbf{B}\bar{\mathbf{u}}^n|^2,$$

and, hence, that

(2.20) $$|\bar{\mathbf{p}}^n|^2 - |\bar{\mathbf{p}}^{n+1}|^2 = -2\rho_n(\bar{\mathbf{p}}^n, \mathbf{B}\bar{\mathbf{u}}^n) - \rho_n^2|\mathbf{B}\bar{\mathbf{u}}^n|^2.$$

It follows from (2.18) that

$$(\mathbf{A}\bar{\mathbf{u}}^n, \bar{\mathbf{u}}^n) + r|\mathbf{B}\bar{\mathbf{u}}^n|^2 = -(\bar{\mathbf{p}}^n, \mathbf{B}\bar{\mathbf{u}}^n),$$

and, hence, by substitution in (2.20), that

(2.21) $$|\bar{\mathbf{p}}^n|^2 - |\bar{\mathbf{p}}^{n+1}|^2 = 2\rho_n(\mathbf{A}\bar{\mathbf{u}}^n, \bar{\mathbf{u}}^n) + \rho_n(2r - \rho_n)|\mathbf{B}\bar{\mathbf{u}}^n|^2.$$

If we then take

(2.22) $$0 < \alpha_0 \le \rho_n \le 2r,$$

the sequence $|\bar{\mathbf{p}}^n|$ can be seen to be decreasing. Being bounded below by 0, it is convergent and, hence, $|\bar{\mathbf{p}}^n|^2-|\bar{\mathbf{p}}^{n+1}|^2\to 0$. Equation (2.21) then implies

$$\lim_{n\to+\infty}(\mathbf{A}\bar{\mathbf{u}}^n,\bar{\mathbf{u}}^n)=0, \tag{2.23}$$

and, since $\mathbf{A}$ is positive-definite, it follows from (2.23) that $\bar{\mathbf{u}}^n\to\mathbf{0}$; also, since $\bar{\mathbf{u}}^n=\mathbf{u}^n-\mathbf{u}$, we have $\lim_{n\to+\infty}\mathbf{u}^n=\mathbf{u}$. □

Remark 2.2. It follows from (2.21) that we actually have convergence of Algorithm (2.12)–(2.14) under the following condition, which is less restrictive than (2.22),

$$0<\alpha_0\le\rho_n\le\alpha_1<2\left(r+\frac{1}{\beta^2}\right), \tag{2.24}$$

where β^2 is the largest eigenvalue of $\mathbf{A}^{-1}\mathbf{B}^t\mathbf{B}$ defined by

$$\beta^2=\max_{\mathbf{v}\neq\mathbf{0}}\frac{|\mathbf{B}\mathbf{v}|^2}{(\mathbf{A}\mathbf{v},\mathbf{v})}. \tag{2.25}$$

2.3.3. Convergence results for $\{\mathbf{p}^n\}_{n\ge 0}$. With a view to studying the behavior of the sequence $\mathbf{p}^n$, it is worth noting that

$$(R(\mathbf{B}))^{\perp}=\ker(\mathbf{B}^t), \tag{2.26}$$

and, hence

$$\mathbb{R}^M=R(\mathbf{B})\oplus\ker(\mathbf{B}^t). \tag{2.27}$$

We thus have, for all $\mathbf{q}\in\mathbb{R}^M$, the unique decomposition

$$\mathbf{q}=\mathbf{q}_1+\mathbf{q}_2,\quad \mathbf{q}_1\in R(\mathbf{B}),\quad \mathbf{q}_2\in\ker(\mathbf{B}^t), \tag{2.28}$$

$$\mathbf{q}_i=P_i(\mathbf{q})\qquad\forall i=1,2, \tag{2.29}$$

P_1 (resp., P_2) denoting the projection of $\mathbb{R}^M$ onto $R(\mathbf{B})$ (resp., $\ker(\mathbf{B}^t)$). Under this notation, it follows from Theorem 2.1 that there exists a unique $\hat{\mathbf{p}}\in R(\mathbf{B})$ such that the Lagrange multipliers of (2.3), (2.6) are of the form

$$\mathbf{p}=\hat{\mathbf{p}}+\mathbf{q},\quad \mathbf{q}\in\ker(\mathbf{B}^t),\quad \text{with } \mathbf{q} \text{ arbitrary}. \tag{2.30}$$

The vector $\hat{\mathbf{p}}$ thus appears as the Lagrange multiplier of (2.3), (2.6) with minimal norm in $\mathbb{R}^M$.

From Theorem 2.2 and the above properties of the Lagrange multipliers we shall now deduce the following theorem.

THEOREM 2.3. *If ρ_n satisfies* (2.24), *then the sequence $\mathbf{p}^n$ defined by Algorithm* (2.12)–(2.14) *converges to $\hat{\mathbf{p}}+\mathbf{p}_2^0$ where $\mathbf{p}_2^0=P_2(\mathbf{p}^0)$ is the component of $\mathbf{p}^0$ in* $\ker(\mathbf{B}^t)$. *In particular, if $\mathbf{p}^0=\mathbf{0}$, then $\lim_{n\to+\infty}\mathbf{p}^n=\hat{\mathbf{p}}$.*

Proof. Relations (2.2), (2.14) immediately imply

$$P_2(\mathbf{p}^{n+1})=P_2(\mathbf{p}^n)=P_2(\mathbf{p}^0)\quad\forall n. \tag{2.31}$$

Under the condition (2.24), we have $\lim_{n\to+\infty} \bar{\mathbf{u}}^n = \mathbf{0}$. It then follows from (2.18) that

$$\lim_{n\to+\infty} \mathbf{B}^t\bar{\mathbf{p}}^n = \mathbf{0}. \tag{2.32}$$

Since $|\mathbf{B}^t\mathbf{q}|$ defines a norm over $R(\mathbf{B})$, (2.32) implies by projection onto $R(\mathbf{B})$ that

$$\lim_{n\to+\infty} P_1(\mathbf{p}^n) = \hat{\mathbf{p}}.$$

Hence, with (2.31), $\lim_{n\to+\infty} \mathbf{p}^n = \hat{\mathbf{p}} + P_2(\mathbf{p}^0)$. □

2.3.4. Interpretation of Algorithm (2.12)–(2.14). Algorithm (2.12)-(2.14) is in fact a gradient-type algorithm applied to the minimization of the dual function $J_r^*: \mathbb{R}^M \to \mathbb{R}$ defined by

$$\begin{cases} J_r^*(\mathbf{q}) = -\min_{\mathbf{v}\in\mathbb{R}^N} \mathscr{L}_r(\mathbf{v}, \mathbf{q}) \\ \quad = \frac{1}{2}(\mathbf{B}\mathbf{A}_r^{-1}\mathbf{B}^t\mathbf{q}, \mathbf{q}) - (\mathbf{B}\mathbf{A}_r^{-1}(\mathbf{b}+r\mathbf{B}^t\mathbf{c}), \mathbf{q}) + (\mathbf{c}, \mathbf{q}) \\ \quad + \frac{1}{2}(\mathbf{A}_r^{-1}(\mathbf{b}+r\mathbf{B}^t\mathbf{c}), (\mathbf{b}+r\mathbf{B}^t\mathbf{c})) - \frac{r}{2}|\mathbf{c}|^2, \\ \text{where } \mathbf{A}_r = \mathbf{A} + r\mathbf{B}^t\mathbf{B}, \end{cases} \tag{2.33}$$

or, equivalently to the solution of the linear system,

$$\mathbf{B}\mathbf{A}_r^{-1}\mathbf{B}^t\mathbf{p} = \mathbf{B}\mathbf{A}_r^{-1}(\mathbf{b}+r\mathbf{B}^t\mathbf{c}) - \mathbf{c}. \tag{2.34}$$

More precisely, by elimination of $\mathbf{u}^n$, Algorithm (2.12)-(2.14) can be re-expressed as

ALGORITHM (2.35)-(2.36).

$$\mathbf{p}^0 \in \mathbb{R}^M, \text{ specified arbitrarily;} \tag{2.35}$$

$$\mathbf{p}^{n+1} = \mathbf{p}^n - \rho_n(\mathbf{B}\mathbf{A}_r^{-1}\mathbf{B}^t\mathbf{p}^n - \mathbf{B}\mathbf{A}_r^{-1}(\mathbf{b}+r\mathbf{B}^t\mathbf{c}) + \mathbf{c}). \tag{2.36}$$

Remark 2.3. The advantage of Algorithm (2.12)-(2.14) compared with Algorithm (2.35)-(2.36) is that we do not have to construct $\mathbf{A}_r^{-1}$ explicitly; in certain applications to partial differential equations this would in practice be unrealizable since $\mathbf{A}_r^{-1}$ would be a full matrix of very large order. Algorithm (2.35)-(2.36) is on the other hand very useful as a theoretical basis for the study of the influence of r and ρ_n on the convergence of Algorithm (2.12)-(2.14).

In a similar fashion, by eliminating $\mathbf{p}^n$, Algorithm (2.12)–(2.14) can be re-expressed as

ALGORITHM (2.37)–(2.38).

$$\mathbf{u}^0 = \mathbf{A}_r^{-1}(\mathbf{b} + r\mathbf{B}^t\mathbf{c} - \mathbf{B}^t\mathbf{p}^0), \tag{2.37}$$

$$\mathbf{u}^{n+1} = \mathbf{u}^n - \rho_n \mathbf{A}_r^{-1}\mathbf{B}^t(\mathbf{B}\mathbf{u}^n - \mathbf{c}), \tag{2.38}$$

and Remark 2.3 holds equally for Algorithm (2.37)–(2.38).

2.3.5. Rate of convergence (for $\rho_n = \rho$). To study the influence of r and ρ_n on the convergence of Algorithm (2.12)–(2.14), we first observe that, if $\bar{\mathbf{p}}^n = \mathbf{p}^n - (\hat{\mathbf{p}} + P_2(\mathbf{p}^0))$, then from (2.2), (2.31), and (2.36) we have

$$\begin{cases} \bar{\mathbf{p}}^n \in R(\mathbf{B}) & \forall n \geq 0, \\ \bar{\mathbf{p}}^{n+1} = \bar{\mathbf{p}}^n - \rho_n \mathbf{B}\mathbf{A}_r^{-1}\mathbf{B}^t\bar{\mathbf{p}}^n. \end{cases} \tag{2.39}$$

From the second relation (2.39), we deduce that

$$\mathbf{A}^{-1}\mathbf{B}^t\bar{\mathbf{p}}^{n+1} = \mathbf{A}^{-1}\mathbf{B}^t\bar{\mathbf{p}}^n - \rho_n \mathbf{A}^{-1}\mathbf{B}^t\mathbf{B}\mathbf{A}_r^{-1}\mathbf{B}^t\bar{\mathbf{p}}^n. \tag{2.40}$$

On the other hand, by inverting the definition of $\mathbf{A}_r$, we have

$$\mathbf{A}_r^{-1} = (\mathbf{I} + r\mathbf{A}^{-1}\mathbf{B}^t\mathbf{B})^{-1}\mathbf{A}^{-1}, \tag{2.41}$$

which, together with (2.40), implies

$$\mathbf{A}^{-1}\mathbf{B}^t\bar{\mathbf{p}}^{n+1} = \mathbf{A}^{-1}\mathbf{B}^t\bar{\mathbf{p}}^n - \rho_n \mathbf{A}^{-1}\mathbf{B}^t\mathbf{B}(\mathbf{I} + r\mathbf{A}^{-1}\mathbf{B}^t\mathbf{B})^{-1}\mathbf{A}^{-1}\mathbf{B}^t\bar{\mathbf{p}}^n. \tag{2.42}$$

By introducing the auxiliary variable $\mathbf{y}^n = \mathbf{A}^{-1}\mathbf{B}^t\bar{\mathbf{p}}^n$, (2.42) finally becomes

$$\mathbf{y}^{n+1} = (\mathbf{I} - \rho_n \mathbf{A}^{-1}\mathbf{B}^t\mathbf{B}(\mathbf{I} + r\mathbf{A}^{-1}\mathbf{B}^t\mathbf{B})^{-1})\mathbf{y}^n. \tag{2.43}$$

Such a recursive relation on the sequence $\{\mathbf{y}^n\}_{n\geq 0}$ is important because it gives an easy way to prove the linear convergence to $\mathbf{0}$ of $\{\mathbf{y}^n\}_{n\geq 0}$, and thus the convergence to $\mathbf{0}$ of $\{\bar{\mathbf{u}}^n\}_{n\geq 0}$ and $\{\bar{\mathbf{p}}^n\}_{n\geq 0}$, which are linearly related to $\{\mathbf{y}^n\}_{n\geq 0}$.

We shall express (2.43) in a basis of eigenvectors of $\mathbf{A}^{-1}\mathbf{B}^t\mathbf{B}$, but, before doing so, we shall first indicate, without proof, several properties of the eigenvectors and eigenvalues of $\mathbf{A}^{-1}\mathbf{B}^t\mathbf{B}$.

PROPOSITION 2.1. *The eigenvalues of* $\mathbf{A}^{-1}\mathbf{B}^t\mathbf{B}$ *are* ≥ 0, *and the eigenvectors corresponding to two distinct eigenvalues are* **A**-*orthogonal, i.e., if*

$$\mathbf{A}^{-1}\mathbf{B}^t\mathbf{B}\mathbf{w}_i = \lambda_i \mathbf{w}_i, \quad \mathbf{A}^{-1}\mathbf{B}^t\mathbf{B}\mathbf{w}_j = \lambda_j \mathbf{w}_j,$$

with $\lambda_i \neq \lambda_j$, *then* $(\mathbf{A}\mathbf{w}_i, \mathbf{w}_j) = 0$. □

PROPOSITION 2.2. *If* 0 *is an eigenvalue of* $\mathbf{A}^{-1}\mathbf{B}^t\mathbf{B}$, *then the corresponding eigen-subspace is* $\ker(\mathbf{B})$, *and* $R(\mathbf{A}^{-1}\mathbf{B}^t)$ *is thus spanned by the eigenvectors of* $\mathbf{A}^{-1}\mathbf{B}^t\mathbf{B}$ *associated with the eigenvalues that are distinct from* 0. □

We have, of course, dim $R(\mathbf{A}^{-1}\mathbf{B}^t) = \operatorname{rank} \mathbf{B}^t = \operatorname{rank} \mathbf{B}$. In the following, we shall denote by N_1 the rank of $\mathbf{B}$ and by λ_m, λ_M, respectively, the smallest nonzero eigenvalue and the largest eigenvalue of $\mathbf{A}^{-1}\mathbf{B}^t\mathbf{B}$. Suppose that $\mathcal{B} = \{\mathbf{w}_i\}_{i=1}^{N_1}$ is a basis of $\mathbf{R}(\mathbf{A}^{-1}\mathbf{B}^t)$, with $\mathbf{w}_i$ the eigenvector of $\mathbf{A}^{-1}\mathbf{B}^t\mathbf{B}$ associated with the eigenvalue λ_i. If $\mathbf{y} \in R(\mathbf{A}^{-1}\mathbf{B}^t\mathbf{B})$, we then have

$$\mathbf{y} = \sum_{i=1}^{N_1} y_i \mathbf{w}_i, \qquad y_i \in \mathbb{R} \qquad \forall i.$$

From the fact that $\mathbf{y}^n \in R(\mathbf{A}^{-1}\mathbf{B}^t\mathbf{B})$ for all $n \geq 0$, and from (2.43), we can deduce that

$$\text{(2.44)} \qquad \begin{cases} y_i^{n+1} = \left(1 - \dfrac{\rho_n \lambda_i}{1 + r\lambda_i}\right) y_i^n = \dfrac{1 + (r - \rho_n)\lambda_i}{1 + r\lambda_i} y_i^n \\ \forall i = 1, \cdots, N_1, \quad n \geq 0. \end{cases}$$

In the remainder of § 2.3.5, we shall assume that $\rho_n = \rho$ for all n. Thus, Algorithm (2.12)-(2.14) reduces to

$$\text{(2.45)} \qquad \boxed{y_i^{n+1} = \frac{1 + (r - \rho)\lambda_i}{1 + r\lambda_i} y_i^n \qquad \forall i = 1, \cdots, N_1.}$$

Using (2.45), we now study the rate of convergence of Algorithm (2.12)-(2.14) for different choices of ρ with r given. This will allow us to draw a number of conclusions concerning the choice of r.

The case $r = 0$.

We have

$$\text{(2.46)} \qquad y_i^{n+1} = (1 - \rho\lambda_i) y_i^n \qquad \forall i = 1, \cdots, N_1, \quad n \geq 0.$$

It follows from (2.46) that the optimal choice for ρ is

$$\text{(2.47)} \qquad \rho_{\text{opt}} = \frac{2}{\lambda_m + \lambda_M}.$$

For $\rho = \rho_{\text{opt}}$, we then have

$$\text{(2.48)} \qquad |y_i^{n+1}| \leq \left(\frac{1 - (\lambda_m/\lambda_M)}{1 + (\lambda_m/\lambda_M)}\right) |y_i^n| \qquad \forall i = 1, \cdots, N_1, \quad n \geq 0.$$

We deduce from (2.48) that, for $\rho = \rho_{\text{opt}}$, Algorithm (2.12)-(2.14) converges linearly with an asymptotic constant R satisfying

$$\text{(2.49)} \qquad R \leq \frac{1 - (\lambda_m/\lambda_M)}{1 + (\lambda_m/\lambda_M)}.$$

Remark 2.4. If $r = 0$, Algorithm (2.12)-(2.14) reduces to Uzawa's algorithm applied to the Lagrangian $\mathcal{L}$ defined by (2.6), namely

$$\text{(2.50)} \qquad \mathbf{p}^0 \in \mathbb{R}^M, \textit{ specified arbitrarily,}$$

$$\mathbf{A}\mathbf{u}^n + \mathbf{B}^t\mathbf{p}^n = \mathbf{b}, \tag{2.51}$$

$$\mathbf{p}^{n+1} = \mathbf{p}^n + \rho_n(\mathbf{B}\mathbf{u}^n - \mathbf{c}). \tag{2.52}$$

The result (2.49) is a standard one in the study of the convergence of fixed-step gradient methods (see, for example, Cea [1971] and Marchouk and Kuznetzov [1974] and the references therein).

The case $\rho = 2r$.

We deduce from (2.45) that

$$y_i^{n+1} = \frac{1 - r\lambda_i}{1 + r\lambda_i} y_i^n \qquad \forall i = 1, \cdots, N_1, \quad n \geq 0. \tag{2.53}$$

As indicated by Fig. 2.1, the optimal choice for r is

$$r_{\text{opt}} = \frac{1}{\sqrt{\lambda_m \lambda_M}}. \tag{2.54}$$

For $r = r_{\text{opt}}$, $\rho = 2r_{\text{opt}}$, we have

$$|y_i^{n+1}| \leq \frac{1 - (\lambda_m/\lambda_M)^{1/2}}{1 + (\lambda_m/\lambda_M)^{1/2}} |y_i^n| \quad \forall i = 1, \cdots, N_1, \quad n \geq 0, \tag{2.55}$$

implying that, for this choice of r and ρ, the asymptotic constant R of the method satisfies

$$R \leq \frac{1 - (\lambda_m/\lambda_M)^{1/2}}{1 + (\lambda_m/\lambda_M)^{1/2}}. \tag{2.56}$$

We deduce from (2.49) and (2.56) as well as from the behavior of the function $\xi \to (1-\xi)/(1+\xi)$ that Algorithm (2.12)–(2.14) with $r = 1/(\lambda_m\lambda_M)^{1/2}$ and $\rho = 2/(\lambda_m\lambda_M)^{1/2}$ is iteratively faster than Algorithm (2.12)–(2.14) with $r = 0$ and $\rho_n = \rho = 2/(\lambda_m + \lambda_M)$.

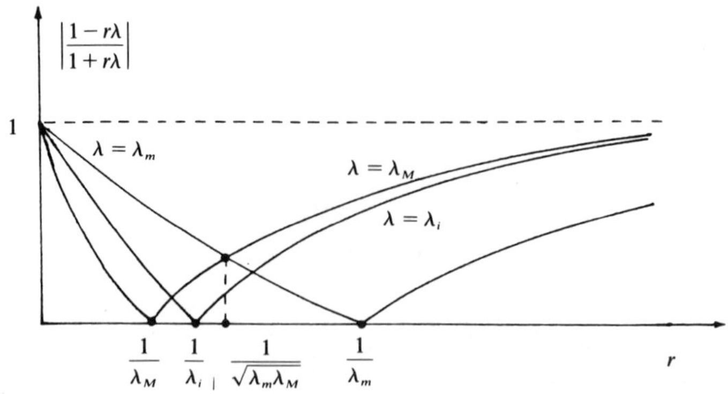

FIG. 2.1

The case $\rho = r$.
This choice is the standard one; in this case, we deduce from (2.45) that

$$y_i^{n+1} = \frac{1}{1+r\lambda_i} y_i^n \qquad \forall i = 1, \cdots, N_1, \quad n \geq 0,$$

and, hence, that

$$|y_i^{n+1}| \leq \frac{1}{1+r\lambda_m} |y_i^n| \qquad \forall i = 1, \cdots, N_1, \quad n \geq 0. \tag{2.57}$$

Therefore, the asymptotic constant R satisfies

$$R \leq \frac{1}{1+r\lambda_m}. \tag{2.58}$$

It follows from (2.58) that, for $\rho_n = \rho = r$, Algorithm (2.12)–(2.14) becomes faster, iteratively, as the value of r gets larger. If, in particular, $r \ll \lambda_m^{-1}$, it follows from (2.58) that Algorithm (2.12)–(2.14) will in general be iteratively slow. We note that, if $r = \lambda_m^{-1}$, then $R \leq \frac{1}{2}$.

The optimal choice of ρ for a given r.

It follows from (2.45) (and Fig. 2.2) that the optimal value of ρ is the solution of the linear equation

$$\frac{1+(r-\rho)\lambda_m}{1+r\lambda_m} = -\frac{1+(r-\rho)\lambda_M}{1+r\lambda_M},$$

and, hence, that

$$\rho_{\text{opt}} = r + \frac{2+r(\lambda_m+\lambda_M)}{2r\lambda_m\lambda_M + (\lambda_m+\lambda_M)}. \tag{2.59}$$

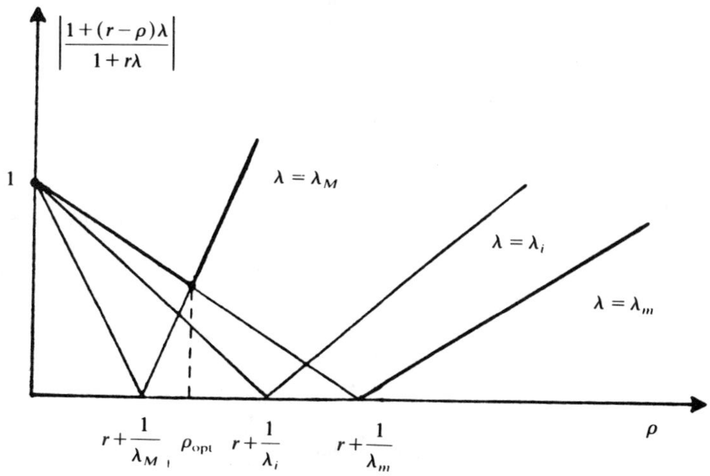

FIG. 2.2

We deduce from (2.59) that, for $\rho=\rho_{\text{opt}}$, we have

$$(2.60)\qquad |y_i^{n+1}|\le\frac{1-(\lambda_m/\lambda_M)}{1+(\lambda_m/\lambda_M)+2r\lambda_m}|y_i^n|\qquad\forall i=1,\cdots,N_1,\quad n\ge 0,$$

and, hence, for the asymptotic constant R, we have

$$(2.61)\qquad R\le\frac{1-(\lambda_m/\lambda_M)}{1+(\lambda_m/\lambda_M)+2r\lambda_m}.$$

Note that $\rho_{\text{opt}}>r$ and that, for the same value of r, Algorithm (2.12)-(2.14) is iteratively faster with ρ given by (2.59) than with $\rho=r$; of course, (2.59) involves λ_m and λ_M, quantities that in general are not known a priori.

2.3.6. On the choice of *r*. Although relation (2.58) appears to indicate it is advantageous to work with $\rho_n=\rho=r$ as large as possible, one must realize that, all other things being equal, the determination of $\mathbf{u}^n$ in (2.13), i.e., the solution of the linear system

$$(2.62)\qquad (\mathbf{A}+r\mathbf{B}^t\mathbf{B})\mathbf{u}^n=\mathbf{b}+r\mathbf{B}^t\mathbf{c}-\mathbf{B}^t\mathbf{p}^n$$

is more costly in computation time and/or memory requirements the larger the value of r is. In fact, as we shall see in the following discussion, the matrix $\mathbf{A}_r=\mathbf{A}+r\mathbf{B}^t\mathbf{B}$ becomes progressively more ill conditioned the larger r becomes.

First, it is appropriate to define some notation. We shall denote by $|\mathbf{v}|$ the standard Euclidian norm on $\mathbb{R}^N$, and for a linear operator $\mathbf{L}$ defined on $\mathbb{R}^N$, we shall denote by $\|\mathbf{L}\|$ the norm associated with $|\cdot|$, namely

$$\|\mathbf{L}\|=\sup_{\mathbf{v}\in\mathbb{R}^N-\{\mathbf{0}\}}\frac{|\mathbf{L}\mathbf{v}|}{|\mathbf{v}|}=\sup_{\mathbf{v}\in S}|\mathbf{L}\mathbf{v}|,$$

where

$$S=\{\mathbf{v}\,|\,\mathbf{v}\in\mathbb{R}^N,\ |\mathbf{v}|=1\}.$$

For the condition number of $\mathbf{A}_r$ $(=\mathbf{A}+r\mathbf{B}^t\mathbf{B})$ when r approaches $+\infty$, we have the following proposition.

PROPOSITION 2.3. *The condition number* $\nu(\mathbf{A}_r)$ $(=\|\mathbf{A}_r\|\,\|\mathbf{A}_r^{-1}\|)$ *of* $\mathbf{A}_r$ *satisfies*

$$(2.63)\qquad \nu(\mathbf{A}_r)\simeq r\frac{\|\mathbf{B}\|^2}{\sigma}\quad\text{when } r \text{ approaches } +\infty,$$

where

$$\sigma=\inf_{\mathbf{v}\in\ker(\mathbf{B})-\{\mathbf{0}\}}\frac{(\mathbf{A}\mathbf{v},\mathbf{v})}{|\mathbf{v}|^2}.$$

Proof. See Fortin and Glowinski ([1982, Chap. 1], [1983]) for a proof. □

Thus, the condition number of $\mathbf{A}_r$ is asymptotically proportional to r. Then, as r increases, it is more difficult, other things being equal, to solve the system

$$\mathbf{A}_r\mathbf{u}^n = \mathbf{b} + r\mathbf{B}^t\mathbf{c} - \mathbf{B}^t\mathbf{p}^n. \tag{2.64}$$

Indeed, if we solve (2.64) by a standard iterative method, the convergence, being related to the condition number, will become slower; it may thus require a large number of iterations to solve (2.64) to an appropriate level of accuracy even if, in the obvious manner, we initialize the calculation of $\mathbf{u}^n$ with $\mathbf{u}^{n-1}$. Furthermore, if we solve (2.64) by a direct method, the sensitivity to round-off errors will be greater when r is large.

For a large number of problems, therefore, a good strategy would seem to be the following.

(i) Work in "double precision."

(ii) With the parameter r having a fixed value as large as possible, carry out once and for all the Cholesky factorization of the matrix $\mathbf{A}_r$, which, we recall, is symmetric and positive-definite.

(iii) Take $\rho_n = \rho = r$ (or use a conjugate-gradient method like the one presented in § 2.4).

Remark 2.5. When an iterative method is used to solve (2.64), we can, in the early stages of Algorithm (2.12)-(2.14), proceed with a low accuracy in the determination of $\mathbf{u}^n$. This effect can be obtained, for example, by choosing to use a fixed (and "small") number of iterations (in the solution of (2.64)); see Fortin and Glowinski [1982], [1983] for a discussion of such algorithms.

Remark 2.6. When solving (2.64) by an iterative method, it may be advantageous to use a parameter r that varies with n, giving, in fact, a sequence $\{r_n\}_n$. Some authors recommend the use of a sequence $\{r_n\}_n$ such that

$$r_0 \geq 0, \quad r_{n+1} \geq r_n \quad \forall n \geq 0 \quad \text{and} \quad \lim_{n\to+\infty} r_n = +\infty.$$

However, the optimal choice for $\{r_n\}_n$ seems to be an open question. The use of such a method combined with a direct solution of (2.64) is usually of little interest, since the factorization $\mathbf{A}_{r_n}$ would need to be carried out every time that $r_n > r_{n-1}$, and this, in general, would be costly.

2.4. Variable step-length algorithms. Conjugate-gradient methods.

2.4.1. General description. We have shown in § 2.3.4 that Algorithm (2.12)-(2.14) can be interpreted as a gradient algorithm applied to the minimization of the dual function J_r^* defined by (2.33). With this interpretation in mind, it is natural to seek to apply the standard iterative methods for minimization of quadratic functionals to the minimization of J_r^* on $\mathbb{R}^M$ (see, for example, Daniel [1970], Cea [1971], [1978], and the review article of Marchouk and Kuznetzov [1974] with its extensive bibliography for a thorough study of these methods). First, however, we will briefly describe these methods as applied to a standard, unconstrained, quadratic minimization problem in finite dimension.

Suppose that $\mathcal{A}$ is an $M \times M$ symmetric, positive-definite matrix, and suppose that $\boldsymbol{\beta} \in \mathbb{R}^M$; we associate to $\mathcal{A}$ and $\boldsymbol{\beta}$ the function $\mathcal{J}$ defined by

$$\mathcal{J}(\mathbf{q}) = \tfrac{1}{2}(\mathcal{A}\mathbf{q}, \mathbf{q}) - (\boldsymbol{\beta}, \mathbf{q}). \tag{2.65}$$

The minimization problem

$$\begin{cases} \mathbf{p} \in \mathbb{R}^M, \\ \mathcal{J}(\mathbf{p}) \le \mathcal{J}(\mathbf{q}) \qquad \forall \mathbf{q} \in \mathbb{R}^M, \end{cases} \tag{2.66}$$

admits a unique solution, which is also the solution of the linear system

$$\mathcal{A}\mathbf{p} = \boldsymbol{\beta}. \tag{2.67}$$

To solve (2.66) and (2.67) we now consider methods of the general type.

ALGORITHM (2.68)–(2.69).

$$\mathbf{p}^0 \in \mathbb{R}^M, \text{ specified arbitrarily;} \tag{2.68}$$

then, for $n \ge 0$, with $\mathbf{p}^n$ known, calculate $\mathbf{p}^{n+1}$ by

$$\mathbf{p}^{n+1} = \mathbf{p}^n - \rho_n \mathbf{w}^n. \tag{2.69}$$

The descent direction $\mathbf{w}^n$ will, in general, be deduced from the direction of the gradient of $\mathcal{J}$ at the point $\mathbf{p}^n$. For a given descent direction, we shall choose ρ_n in such a way as to optimize a criterion related to the problem. In practice, we shall confine ourselves to the following methods.

Steepest descent method. The descent is made in the direction opposite to the gradient; hence,

$$\boxed{\mathbf{w}^n = \mathbf{g}^n = \nabla \mathcal{J}(\mathbf{p}^n) = \mathcal{A}\mathbf{p}^n - \boldsymbol{\beta}.} \tag{2.70}$$

The choice of ρ_n is made by minimizing, with respect to ρ, the function

$$\rho \to \mathcal{J}(\mathbf{p}^n - \rho \mathbf{g}^n). \tag{2.71}$$

Since we have

$$\mathcal{J}(\mathbf{p}^n - \rho \mathbf{g}^n) = \mathcal{J}(\mathbf{p}^n) - \rho |\mathbf{g}^n|^2 + \frac{\rho^2}{2}(\mathcal{A}\mathbf{g}^n, \mathbf{g}^n), \tag{2.72}$$

ρ_n is thus given by

(2.73) $$\rho_n = \frac{|\mathbf{g}^n|^2}{(\mathscr{A}\mathbf{g}^n, \mathbf{g}^n)}.$$

We observe that (2.69) and (2.70) imply

(2.74) $$\mathbf{g}^{n+1} = \mathbf{g}^n - \rho_n \mathscr{A}\mathbf{g}^n \qquad \forall n \geq 0,$$

which is most important in practice. Indeed, due to (2.74), we can save one matrix vector multiplication in the computation of $\mathbf{g}^{n+1}$.

Minimum residual method. The descent is made in the gradient direction, and, hence, $\mathbf{w}^n = \mathbf{g}^n$ in (2.69). We choose ρ_n so as to minimize, with respect to ρ, the residual $|\mathscr{A}(\mathbf{p}^n - \rho\mathbf{g}^n) - \boldsymbol{\beta}|$.

By a direct algebraic development, we have

(2.75) $$|\mathscr{A}(\mathbf{p}^n - \rho\mathbf{g}^n) - \boldsymbol{\beta}|^2 = |\mathbf{g}^n - \rho\mathscr{A}\mathbf{g}^n|^2 = |\mathbf{g}^n|^2 - 2\rho(\mathscr{A}\mathbf{g}^n, \mathbf{g}^n) + \rho^2|\mathscr{A}\mathbf{g}^n|^2;$$

the optimal ρ is, therefore, given by

(2.76) $$\boxed{\rho_n = \frac{(\mathscr{A}\mathbf{g}^n, \mathbf{g}^n)}{|\mathscr{A}\mathbf{g}^n|^2}.}$$

The equations characterizing the steepest descent method also characterize this method with the exception of (2.73), which has to be replaced by (2.76).

Conjugate-gradient method. The conjugate-gradient method is especially attractive for solving quadratic problems because, theoretically (i.e., ignoring round-off errors), it converges in a finite number of iterations ($\leq M$); moreover, in the general case it leads to a fairly fast convergence (for M "large," this fast convergence is more attractive than is convergence in a finite number of iterations). For detailed studies of the convergence of this method, we refer the reader to Daniel [1970], Polak [1971], Concus and Golub [1976], Hestenes [1980], Golub and Meurant [1983], etc.

For convenience we shall split the description of the algorithm into three steps.

ALGORITHM (2.77)–(2.84).

Step 0: Initialization

(2.77) $$\mathbf{p}^0 \in \mathbb{R}^M, \text{ given arbitrarily,}$$

(2.78) $$\mathbf{g}^0 = \mathscr{A}\mathbf{p}^0 - \boldsymbol{\beta},$$

(2.79) $$\mathbf{w}^0 = \mathbf{g}^0.$$

□

Then, for $n \geq 0$, with $\mathbf{p}^n$, $\mathbf{g}^n$, $\mathbf{w}^n$ known, we compute $\mathbf{p}^{n+1}$, $\mathbf{g}^{n+1}$, $\mathbf{w}^{n+1}$ as follows.

Step 1: Descent

$$\begin{cases} \text{Find } \rho_n \in \mathbb{R} \text{ such that} \\ \mathcal{J}(\mathbf{p}^n - \rho_n \mathbf{w}^n) \leq \mathcal{J}(\mathbf{p}^n - \rho \mathbf{w}^n) \qquad \forall \rho \in \mathbb{R}, \end{cases} \tag{2.80}$$

$$\mathbf{p}^{n+1} = \mathbf{p}^n - \rho_n \mathbf{w}^n, \tag{2.81}$$

$$\mathbf{g}^{n+1} = \mathcal{A}\mathbf{p}^{n+1} - \boldsymbol{\beta}. \tag{2.82}$$

Step 2: New Descent Direction

$$\mathbf{w}^{n+1} = \mathbf{g}^{n+1} + \gamma_n \mathbf{w}^n, \tag{2.83}$$

where γ_n is such that

$$(\mathcal{A}\mathbf{w}^{n+1}, \mathbf{w}^n) = 0. \tag{2.84}$$

□

Do $n = n+1$ and go to (2.80).

Combining (2.81) and (2.82), we have

$$\boxed{\mathbf{g}^{n+1} = \mathbf{g}^n - \rho_n \mathcal{A}\mathbf{w}^n,} \tag{2.85}$$

which, in practice, should be used instead of (2.82) since it saves one matrix vector multiplication. On the other hand, it follows from (2.83), (2.84) that we have

$$\gamma_n = -\frac{(\mathcal{A}\mathbf{g}^{n+1}, \mathbf{w}^n)}{(\mathcal{A}\mathbf{w}^n, \mathbf{w}^n)}. \tag{2.86}$$

We can further prove the orthogonality relations

$$\begin{cases} (\mathbf{g}^i, \mathbf{g}^j) = 0 & \text{if } i \neq j, \\ (\mathbf{g}^i, \mathbf{w}^j) = 0 & \text{if } i > j, \\ (\mathcal{A}\mathbf{w}^i, \mathbf{w}^j) = 0 & \text{if } i \neq j. \end{cases} \tag{2.87}$$

By virtue of these relations, (2.86) can be reduced by elementary manipulations to

$$\boxed{\gamma_n = \frac{|\mathbf{g}^{n+1}|^2}{|\mathbf{g}^n|^2}.} \tag{2.88}$$

By a calculation analogous to that carried out in the steepest descent method, we obtain

$$\rho_n = \frac{(\mathbf{g}^n, \mathbf{w}^n)}{(\mathcal{A}\mathbf{w}^n, \mathbf{w}^n)} \left(= \frac{|\mathbf{g}^n|^2}{(\mathcal{A}\mathbf{w}^n, \mathbf{w}^n)} \right) \tag{2.89}$$

It can be observed from (2.89) that the determination of ρ_n requires the calculation of $\mathcal{A}\mathbf{w}^n$, which is then used in (2.85). Thus, using (2.85) instead of (2.82) saves one multiplication by $\mathcal{A}$, which can be very substantial in many applications.

Remark 2.7. In all three algorithms, we should stop iterating as soon as $|\mathbf{g}^n| \le \varepsilon$, where ε is a "small" positive number given in advance (in fact, $|\mathbf{g}^n|/|\mathbf{g}^0| \le \varepsilon$ or $|\mathbf{g}^n|/|\boldsymbol{\beta}| \le \varepsilon$ are more convenient tests). Concerning the conjugate-gradient Algorithm (2.77)–(2.84), it follows from the orthogonality relations (2.87) that $\mathbf{w}^n = \mathbf{0}$ implies $\mathbf{g}^n = \mathbf{0}$ and thus $\mathbf{p}^n = \mathbf{p}$.

Remark 2.8. We have supposed at the beginning of § 2.4.1 that $\mathcal{A}$ is a positive-definite matrix; actually, the convergence properties of the above algorithms still hold if we suppose that $\mathcal{A}$ is only positive-semidefinite with $\boldsymbol{\beta} \in R(\mathcal{A})$. When this is the case, the algorithms converge to the solution $\mathbf{p}$ of (2.67) such that

$$\mathbf{p} = \hat{\mathbf{p}} + \mathbf{p}_2^0, \tag{2.90}$$

where $\hat{\mathbf{p}}$ is the unique solution of (2.67) in $R(\mathcal{A})$ and $\mathbf{p}_2^0$ is the component of $\mathbf{p}^0$ in $\ker(\mathcal{A})$ in the decomposition

$$\mathbb{R}^M = R(\mathcal{A}) \oplus \ker(\mathcal{A}).$$

Remark 2.9. The minimum residual algorithm still makes sense if $\mathcal{A}$ is only positive-semidefinite without being symmetric (with $\boldsymbol{\beta} \in R(\mathcal{A})$ as above).

2.4.2. Application to the minimization of J_r^*. We shall now show how the methods described in § 2.4.1 can be applied to the minimization of J_r^* and thus to the solution of problem (2.3). We recall from § 2.3.4 that we have

$$\begin{cases} J_r^*(\mathbf{q}) = \frac{1}{2}(\mathbf{B}\mathbf{A}_r^{-1}\mathbf{B}^t\mathbf{q}, \mathbf{q}) - (\mathbf{B}\mathbf{A}_r^{-1}(\mathbf{b} + r\mathbf{B}^t\mathbf{c}), \mathbf{q}) + (\mathbf{c}, \mathbf{q}) \\ \qquad + \frac{1}{2}(\mathbf{A}_r^{-1}(\mathbf{b} + r\mathbf{B}^t\mathbf{c}), \mathbf{b} + r\mathbf{B}^t\mathbf{c}) - \frac{r}{2}|\mathbf{c}|^2. \end{cases} \tag{2.91}$$

Obviously, the constant term plays no part in the minimization. We thus have, using the notation of § 2.4.1,

$$\mathcal{A} = \mathbf{B}\mathbf{A}_r^{-1}\mathbf{B}^t, \tag{2.92}$$

$$\boldsymbol{\beta} = \mathbf{B}\mathbf{A}_r^{-1}(\mathbf{b} + r\mathbf{B}^t\mathbf{c}) - \mathbf{c}. \tag{2.93}$$

The matrix $\mathscr{A}$ given by (2.92) is only positive-semidefinite, but, as stated in Remark 2.8, this does not affect the three algorithms considered in § 2.4.1. We can prove, as in Theorem 2.3, that the component of $\mathbf{p}^n$ in ker $(\mathbf{B}^t)$ is in fact constant and therefore equal to that of $\mathbf{p}^0$. Under these conditions, the three algorithms considered in § 2.4.1 become as follows (we give directly practical descriptions).

Steepest descent algorithm.

ALGORITHM (2.94)–(2.101).

(2.94) $$\mathbf{p}^0 \in \mathbb{R}^M \text{ is given,}$$

(2.95) $$\mathbf{u}^0 = \mathbf{A}_r^{-1}(\mathbf{b} + r\mathbf{B}^t\mathbf{c} - \mathbf{B}^t\mathbf{p}^0),$$

(2.96) $$\mathbf{g}^0 = \mathbf{c} - \mathbf{B}\mathbf{u}^0.$$

Then, for $n \geq 0$, assuming $\mathbf{p}^n$, $\mathbf{u}^n$, $\mathbf{g}^n$ known, we compute $\mathbf{p}^{n+1}$, $\mathbf{u}^{n+1}$, $\mathbf{g}^{n+1}$ as follows.

Solve the linear problem

(2.97) $$\mathbf{A}_r\mathbf{z}^n = \mathbf{B}^t\mathbf{g}^n,$$

then

(2.98) $$\rho_n = \frac{|\mathbf{g}^n|^2}{(\mathbf{B}\mathbf{z}^n, \mathbf{g}^n)},$$

(2.99) $$\mathbf{p}^{n+1} = \mathbf{p}^n - \rho_n\mathbf{g}^n,$$

(2.100) $$\mathbf{u}^{n+1} = \mathbf{u}^n + \rho_n\mathbf{z}^n,$$

(2.101) $$\mathbf{g}^{n+1} = \mathbf{g}^n - \rho_n\mathbf{B}\mathbf{z}^n.$$

We stop iterating as soon as $|\mathbf{g}^n|/|\mathbf{g}^0|$ is sufficiently small.

Minimum residual algorithm. This is the same algorithm as Algorithm (2.94)–(2.101) except that (2.98) has to be replaced by

(2.102) $$\rho_n = \frac{(\mathbf{B}\mathbf{z}^n, \mathbf{g}^n)}{|\mathbf{B}\mathbf{z}^n|^2}.$$

Conjugate-gradient algorithm.

ALGORITHM (2.103)–(2.113).

Step 0: Initialization

(2.103) $$\mathbf{p}^0 \in \mathbb{R}^M \text{ is given,}$$

(2.104) $$\mathbf{u}^0=\mathbf{A}_r^{-1}(\mathbf{b}+r\mathbf{B}^t\mathbf{c}-\mathbf{B}^t\mathbf{p}^0),$$

(2.105) $$\mathbf{g}^0=\mathbf{c}-\mathbf{B}\mathbf{u}^0,$$

(2.106) $$\mathbf{w}^0=\mathbf{g}^0. \qquad \square$$

Then, for $n\geq 0$, with $\mathbf{p}^n$, $\mathbf{u}^n$, $\mathbf{g}^n$, $\mathbf{w}^n$ known, we compute $\mathbf{p}^{n+1}$, $\mathbf{u}^{n+1}$, $\mathbf{g}^{n+1}$, $\mathbf{w}^{n+1}$ as follows.

Step 1: Descent

Solve

(2.107) $$\mathbf{A}_r\mathbf{z}^n=\mathbf{B}^t\mathbf{w}^n,$$

then

(2.108) $$\rho_n=\frac{|\mathbf{g}^n|^2}{(\mathbf{B}\mathbf{z}^n,\mathbf{w}^n)},$$

(2.109) $$\mathbf{p}^{n+1}=\mathbf{p}^n-\rho_n\mathbf{w}^n,$$

(2.110) $$\mathbf{u}^{n+1}=\mathbf{u}^n+\rho_n\mathbf{z}^n.$$

Step 2: New Descent Direction

(2.111) $$\mathbf{g}^{n+1}=\mathbf{g}^n-\rho_n\mathbf{B}\mathbf{z}^n,$$

(2.112) $$\gamma_n=\frac{|\mathbf{g}^{n+1}|^2}{|\mathbf{g}^n|^2},$$

(2.113) $$\mathbf{w}^{n+1}=\mathbf{g}^{n+1}+\gamma_n\mathbf{w}^n. \qquad \square$$

Do $n=n+1$ and go to (2.107).

Remark 2.10. In the three algorithms described above, we have to solve at each iteration only a single linear system, with matrix $\mathbf{A}_r$. Compared with Algorithm (2.12)-(2.14) used with a fixed ρ, these algorithms require the presence of additional vectors in memory. This requirement for increased memory will be justified if the automatic determination of the step-length ρ_n leads to a very clear improvement in the speed of convergence compared with Algorithm (2.12)-(2.14) used with $\rho_n=\rho=r$. This is very often, but not always, the case.

Remark 2.11. The conjugate-gradient Algorithm (2.103)-(2.113) converges theoretically in N_1 (=rank $\mathbf{B}$)-iterations at most. Since round-off errors are present, this result does not hold in practice. Furthermore, when considering the large size of problems arising from the discretization of partial differential equations, it is desirable that, for these problems, convergence should be obtained in a number of iterations considerably less than N_1, with an adequate termination test. This essentially can be achieved if the condition number of

$\mathbf{BA}_r^{-1}\mathbf{B}^t$ restricted to $R(\mathbf{B})$ is small, this quantity being henceforth denoted by $\nu(\mathbf{BA}_r^{-1}\mathbf{B}^t)_{R(\mathbf{B})}$. It can be shown that

$$\lim_{r\to+\infty} \nu(\mathbf{BA}_r^{-1}\mathbf{B}^t)_{R(\mathbf{B})} = 1. \tag{2.114}$$

We note that

$$\mathbf{BA}_r^{-1}\mathbf{B}^t = \mathbf{B}(\mathbf{I}+r\mathbf{A}^{-1}\mathbf{B}^t\mathbf{B})^{-1}\mathbf{A}^{-1}\mathbf{B}^t, \tag{2.115}$$

and that the matrix corresponding to the case $r=0$ is $\mathbf{BA}^{-1}\mathbf{B}^t$. It follows from these properties that the replacement of the Lagrangian $\mathcal{L}$, defined by (2.6), by the augmented Lagrangian $\mathcal{L}_r$, defined by (2.10), may be considered a method of preconditioning in a sense close to that of Axelsson [1976], the preconditioning matrix being $(\mathbf{I}+r\mathbf{A}^{-1}\mathbf{B}^t\mathbf{B})$. This remark is true not only for the conjugate-gradient algorithm but also for the other algorithms studied in the preceding sections. In particular, in view of this preconditioning, it would seem unnecessary to carry out a reinitialization (of the type $\mathbf{w}^n=\mathbf{g}^n$) in the conjugate-gradient algorithm to counteract the accumulation of rounding errors.

2.5. Further comments.

Remark 2.12. Suppose that $\mathbf{A}\in\mathcal{L}(\mathbb{R}^N,\mathbb{R}^N)$ is positive-definite and not necessarily symmetric, and consider the solution of problem

$$\mathbf{Au}+\mathbf{B}^t\mathbf{p}=\mathbf{b}, \qquad \mathbf{Bu}=\mathbf{c}. \tag{2.116}$$

This problem has a unique solution in $\mathbb{R}^N\times(\mathbb{R}^M/\ker\mathbf{B}^t)$. For its solution, consider

ALGORITHM (2.117)–(2.119).

$$\mathbf{p}^0\in\mathbb{R}^M \text{ specified arbitrarily;} \tag{2.117}$$

with $\mathbf{p}^n$ known, compute $\mathbf{u}^n$, then $\mathbf{p}^{n+1}$, by

$$(\mathbf{A}+r\mathbf{B}^t\mathbf{B})\mathbf{u}^n=\mathbf{b}-\mathbf{B}^t\mathbf{p}^n+r\mathbf{B}^t\mathbf{c}, \tag{2.118}$$

$$\mathbf{p}^{n+1}=\mathbf{p}^n+\rho_n(\mathbf{Bu}^n-\mathbf{c}), \qquad \rho_n>0. \tag{2.119}$$

Algorithm (2.117)–(2.119) coincides with Algorithm (2.12)–(2.14) when $\mathbf{A}$ is symmetric.

By proceeding as for Theorem 2.2, it can easily be shown that Algorithm (2.117)–(2.119) converges whatever the value of $\mathbf{p}^0$ is, subject to the condition that

$$0<\alpha_0\le\rho_n\le\alpha_1<2\left(r+\frac{1}{\beta^2}\right), \tag{2.120}$$

where β^2 is defined by

$$\beta^2=\max_{\mathbf{v}\neq\mathbf{0}}\frac{|\mathbf{Bv}|^2}{(\mathbf{A}_\sigma\mathbf{v},\mathbf{v})}, \tag{2.121}$$

and where, in (2.121), $\mathbf{A}_\sigma$ is the symmetric component of $\mathbf{A}$, i.e., $\mathbf{A}_\sigma = \frac{1}{2}(\mathbf{A}+\mathbf{A}^t)$. By contrast, a "finely detailed" study of convergence rates seems more difficult, since the spectral methods of § 2.3 cannot then be used. Likewise, the extension to problem (2.116) of the variable step-length and conjugate-gradient methods of § 2.4 may pose difficulties; this applies particularly to the conjugate-gradient method.

In Fortin and Glowinski ([1982, Chap. 2], [1983]), the authors use algorithms of the Algorithm (2.117)-(2.119) type, with $\mathbf{A}$ nonsymmetric, for the solution of the Navier-Stokes equations.

Remark 2.13. We complete the above sections—and also Remark 2.12—by considering the case of (2.116) where $\mathbf{c} \notin R(\mathbf{B})$. In this case, (2.116) has no solution. Apply, however, Algorithm (2.117)-(2.119) to the solution of that ill-posed problem. We can show that, under condition (2.120), we have

$$\lim_{n\to+\infty} \mathbf{u}^n = \mathbf{u}^*, \tag{2.122}$$

where $\mathbf{u}^*$ is the solution of the problem

$$\begin{cases} \mathbf{u}^* \in H^* = \{\mathbf{v} \mid \mathbf{v} \in \mathbb{R}^N,\ \mathbf{B}^t(\mathbf{B}\mathbf{v}-\mathbf{c}) = \mathbf{0}\}, \\ (\mathbf{A}\mathbf{u}^*, \mathbf{v}) = (\mathbf{b}, \mathbf{v}) \qquad \forall \mathbf{v} \in \ker \mathbf{B}^t\mathbf{B}. \end{cases} \tag{2.123}$$

Problem (2.123) is equivalent to the linear system

$$\begin{cases} \mathbf{A}\mathbf{u} + \mathbf{B}^t\mathbf{p} = \mathbf{b}, \\ \mathbf{B}^t\mathbf{B}\mathbf{u} = \mathbf{B}^t\mathbf{c}. \end{cases} \tag{2.124}$$

We know that H^* ($\neq\emptyset$) is the set of the solutions of the normal equation

$$\mathbf{B}^t\mathbf{B}\mathbf{z} = \mathbf{B}^t\mathbf{c}. \tag{2.125}$$

We can likewise show that the convergence of $\mathbf{u}^n$ to $\mathbf{u}$ is linear. Regarding the sequence $\{\mathbf{p}^n\}_{n\geq 0}$, it follows from (2.119) and from the fact that $\mathbf{c} \notin R(\mathbf{B})$, that this sequence diverges like an arithmetic progression. This divergence is "much slower" than the convergence of $\{\mathbf{u}^n\}_{n\geq 0}$, which means that, in practice, there will be no risk of "overflow."

The convergence result stated above shows the robustness of the methods described in this section, particularly in the presence of round-off errors. In actual fact, the condition $\mathbf{c} \in R(\mathbf{B})$ can no longer be satisfied exactly because of these errors; nonetheless, the above convergence results show that the Lagrangian methods remain usable and provide the best possible result (in the least-squares sense) in this "noisy" environment.

Remark 2.14. In the particular case where $\rho = r$, we can easily eliminate $\mathbf{p}^n$ in Algorithm (2.12)-(2.14). We then obtain

$$(\mathbf{A} + r\mathbf{B}^t\mathbf{B})\mathbf{u}^{n+1} = \mathbf{A}\mathbf{u}^n + r\mathbf{B}^t\mathbf{c}.$$

This formulation of Algorithm (2.12)–(2.14) has been discussed by Gunzburger, Liu, and Nicolaides [1983] for the solution of the Stokes problem. Actually, in the case of the Stokes problem, it has been proved in Fortin and Glowinski ([1981, Chap 2], [1983]) that this algorithm is less efficient than the conjugate-gradient Algorithm (2.103)–(2.113).

Remark 2.15. Linear systems such as

$$\begin{cases} \mathbf{A}\mathbf{u}+\mathbf{B}^t\mathbf{p}=\mathbf{b}, \\ \mathbf{B}\mathbf{u}=\mathbf{c}, \end{cases}$$

occur in many applications such as mechanics, statistics, etc. The practical solution of such systems has motivated several papers; among others, we shall mention Gill and Murray [1974], Luenberger [1970], Paige and Saunders [1975], Dyn and Ferguson [1983], and also Fortin and Glowinski ([1982, Chap. 1], [1983]).

3. Application to the Stokes problem.

3.1. Physical motivation and formulation of the problem. In § 5 of Chapter 1, we discussed formulations of flows of various viscous media. In this section, we wish to apply the concepts introduced in § 2 of this chapter—which, so far, we have applied to finite-dimensional problems only—to the solution of the Stokes problem, which plays an important role in computational fluid dynamics.

Our starting point will be the following time-dependent Navier–Stokes equations modeling the unsteady flows of Newtonian incompressible viscous fluids (see Chap. 1, § 5.3 for their derivation).

$$\frac{\partial \mathbf{u}}{\partial t}-\nu\Delta\mathbf{u}+(\mathbf{u}\cdot\nabla)\mathbf{u}+\nabla p=\mathbf{f}, \tag{3.1}$$

$$\nabla\cdot\mathbf{u}=0 \qquad \text{(incompressibility condition)}, \tag{3.2}$$

where $\mathbf{u}=\{u_i\}_{i=1}^N$ is the velocity vector, p is the pressure, $\mathbf{f}$ is the density of external forces, ν is a viscosity parameter, $\Delta\mathbf{u}$ denotes the Laplace operator $\sum_{i=1}^N \partial^2/\partial x_i^2$ operating on $\mathbf{u}$, $\nabla\cdot\mathbf{u}$ is the divergence of the vector field $\mathbf{u}$, and $(\mathbf{u}\cdot\nabla)\mathbf{u}$ is a symbolic notation for the nonlinear operator defined by

$$(\mathbf{u}\cdot\nabla)\mathbf{u}=\left\{\sum_{j=1}^N u_j\frac{\partial u_i}{\partial x_j}\right\}_{i=1}^N.$$

Some of the notation used in this section differs from that of Chapters 1 and 2; in fact, since this section on the Stokes problem can be read independently of the other chapters, we have used notation that is classical in the Navier–Stokes context, such as can be found in Lions [1969], Temam [1977], and Girault and Raviart [1986].

Boundary and initial-value conditions have to be prescribed; if $\Omega \subset \mathbb{R}^N$ ($N = 2, 3$ in practice) is the flow domain, and if Γ is its boundary, we shall suppose that

$$\mathbf{u}(\mathbf{x}, 0) = \mathbf{u}_0(\mathbf{x}) \quad \text{a.e. on } \Omega, \tag{3.3}$$

$$\mathbf{u} = \mathbf{u}_1 \quad \text{on } \Gamma. \tag{3.4}$$

Note that, from the incompressibility condition (3.2), the function $\mathbf{u}_1$ in (3.4) has to satisfy

$$\int_\Gamma \mathbf{u}_1 \cdot \mathbf{n} \, d\Gamma = 0, \tag{3.5}$$

where $\mathbf{n}$ is the unit vector of the outward normal at Γ.

More complicated boundary conditions than (3.4) can be associated with (3.1) and (3.2); see, for example, Glowinski [1984] for some examples and further references.

If we consider steady flows only, and if we neglect the nonlinear terms (the assumption of highly viscous flows), (3.1), (3.2), and (3.4) reduce to the following Stokes problem:

$$-\nu \Delta \mathbf{u} + \nabla p = \mathbf{f} \quad \text{in } \Omega, \tag{3.6}$$

$$\nabla \cdot \mathbf{u} = 0 \quad \text{in } \Omega, \tag{3.7}$$

$$\mathbf{u} = \mathbf{u}_1 \quad \text{on } \Gamma. \tag{3.8}$$

Linear problems closely related to (3.6)–(3.8) are also obtained when one discretizes the time-dependent problem by operator-splitting methods such as the following (with Δt (>0) denoting a time-discretization step).

$$\mathbf{u}^0 = \mathbf{u}_0, \tag{3.9}$$

then, for $n \geq 0$, with $\mathbf{u}^n$ known, compute $\{\mathbf{u}^{n+1/2}, p^{n+1/2}\}$ and $\mathbf{u}^{n+1}$ as follows.

$$\begin{cases} \dfrac{\mathbf{u}^{n+1/2} - \mathbf{u}^n}{\Delta t/2} - \dfrac{\nu}{2} \Delta \mathbf{u}^{n+1/2} + \nabla p^{n+1/2} = \mathbf{f}^{n+1/2} + \dfrac{\nu}{2} \Delta \mathbf{u}^n - (\mathbf{u}^n \cdot \nabla) \mathbf{u}^n, \\ \nabla \cdot \mathbf{u}^{n+1/2} = 0, \qquad \mathbf{u}^{n+1/2} = \mathbf{u}_1^{n+1/2} \quad \text{on } \Gamma, \end{cases} \tag{3.10}$$

$$\begin{cases} \dfrac{\mathbf{u}^{n+1} - \mathbf{u}^{n+1/2}}{\Delta t/2} - \dfrac{\nu}{2} \Delta \mathbf{u}^{n+1} + (\mathbf{u}^{n+1} \cdot \nabla) \mathbf{u}^{n+1} = \mathbf{f}^{n+1} + \dfrac{\nu}{2} \Delta \mathbf{u}^{n+1/2} - \nabla p^{n+1/2}, \\ \mathbf{u}^{n+1} = \mathbf{u}_1^{n+1} \quad \text{on } \Gamma. \end{cases} \tag{3.11}$$

In (3.10) and (3.11), $\mathbf{u}^\alpha$ (resp., p^α) approximates $\mathbf{u}$ (resp., p) at time $\alpha \Delta t$ (provided the above scheme converges); similarly, $\mathbf{f}^\alpha(\mathbf{x}) = \mathbf{f}(\mathbf{x}, \alpha \Delta t)$, $\mathbf{u}_1^\alpha(\mathbf{x}) = \mathbf{u}_1(\mathbf{x}, \alpha \Delta t)$. The above scheme is derived from the well-known Peaceman–Rachford alternating-direction scheme (see Glowinski [1984] for more details, as well as for other schemes using operator-splitting and for further references).

The solution of nonlinear elliptic problems such as (3.11) is discussed in this same reference; in this monograph we shall concentrate on the solution of linear problems such as (3.10), which all belong to the following family.

$$\alpha \mathbf{u} - \nu \Delta \mathbf{u} + \nabla p = \mathbf{f} \quad \text{in } \Omega, \tag{3.12}$$

$$\nabla \cdot \mathbf{u} = 0 \quad \text{in } \Omega, \tag{3.13}$$

$$\mathbf{u} = \mathbf{u}_1 \quad \text{on } \Gamma \quad \left(\text{with} \int_\Gamma \mathbf{u}_1 \cdot \mathbf{n}\, d\Gamma = 0\right), \tag{3.14}$$

where the constants α and ν satisfy $\alpha \geq 0$, $\nu > 0$.

3.2. Mathematical properties of problem (3.12)–(3.14). Until now, (3.12)-(3.14) are formal differential equations only that can be considered either in a strong (Frechet) or in a weak (distributional) sense. This section studies the mathematical properties of (3.12)-(3.14) when they are considered as weak equations set in $\mathbf{H}^{-1}(\Omega)$, i.e., in the topological dual of the Hilbert space

$$\mathbf{H}_0^1(\Omega) = \{\mathbf{v} \in \mathbf{H}^1(\Omega), \mathbf{v} = \mathbf{0} \text{ on } \Gamma\}.$$

By definition, the following is the weak formulation of (3.12)-(3.14) in $\mathbf{H}^{-1}(\Omega)$.

$$\begin{cases} \text{Find } \{\mathbf{u}, p\} \text{ in } \mathbf{H}^1(\Omega) \times L^2(\Omega) \text{ such that} \\ \alpha\langle \mathbf{u}, \mathbf{v}\rangle - \nu\langle \Delta \mathbf{u}, \mathbf{v}\rangle + \langle \nabla p, \mathbf{v}\rangle = \langle \mathbf{f}, \mathbf{v}\rangle \qquad \forall \mathbf{v} \in \mathbf{H}_0^1(\Omega), \\ \mathbf{u} = \mathbf{u}_1 \quad \text{on } \Gamma, \qquad \nabla \cdot \mathbf{u} = 0, \end{cases} \tag{3.15}$$

$\langle \cdot, \cdot \rangle$ denoting the duality pairing between $\mathbf{H}_0^1(\Omega)$ and $\mathbf{H}^{-1}(\Omega)$.

From the definition of a distributional derivative (Chap. 1, § 1.2), and since $(\nabla \mathbf{u})_{ij}$ and p belong to $L^2(\Omega)$, we have

$$\begin{aligned} &\alpha\langle \mathbf{u}, \mathbf{v}\rangle - \nu\langle \Delta \mathbf{u}, \mathbf{v}\rangle + \langle \nabla p, \mathbf{v}\rangle \\ &= \alpha \int_\Omega \mathbf{u} \cdot \mathbf{v}\, d\mathbf{x} + \nu \int_\Omega \nabla \mathbf{u} \cdot \nabla \mathbf{v}\, d\mathbf{x} - \int_\Omega p \nabla \cdot \mathbf{v}\, d\mathbf{x} \qquad \forall \mathbf{v} \in (\mathcal{D}(\Omega))^N. \end{aligned} \tag{3.16}$$

By density, (3.16) extends to any $\mathbf{v}$ in $\mathbf{H}_0^1(\Omega)$. Therefore, the weak formulation of (3.12)-(3.14) is equivalent to the following variational problem.

$$\begin{aligned} &\text{Find } \{\mathbf{u}, p\} \in \mathbf{H}^1(\Omega) \times L^2(\Omega) \text{ such that} \\ &\int_\Omega \{\alpha \mathbf{u} \cdot \mathbf{v} + \nu \nabla \mathbf{u} \cdot \nabla \mathbf{v} - p \nabla \cdot \mathbf{v}\}\, d\mathbf{x} = \langle \mathbf{f}, \mathbf{v}\rangle \qquad \forall \mathbf{v} \in \mathbf{H}_0^1(\Omega), \\ &\mathbf{u} = \mathbf{u}_1 \quad \text{on } \Gamma, \qquad \nabla \cdot \mathbf{u} = 0. \end{aligned} \tag{3.17}$$

In Remark 2.4 of Chapter 2, it was proved that the variational problem (3.17) has a unique solution $\{\mathbf{u}, p\}$ in $\mathbf{H}^1(\Omega) \times L_0^2(\Omega)$. This result uses the techniques of convex analysis introduced in Theorem 1.10 of Chapter 2 and involves the following functional spaces:

$$L_0^2(\Omega)=\left\{q\in L^2(\Omega),\int_\Omega q\,d\mathbf{x}=0\right\}, \tag{3.18}$$

$$X_1=\{\mathbf{v}\in\mathbf{H}^1(\Omega),\mathbf{v}=\mathbf{u}_1 \text{ on } \Gamma, \nabla\cdot\mathbf{v}=0\}, \tag{3.19}$$

$$X=\{\mathbf{v}\in\mathbf{H}_0^1(\Omega),\nabla\cdot\mathbf{v}=0\}, \tag{3.20}$$

$$V_1=\{\mathbf{v}\in\mathbf{H}^1(\Omega),\mathbf{v}=\mathbf{u}_1 \text{ on } \Gamma\}. \tag{3.21}$$

These sets also play an important role in the numerical treatment of (3.17) by augmented Lagrangian methods.

3.3. Lagrangian formulations of problem (3.12)–(3.14). Consider now the Lagrangian function $\mathcal{L}:\mathbf{H}^1(\Omega)\times L^2(\Omega)\to\mathbb{R}$ defined by

$$\mathcal{L}(\mathbf{v},q)=\frac{\alpha}{2}\int_\Omega|\mathbf{v}|^2\,d\mathbf{x}+\frac{\nu}{2}\int_\Omega|\nabla\mathbf{v}|^2\,d\mathbf{x}-\int_\Omega q\nabla\cdot\mathbf{v}\,d\mathbf{x}-\langle\mathbf{f},\mathbf{v}\rangle. \tag{3.22}$$

We have the following proposition.

PROPOSITION 3.1. *Any saddle-point* $\{\mathbf{u},p\}$ *of* $\mathcal{L}$ *over* $V_1\times L^2(\Omega)$ *is a solution of* (3.12)–(3.14) *and conversely. The same result holds for the augmented Lagrangian* $\mathcal{L}_r$ *defined by*

$$\mathcal{L}_r(\mathbf{v},q)=\mathcal{L}(\mathbf{v},q)+\frac{r}{2}\int_\Omega|\nabla\cdot\mathbf{v}|^2\,d\mathbf{x}, \tag{3.23}$$

where r is any positive constant.

Proof. Let $\{\mathbf{u},p\}$ be a saddle-point of $\mathcal{L}_r$ $(r\geq 0)$ over $V_1\times L^2(\Omega)$. Then, from the definition of a saddle-point, we have

$$\mathcal{L}_r(\mathbf{u},q)\leq\mathcal{L}_r(\mathbf{u},p)\qquad\forall q\in L^2(\Omega), \tag{3.24}$$

$$\mathcal{L}_r(\mathbf{u},p)\leq\mathcal{L}_r(\mathbf{v},p)\qquad\forall\mathbf{v}\in V_1. \tag{3.25}$$

It follows from (3.22), (3.23), and (3.24) that

$$\int_\Omega(p-q)\nabla\cdot\mathbf{u}\,d\mathbf{x}\leq 0\qquad\forall q\in L^2(\Omega),$$

which clearly implies

$$\nabla\cdot\mathbf{u}=0. \tag{3.26}$$

On the other hand, $\mathbf{u}$ being a minimizer of $\mathcal{L}_r(\cdot,p)$ over V_1, the gradient of

$\mathcal{L}_r$ at $\mathbf{u}$ vanishes on $\mathbf{H}_0^1(\Omega)$, and thus

$$(3.27)\quad \begin{cases} \alpha \int_\Omega \mathbf{u}\cdot\mathbf{v}\,d\mathbf{x} + \nu\int_\Omega \nabla\mathbf{u}\cdot\nabla\mathbf{v}\,d\mathbf{x} \\ \qquad + r\int_\Omega \nabla\cdot\mathbf{u}\nabla\cdot\mathbf{v}\,d\mathbf{x} - \int_\Omega p\nabla\cdot\mathbf{v}\,d\mathbf{x} = \langle\mathbf{f},\mathbf{v}\rangle \qquad \forall\mathbf{v}\in\mathbf{H}_0^1(\Omega). \end{cases}$$

From (3.26) and (3.27), the pair $\{\mathbf{u}, p\}$ can be seen to be the solution of the variational formulation (3.17) of (3.12)–(3.14).

Conversely, let $\{\mathbf{u}, p\}$ be a solution of (3.12)–(3.14) and , therefore, of (3.17). Since $\nabla\cdot\mathbf{u}=0$, we obviously have (3.24). Moreover, a direct computation yields

$$\begin{aligned} \mathcal{L}_r(\mathbf{v},p)-\mathcal{L}_r(\mathbf{u},p) &= \frac{\alpha}{2}\int_\Omega |\mathbf{v}-\mathbf{u}|^2\,d\mathbf{x} + \frac{\nu}{2}\int_\Omega |\nabla(\mathbf{v}-\mathbf{u})|^2\,d\mathbf{x} \\ &\quad + \frac{r}{2}\int_\Omega |\nabla\cdot(\mathbf{v}-\mathbf{u})|^2\,d\mathbf{x} + \alpha\int_\Omega \mathbf{u}\cdot(\mathbf{v}-\mathbf{u})\,d\mathbf{x} \\ &\quad + \nu\int_\Omega \nabla\mathbf{u}\cdot\nabla(\mathbf{v}-\mathbf{u})\,d\mathbf{x} \\ &\quad + r\int_\Omega \nabla\cdot\mathbf{u}\nabla\cdot(\mathbf{v}-\mathbf{u})\,d\mathbf{x} \\ &\quad - \int_\Omega p\nabla\cdot(\mathbf{v}-\mathbf{u})\,d\mathbf{x} - \langle\mathbf{f},\mathbf{v}-\mathbf{u}\rangle \qquad \forall\mathbf{v}\in V_1, \end{aligned}$$

the right-hand side of which is ≥ 0 from the variational equation (3.17). □

The augmented Lagrangian formulation of (3.12)–(3.14) is particularly interesting because it leads naturally to the following dual formulation

$$(3.28)\qquad \sup_{q\in L^2(\Omega)}\ \inf_{\mathbf{v}\in V_1} \mathcal{L}_r(\mathbf{v},q).$$

This dual formulation appears to be the right framework by which to introduce and analyze, along the lines of § 2.4, several efficient descent algorithms for solving (3.12)–(3.14).

In the following chapters we will systematically construct such augmented Lagrangian formulations to derive numerical methods that take advantage of the particular structure of the variational problems to be considered.

3.4. Decomposition properties of the Stokes problem. First, let us state the following.

LEMMA 3.1. *On $L_0^2(\Omega)$, there is equivalence between $\|\nabla q\|_{\mathbf{H}^{-1}(\Omega)}$ and $\|q\|_{L^2(\Omega)}$.*

Proof. For the proof, see, for example, Ladyshenkaya [1969]. □

Hereafter we shall use the notation $\|\mathbf{f}\|_{-1}$ for $\|\mathbf{f}\|_{\mathbf{H}^{-1}(\Omega)}$. For $r\geq 0$, we define $\mathcal{A}_r: L^2(\Omega)\to L^2(\Omega)$ as follows:

$$(3.29)\qquad \mathcal{A}_r q = \nabla\cdot\mathbf{u}(q),$$

where $\mathbf{u}(q)$ is the unique solution in $\mathbf{H}_0^1(\Omega)$ of the following elliptic system:

$$\alpha\mathbf{u}(q)-\nu\Delta\mathbf{u}(q)-r\nabla(\nabla\cdot\mathbf{u}(q))=-\nabla q,\qquad \mathbf{u}(q)=\mathbf{0}\quad\text{on }\Gamma. \tag{3.30}$$

We shall use the notation $\mathcal{A}$ for $\mathcal{A}_0$. We then have the following proposition.

PROPOSITION 3.2. *For any $r\geq 0$, $\mathcal{A}_r$ is a self-adjoint, strongly elliptic isomorphism from $L_0^2(\Omega)$ onto itself.*

Proof. Operator $\mathcal{A}_r$ is clearly linear from $L^2(\Omega)$ to $L^2(\Omega)$. Moreover, since

$$\int_\Omega \mathcal{A}_r q\,d\mathbf{x}=\int_\Omega \nabla\cdot\mathbf{u}(q)\,d\mathbf{x}=\int_\Gamma \mathbf{u}(q)\cdot\mathbf{n}\,d\Gamma=0,$$

we have that $\mathcal{A}_r q\in L_0^2(\Omega)$ for all $q\in L^2(\Omega)$.

Let us now consider the bilinear form associated with $\mathcal{A}_r$; we have, for all $q, q'\in L^2(\Omega)$,

$$\begin{cases}\displaystyle\int_\Omega (\mathcal{A}_r q)q'\,d\mathbf{x}=\int_\Omega \nabla\cdot\mathbf{u}(q)q'\,d\mathbf{x}=-\langle\mathbf{u}(q),\nabla q'\rangle\\[2mm] \displaystyle\qquad=\alpha\int_\Omega \mathbf{u}(q)\cdot\mathbf{u}(q')\,d\mathbf{x}+\nu\int_\Omega \nabla\mathbf{u}(q)\cdot\nabla\mathbf{u}(q')\,d\mathbf{x}\\[2mm] \displaystyle\qquad\quad+r\int_\Omega (\nabla\cdot\mathbf{u}(q))(\nabla\cdot\mathbf{u}(q'))\,d\mathbf{x}.\end{cases} \tag{3.31}$$

The above bilinear form is clearly symmetric, implying that $\mathcal{A}_r$ is self-adjoint. To prove that $\mathcal{A}_r$ is a strongly elliptic isomorphism from $L_0^2(\Omega)$ onto $L_0^2(\Omega)$, it suffices to prove that the bilinear form in (3.31) is $L_0^2(\Omega)$-elliptic, i.e., there exists $\beta>0$ such that

$$\int_\Omega (\mathcal{A}_r q)q\,d\mathbf{x}\geq\beta\|q\|^2_{L^2(\Omega)}\qquad \forall q\in L_0^2(\Omega).$$

We have, from (3.31),

$$\int_\Omega (\mathcal{A}_r q)q\,d\mathbf{x}=\alpha\int_\Omega |\mathbf{u}(q)|^2\,d\mathbf{x}+\nu\int_\Omega |\nabla\mathbf{u}(q)|^2\,d\mathbf{x}+r\int_\Omega |\nabla\cdot\mathbf{u}(q)|^2\,d\mathbf{x}. \tag{3.32}$$

Since the operator

$$\mathbf{v}\rightarrow\alpha\mathbf{v}-\nu\Delta\mathbf{v}-r\nabla(\nabla\cdot\mathbf{v})$$

is, for all $r\geq 0$, an isomorphism from $\mathbf{H}_0^1(\Omega)$ onto $\mathbf{H}^{-1}(\Omega)$, the ellipticity of the above bilinear form follows from (3.30), (3.32), and from Lemma 3.1. □

Remark 3.1. We observe that we have for $\mathbf{u}(q)$ the following relation.

$$\alpha\mathbf{u}(q)-\nu\Delta\mathbf{u}(q)=-\nabla(q-r\nabla\cdot\mathbf{u}(q)),\qquad \mathbf{u}(q)=\mathbf{0}\quad\text{on }\Gamma, \tag{3.33}$$

implying, from the definition of $\mathcal{A}$ $(=\mathcal{A}_0)$ and $\mathcal{A}_r$, that

$$\mathcal{A}_r q = \nabla \cdot \mathbf{u}(q) = \mathcal{A}(q - r\mathcal{A}_r q),$$

i.e.,

$$\mathcal{A}_r = (\mathbf{I} + r\mathcal{A})^{-1}\mathcal{A}. \tag{3.34}$$

We state now our main result.

PROPOSITION 3.3. *Let p be that pressure solution of the Stokes problem* (3.12)–(3.14) *belonging to $L^2_0(\Omega)$. We have then*

$$\mathcal{A}_r p = -\nabla \cdot \mathbf{u}_0, \tag{3.35}$$

where $\mathbf{u}_0$ is the solution of

$$\begin{cases} \alpha \mathbf{u}_0 - \nu \Delta \mathbf{u}_0 - r\nabla(\nabla \cdot \mathbf{u}_0) = \mathbf{f}, \\ \mathbf{u}_0 = \mathbf{u}_1 \quad \text{on } \Gamma. \end{cases} \tag{3.36}$$

Proof. Subtracting (3.36) from (3.12)–(3.14), we obtain (3.35) from the definition of $\mathcal{A}_r$ and from $\nabla \cdot \mathbf{u} = 0$. □

Remark 3.2. We observe that $\nabla \cdot \mathbf{u}_0 \in L^2_0(\Omega)$ since

$$\int_\Omega \nabla \cdot \mathbf{u}_0 \, d\mathbf{x} = \int_\Gamma \mathbf{u}_1 \cdot \mathbf{n} \, d\Gamma = 0.$$

Remark 3.3. In fact, problem (3.35) is a dual formulation of the Stokes problem (3.12)–(3.14). Indeed, $\mathcal{A}_r q + \nabla \cdot \mathbf{u}_0$ is the derivative at q of the dual functional J_r^* (in the sense of § 2.3.4) defined by

$$J_r^*(q) = -\min_{\mathbf{v} \in V_1} \mathcal{L}_r(\mathbf{v}, q),$$

with V_1 and $\mathcal{L}_r$ defined by (3.21) and (3.22), (3.23), respectively.

To conclude the present section, we observe that, from Proposition 3.3, we can solve the Stokes problem (3.12)–(3.14) in theory as follows:

(i) Solve (3.36) to obtain $\mathbf{u}_0$ from $\mathbf{f}$ and $\mathbf{u}_1$;

(ii) solve (3.35) to obtain p;

(iii) once p is known, compute $\mathbf{u}$ through the solution of either

$$\alpha \mathbf{u} - \nu \Delta \mathbf{u} = \mathbf{f} - \nabla p, \qquad \mathbf{u} = \mathbf{u}_1 \quad \text{on } \Gamma,$$

or, if one wants to use the same solver as for (3.36),

$$\alpha \mathbf{u} - \nu \Delta \mathbf{u} - r\nabla(\nabla \cdot \mathbf{u}) = \mathbf{f} - \nabla p, \qquad \mathbf{u} = \mathbf{u}_1 \quad \text{on } \Gamma.$$

In practice, $\mathcal{A}_r$ is not known in (3.35); however, by generalizing the descent methods of §§ 2.3 and 2.4, the dual problem (3.35) can be solved without the explicit knowledge of $\mathcal{A}_r$. These methods will be discussed in § 3.5 within a general Hilbert space framework and applied in § 3.6 to problem (3.35).

3.5. Iterative methods for symmetric and strongly elliptic linear variational problems in Hilbert spaces. Let V be a real Hilbert space for the scalar product $(\cdot,\cdot)$ and the associated norm $\|\cdot\|$. We consider in V the linear variational problem

$$(P_v)\qquad \begin{cases}\text{Find } u\in V, \text{ such that}\\ a(u,v)=L(v)\quad \forall v\in V,\end{cases}$$

where $L(\cdot)$ is linear and continuous over V, and where $a(\cdot,\cdot)$ is bilinear, continuous over $V\times V$, and V-elliptic, the latter meaning there exists $\alpha>0$ such that $a(v,v)\geq\alpha\|v\|^2$ for all $v\in V$. It follows from the Lax–Milgram lemma (Chap. 2, Thm. 1.7) that (P_v) has a unique solution (see, for example, Ciarlet [1978] or Glowinski [1984] for a proof of this classical result). Moreover, if $a(\cdot,\cdot)$ is symmetric, i.e., $a(v,w)=a(w,v)$ for all $v, w\in V$, one can easily prove that (P_v) is equivalent to the following minimization problem:

$$(\Pi)\qquad \begin{cases}\text{Find } u\in V, \text{ such that}\\ J(u)\leq J(v)\qquad \forall v\in V,\end{cases}$$

where

$$J(v)=\tfrac{1}{2}a(v,v)-L(v). \tag{3.37}$$

From the Riesz representation theorem (see, for example, Yoshida [1968]), (P_v) can also be expressed as

$$Au=f, \tag{3.38}$$

where f is the unique element of V such that $L(v)=(f,v)$ for all $v\in V$, and where A is the unique linear and continuous operator from V to V such that

$$(Av,w)=a(v,w)\qquad \forall v,w\in V. \tag{3.39}$$

Actually, A is an isomorphism from V onto V. The symmetry of A is equivalent to that of $a(\cdot,\cdot)$, and, from (3.39), we clearly have that $\|A\|\geq\alpha$ where α is the ellipticity constant of $a(\cdot,\cdot)$.

3.5.1. A first iterative method for solving (P_v). In operator form it is described by

ALGORITHM (3.40)–(3.41).

$$u^0\in V \text{ is given;} \tag{3.40}$$

then, for $n\geq 0$, we compute u^{n+1} from u^n by

$$u^{n+1}=u^n-\rho(Au^n-f),\qquad \rho>0. \tag{3.41a}$$

In variational form we should use, instead of (3.41a), the equivalent relation

$$\boxed{\begin{cases} u^{n+1} \in V, \\ (u^{n+1}, v) = (u^n, v) - \rho\{a(u^n, v) - L(v)\} \qquad \forall v \in V. \end{cases}} \tag{3.41b}$$

We have the following classical result concerning the convergence of Algorithm (3.40)–(3.41).

PROPOSITION 3.4. *Suppose that the above hypotheses about V, $L(\cdot)$ and $a(\cdot, \cdot)$ hold, and also suppose that*

$$\boxed{0 < \rho < 2\frac{\alpha}{\|A\|^2}.} \tag{3.42}$$

Then, for all $u^0 \in V$, we have

$$\lim_{n \to +\infty} \|u^n - u\| = 0, \tag{3.43}$$

where u is the solution of (P_v) and where $\{u^n\}_{n \geq 0}$ is defined by Algorithm (3.40)–(3.41).

Proof. Denote $u^n - u$ by $\bar{u}^n$. Since $f = Au$, we clearly have from (3.41) that

$$\bar{u}^{n+1} = \bar{u}^n - \rho A \bar{u}^n. \tag{3.44}$$

We have, in turn, from (3.44), (3.39), and from the V-ellipticity of $a(\cdot, \cdot)$, that

$$\begin{cases} \|\bar{u}^{n+1}\|^2 = \|\bar{u}^n\|^2 - 2\rho a(\bar{u}^n, \bar{u}^n) + \rho^2 \|A\bar{u}^n\|^2 \\ \qquad\qquad \leq (1 - 2\rho\alpha + \rho^2 \|A\|^2) \|\bar{u}^n\|^2. \end{cases} \tag{3.45}$$

If (3.42) holds, it follows clearly from (3.45) that $\lim_{n \to +\infty} \|\bar{u}^n\| = 0$, i.e., (3.43) has been proved. Moreover, we observe that the convergence is linear, i.e., $\|u^n - u\|$ approaches 0 as fast as a converging geometric sequence. □

Remark 3.4. The above result does not require the symmetry of $a(\cdot, \cdot)$. If $a(\cdot, \cdot)$ is symmetric, and if V has been identified to its dual space, we clearly have for the function J, defined by (3.37),

$$J'(v) = Av - f. \tag{3.46}$$

Therefore, since (3.41) and (3.46) imply

$$u^{n+1} = u^n - \rho J'(u^n), \tag{3.47}$$

when $a(\cdot, \cdot)$ is symmetric, Algorithm (3.40)–(3.41) appears as a gradient algorithm with constant step ρ.

Remark 3.5. A main drawback of the above algorithm is that it requires some knowledge of α and $\|A\|$. Actually, this difficulty can be overcome by replacing ρ by a sequence $\{\rho_n\}_{n\geq 0}$ automatically generated by the algorithm itself according to some criterion (as shown in § 2.4 for finite-dimensional problems). The conjugate-gradient algorithm below satisfies such properties.

3.5.2. A conjugate-gradient method for solving (P_v) if $a(\cdot,\cdot)$ is symmetric. Without going into too much detail, the conjugate-gradient algorithm described in § 2.4.1 for the solution of finite-dimensional problems can be generalized as follows.

ALGORITHM (3.48)–(3.55).

Step 0: Initialization

(3.48) $$u^0 \in V \text{ is given;}$$

solve

(3.49) $$\begin{cases} g^0 \in V, \\ (g^0, v) = a(u^0, v) - L(v) \qquad \forall v \in V, \end{cases}$$

and set (if $g^0 \neq 0$),

(3.50) $$w^0 = g^0. \qquad \square$$

Then, for $n \geq 0$, assuming u^n, g^n, w^n known with $w^n \neq 0$, compute u^{n+1}, g^{n+1}, w^{n+1} as follows.

Step 1: Descent

Compute

(3.51) $$\rho_n = \frac{(g^n, w^n)}{a(w^n, w^n)},$$

then set

(3.52) $$u^{n+1} = u^n - \rho_n w^n,$$

and solve

(3.53) $$\begin{cases} g^{n+1} \in V, \\ (g^{n+1}, v) = (g^n, v) - \rho_n a(w^n, v) \qquad \forall v \in V. \end{cases}$$

If $g^{n+1} = 0$ (or is small, in practice) then $u^{n+1} = u$; if this is not the case, go to Step 2.

Step 2: New Descent Direction

Compute

(3.54) $$\gamma_n = \frac{\|g^{n+1}\|^2}{\|g^n\|^2}$$

and set

$$w^{n+1}=g^{n+1}+\gamma_n w^n. \qquad (3.55)$$

□

Do $n=n+1$ and go to (3.51).

Remark 2.7 of § 2.4.1 still holds for Algorithm (3.48)–(3.55).

Remark 3.6. Relation (3.51) implies that ρ_n is a minimizer over $\mathbb{R}$ of the real function $\rho \to J(u^n - \rho w^n)$. If γ_n in particular is forced to zero in (3.55), we recover the steepest descent method for the solution of (Π).

Remark 3.7. The convergence of algorithms like Algorithm (3.48)–(3.55) is discussed in Daniel [1970], where it is proved that $\|u^n - u\|$ converges to zero at least as fast as

$$\left(\frac{\sqrt{\|A\|/\alpha}-1}{\sqrt{\|A\|/\alpha}+1}\right)^n.$$

3.6. Application to the solution of the Stokes problem via (3.35). From the properties of $\mathcal{A}_r$ that were proved in § 3.4, problem (3.35) falls into the class of linear problems discussed in § 3.5. Therefore, it can be solved either by the fixed-step Algorithm (3.40)–(3.41) or by the conjugate-gradient Algorithm (3.48)–(3.55).

3.6.1. Application of Algorithm (3.40)–(3.41) to the solution of problem (3.35). In this case, Algorithm (3.40)–(3.41) becomes

ALGORITHM (3.56)–(3.57).

$$p^0 \in L_0^2(\Omega), \text{ specified arbitrarily}; \qquad (3.56)$$

then, for $n \geq 0$, we compute p^{n+1} from p^n by

$$p^{n+1}=p^n-\rho(\mathcal{A}_r p^n+\nabla\cdot\mathbf{u}_0). \qquad (3.57)$$

From (3.36) and the definition of $\mathcal{A}_r$, this algorithm takes the following practical form.

ALGORITHM (3.58)–(3.60).

$$p^0 \in L^2(\Omega), \text{ specified arbitrarily}; \qquad (3.58)$$

then, for $n \geq 0$, we compute $\mathbf{u}^n$ and p^{n+1} from p^n as follows.

$$\begin{cases}\alpha\mathbf{u}^n-\nu\Delta\mathbf{u}^n-r\nabla(\nabla\cdot\mathbf{u}^n)=\mathbf{f}-\nabla p^n,\\ \mathbf{u}^n=\mathbf{u}_1 \quad \text{on } \Gamma,\end{cases} \qquad (3.59)$$

$$p^{n+1}=p^n-\rho\nabla\cdot\mathbf{u}^n. \qquad (3.60)$$

Remark 3.8. The implementation of Algorithm (3.58)–(3.60) does not involve either the operator $\mathcal{A}_r$ or the vector field u_0. For further references see Segal [1979] and Fortin and Glowinski ([1982, Chap. 2], [1983]).

Remark 3.9. We recognize in Algorithm (3.58)-(3.60) Uzawa's algorithm (2.12)-(2.14) applied to the search of a saddle-point $\{\mathbf{u}, p\}$ of the augmented Lagrangian $\mathscr{L}_r$ defined in (3.22).

The convergence of Algorithm (3.58)-(3.60) follows from Proposition 3.4; if ρ verifies $0<\rho<\rho_{\max}$, the pair $\{\mathbf{u}^n, p^n\}$ defined by Algorithm (3.58)-(3.60) converges strongly in $\mathbf{H}^1(\Omega)\times L_0^2(\Omega)$ toward the solution $\{\mathbf{u}, p\}$ of the Stokes problem (3.12)-(3.14). Moreover, as proved in Glowinski [1984], we have $\rho_{\max}\geq 2(r+(\nu/N))$ and the convergence is linear; in the same reference it is also proved that, if $\rho=r$, then $\|\{\mathbf{u}^n-\mathbf{u}, p^n-p\}\|_{\mathbf{H}^1\times L^2}$ converges to zero as r^{-n}

3.6.2. Application of the conjugate-gradient Algorithm (3.48)–(3.55) to the solution of problem (3.35). Let us endow $L_0^2(\Omega)$ with the scalar product

$$(q_1, q_2)=\int_\Omega (\mathscr{B}^{-1}q_1)q_2\,d\mathbf{x}. \tag{3.61}$$

In (3.61), $\mathscr{B}$ is a strongly elliptic isomorphism from $L_0^2(\Omega)$ onto itself defined by

$$\mathscr{B}q=\nu q+\alpha\phi(q)\qquad \forall q\in L_0^2(\Omega), \tag{3.62}$$

where $\phi(q)$ is the unique solution in $H^1(\Omega)\cap L_0^2(\Omega)$ of the Neumann problem

$$\begin{cases}-\Delta\phi(q)=q & \text{in } \Omega,\\ \dfrac{\partial\phi(q)}{\partial n}=0 & \text{on } \Gamma.\end{cases} \tag{3.63}$$

Let us also introduce

$$a_r(q_1, q_2)=\int_\Omega \nabla\cdot\mathbf{u}(q_1)q_2\,d\mathbf{x}, \tag{3.64}$$

$$L(q)=-\int_\Omega \nabla\cdot\mathbf{u}_0 q\,d\mathbf{x}, \tag{3.65}$$

where $\mathbf{u}(q)$ and $\mathbf{u}_0$ are defined by (3.30) and (3.36), respectively.

Using this notation, problem (3.35) takes the form

$$\begin{cases}\text{Find } p\in L_0^2(\Omega) \text{ such that}\\ a_r(p, q)=L(q)\qquad \forall q\in L_0^2(\Omega),\end{cases} \tag{3.66}$$

and can therefore be solved by the conjugate-gradient Algorithm (3.48)-(3.55).

We then obtain the following algorithm.

ALGORITHM (3.67)–(3.79).

Step 0: Initialization

(3.67) $$p^0 \in L_0^2(\Omega), \quad \text{arbitrarily given;}$$

solve successively

(3.68) $$\begin{cases} \alpha \int_\Omega \mathbf{u}^0 \cdot \mathbf{v}\, d\mathbf{x} + \nu \int_\Omega \nabla \mathbf{u}^0 \cdot \nabla \mathbf{v}\, d\mathbf{x} + r \int_\Omega \nabla \cdot \mathbf{u}^0 \nabla \cdot \mathbf{v}\, d\mathbf{x} \\ \quad = \int_\Omega p^0 \nabla \cdot \mathbf{v}\, d\mathbf{x} + \langle \mathbf{f}, \mathbf{v} \rangle \qquad \forall \mathbf{v} \in \mathbf{H}_0^1(\Omega), \quad \mathbf{u}^0 - \mathbf{u}_1 \in \mathbf{H}_0^1(\Omega), \end{cases}$$

(3.69) $$\int_\Omega \nabla \psi^0 \cdot \nabla \phi\, d\mathbf{x} = \int_\Omega \nabla \cdot \mathbf{u}^0 \phi\, d\mathbf{x} \qquad \forall \phi \in H^1(\Omega), \quad \psi^0 \in H^1(\Omega) \cap L_0^2(\Omega),$$

and

(3.70) $$\int_\Omega g^0 q\, d\mathbf{x} = \int_\Omega (\nu \nabla \cdot \mathbf{u}^0 + \alpha \psi^0) q\, d\mathbf{x} \qquad \forall q \in L_0^2(\Omega), \quad g^0 \in L_0^2(\Omega);$$

then set

(3.71) $$w^0 = g^0. \qquad \square$$

Then, for $n \geq 0$, assuming p^n, $\mathbf{u}^n$, g^n, w^n are known, we obtain p^{n+1}, $\mathbf{u}^{n+1}$, g^{n+1}, w^{n+1} as follows.

Step 1: Descent

Solve

(3.72) $$\begin{cases} \alpha \int_\Omega \mathbf{z}^n \cdot \mathbf{v}\, d\mathbf{x} + \nu \int_\Omega \nabla \mathbf{z}^n \cdot \nabla \mathbf{v}\, d\mathbf{x} + r \int_\Omega \nabla \cdot \mathbf{z}^n \nabla \cdot \mathbf{v}\, d\mathbf{x} \\ \quad = \int_\Omega w^n\, \nabla \cdot \mathbf{v}\, d\mathbf{x} \qquad \forall \mathbf{v} \in \mathbf{H}_0^1(\Omega), \quad \mathbf{z}^n \in \mathbf{H}_0^1(\Omega), \end{cases}$$

(3.73) $$\int_\Omega \nabla \psi^n \cdot \nabla \phi\, d\mathbf{x} = \int_\Omega \nabla \cdot \mathbf{z}^n \phi\, d\mathbf{x} \qquad \forall \phi \in H^1(\Omega), \quad \psi^n \in H^1(\Omega) \cap L_0^2(\Omega),$$

and compute

(3.74) $$\rho_n = \frac{\int_\Omega \nabla \cdot \mathbf{u}^n w^n\, d\mathbf{x}}{\int_\Omega \nabla \cdot \mathbf{z}^n w^n\, d\mathbf{x}} \left(= \frac{\int_\Omega \nabla \cdot \mathbf{u}^n g^n\, d\mathbf{x}}{\int_\Omega \nabla \cdot \mathbf{z}^n w^n\, d\mathbf{x}} \right),$$

(3.75) $$p^{n+1} = p^n - \rho_n w^n,$$

(3.76) $$\mathbf{u}^{n+1} = \mathbf{u}^n - \rho_n \mathbf{z}^n.$$

Step 2: New Descent Direction

Solve

$$\int_\Omega g^{n+1} q\, d\mathbf{x} = \int_\Omega (g^n - \rho_n(\nu \nabla \cdot \mathbf{z}^n + \alpha \psi^n)) q\, d\mathbf{x} \tag{3.77}$$

$$\forall q \in L_0^2(\Omega), \quad g^{n+1} \in L_0^2(\Omega),$$

and compute

$$\gamma_n = \frac{\int_\Omega g^{n+1} \nabla \cdot \mathbf{u}^{n+1}\, d\mathbf{x}}{\int_\Omega g^n \nabla \cdot \mathbf{u}^n\, d\mathbf{x}}, \tag{3.78}$$

$$w^{n+1} = g^{n+1} + \gamma_n w^n. \tag{3.79}$$

□

Do $n = n+1$ and go to (3.72).

From various numerical experiments, the above algorithm appears to be quite efficient even if $r = 0$. The preconditioning defined by (3.61)–(3.63) was introduced in a slightly different form in Cahouet and Chabard [1986] and Cahouet and Hauguel [1986], and has since been generalized for more complicated boundary conditions in Glowinski, Goussebaile, and Labadie [1990]. This preconditioning significantly improves the performances of the conjugate-gradient Algorithm (3.67)–(3.79), especially for large values of α/ν.

Remark 3.10. Conjugate-gradient Algorithm (3.67)–(3.79), which is analogous to the finite-dimensional Algorithm (2.103)–(2.113) when $B\mathbf{v} = -\nabla \cdot \mathbf{v}$, is no more costly than the fixed-step Algorithm (3.58)–(3.60) but has, in general, a faster convergence.

Remark 3.11. The elliptic operator

$$\mathbf{v} \to \alpha \mathbf{v} - \nu \Delta \mathbf{v} - r \nabla(\nabla \cdot \mathbf{v})$$

is similar to the linear elasticity operator for $r > 0$ and reduces to a Laplacian-like operator for $r = 0$.

4. Decomposition of variational problems by augmented Lagrangian techniques. The aim of this section is to show that a large class of variational problems can be transformed into saddle-point problems using decomposition principles. These problems can then be solved using augmented Lagrangian techniques, generalizing those of § 2. This approach will be systematically used to solve the problems discussed in the following chapters.

4.1. A family of variational problems. We shall restrict our attention to real Hilbert spaces; thus, let V and H be two such spaces equipped with the norms and inner products

$$\|\cdot\|, ((\cdot,\cdot)) \quad \text{and} \quad |\cdot|, (\cdot,\cdot),$$

respectively. Let $B \in \mathcal{L}(V, H)$ and let F and G be two convex, proper, lower semicontinuous functionals from H and V into $\mathbb{R} \cup \{+\infty\}$, respectively. We assume that

$$\text{dom}(G) \cap \text{dom}(F \circ B) \neq \emptyset, \tag{4.1}$$

where

$$\text{dom}(G) = \{v \mid v \in V, -\infty < G(v) < +\infty\},$$

with a similar definition for dom $(F \circ B)$. We associate with V, H, B, F, and G the minimization problem

$$(P) \qquad \boxed{\begin{cases} \text{Find } u \in V \text{ such that} \\ J(u) \leq J(v) \qquad \forall v \in V, \end{cases}}$$

where $J : V \to \bar{\mathbb{R}}$ is defined by

$$J(v) = F(Bv) + G(v). \tag{4.2}$$

If we assume that, in addition to (4.1), we also have

$$\lim_{\|v\| \to +\infty} J(v) = +\infty,$$

then Problem (P) admits a solution (which is unique if J is strictly convex); this follows, for example, from a direct application of the Weierstrass theorem (Chap. 2, Thm. 1.9).

Example. Let us consider a Bingham viscoplastic fluid flowing in a cylindrical pipe under the action of a constant gradient of pressure. In the case of a steady flow, the variational inequality (3.2) of Chapter 2 (which corresponds to the formulation of the time-dependent flow of an incompressible viscoplastic fluid) reduces to the minimization problem

$$\begin{cases} \text{Find } u \in H_0^1(\Omega), \text{ such that} \\ J(u) \leq J(v) \qquad \forall v \in H_0^1(\Omega), \end{cases}$$

where

$$J(v) = \frac{\nu}{2} \int_\Omega |\nabla v|^2 \, d\mathbf{x} + g \int_\Omega |\nabla v| \, d\mathbf{x} - c \int_\Omega v \, d\mathbf{x},$$

and where Ω is the cross-section of the pipe, u is the axial velocity of the fluid, ν and g are real positive constants representing the viscosity and rigidity of the fluid, and c is the (constant) gradient of pressure along the axial direction.

Obviously the above flow problem can be seen to be a particular case of problem (P) if we assume that

$$B = \nabla, \qquad H = L^2(\Omega) \times L^2(\Omega),$$

and if F and G are defined by

$$F(q)=\frac{\nu}{2}\int_\Omega |q|^2\,d\mathbf{x}+g\int_\Omega |q|\,d\mathbf{x} \qquad (|q|=\sqrt{q_1^2+q_2^2}),$$

$$G(v)=-c\int_\Omega v\,d\mathbf{x}.$$

An alternative choice for F and G is given by

$$F(q)=g\int_\Omega |q|\,d\mathbf{x},$$

$$G(v)=\frac{\nu}{2}\int_\Omega |\nabla v|^2\,d\mathbf{x}-c\int_\Omega v\,d\mathbf{x}.$$

The above functions F and G are convex and continuous.

The function $J(\cdot)$ and problem (P) have a very special structure, and thus it is natural to take advantage of this structure when considering methods of solution.

Remark 4.1. Most of the considerations that follow can be applied to nonconvex minimization problems like the eigenvalue problems discussed in § 6 as well as the nonlinear elasticity problems discussed in Chapters 7 and 8. They also apply to variational problems that are not equivalent to minimization problems, as shown, for example, in Lions and Mercier [1979] and Gabay [1979].

4.2. A decomposition principle. Let us define $W\subset V\times H$ by

(4.3) $$W=\{\{v,q\}\,|\,\{v,q\}\in V\times H,\ Bv-q=0\}.$$

Then problem (P) is clearly equivalent to

(Π) $$\begin{cases}\text{Find } \{u,p\}\in W \text{ such that}\\ j(u,p)\le j(v,q) \qquad \forall\{v,q\}\in W,\end{cases}$$

with

(4.4) $$j(v,q)=F(q)+G(v).$$

Although problems (P) and (Π) are equivalent, by considering (Π) we have in some ways simplified the nonlinear structure of (P), although at the cost of a new variable q and of the linear relation

(4.5) $$Bv-q=0.$$

In fact, problem (Π) looks very much like problem (2.3) of § 2.1, and, therefore, we can think of generalizing the augmented Lagrangian techniques of § 2 for its solution.

4.3. An augmented Lagrangian associated with (Π). Let $r>0$; by analogy with § 2, we define the augmented Lagrangian $\mathscr{L}_r: V\times H\times H\to\bar{\mathbb{R}}$ by

$$\mathscr{L}_r(v, q; \mu)=F(q)+G(v)+\frac{r}{2}|Bv-q|^2+(\mu, Bv-q), \tag{4.6}$$

and introduce the following saddle-point problem:

$$\begin{cases}\text{Find } \{u, p; \lambda\}\in V\times H\times H \text{ such that}\\ \mathscr{L}_r(u, p; \mu)\leq \mathscr{L}_r(u, p; \lambda)\leq \mathscr{L}_r(v, q; \lambda) \qquad \forall\{v, q; \mu\}\in V\times H\times H.\end{cases} \tag{4.7}$$

In general, the existence of a solution for problem (Π) (and (P)) does not imply that the saddle-point problem (4.7) has a solution; such an existence result has to be proved in each specific case. However, the converse holds.

THEOREM 4.1. *Suppose that $\{u, p; \lambda\}$ is a saddle-point of $\mathscr{L}_r$. Then it is also a saddle-point of $\mathscr{L}_{r'}$ for any $r'\geq 0$, and $\{u, p\}$ is a solution of problem* (Π).

Proof. We follow Fortin and Glowinski [1982], [1983]. Let $\{u, p; \lambda\}$ be a saddle-point of $\mathscr{L}_r$ over $V\times H\times H$. We thus have

$$\mathscr{L}_r(u, p; \mu)\leq \mathscr{L}_r(u, p; \lambda)\leq \mathscr{L}_r(v, q; \lambda) \qquad \forall\{v, q\}\in V\times H, \quad \mu\in H. \tag{4.8}$$

From the first inequality in (4.8) we deduce

$$(\mu, Bu-p)\leq(\lambda, Bu-p) \quad \forall\mu\in H,$$

and, hence,

$$Bu=p. \tag{4.9}$$

From the second inequality in (4.8) we deduce

$$F(Bu)+G(u)=\mathscr{L}_r(u, p; \lambda)\leq \mathscr{L}_r(v, q; \lambda)=F(q)+G(v) \qquad \forall\{v, q\}\in W,$$

and, hence, $\{u, p\}$ is a solution of (Π).

On the other hand, in view of (4.9), we immediately have, for all $r'\geq 0$,

$$\mathscr{L}_{r'}(u, p; \mu)=\mathscr{L}_{r'}(u, p; \lambda) \qquad \forall\mu\in H. \tag{4.10}$$

Moreover, if we employ the second inequality of (4.8), requiring that $v=u+t(w-u)$, with $0<t\leq 1$ and $q=p=Bu$, we have

$$G(u+t(w-u))-G(u)+\frac{r}{2}t^2|B(w-u)|^2+t(\lambda, B(w-u))\geq 0 \qquad \forall w\in V,$$

which, from the convexity of G, implies

$$t(G(w)-G(u))+\frac{r}{2}t^2|B(w-u)|^2+t(\lambda, B(w-u))\geq 0 \qquad \forall w\in V. \tag{4.11}$$

Dividing by t in (4.11) and making t approach 0, we finally obtain

$$G(w)-G(u)+(\lambda, B(w-u))\geq 0 \qquad \forall w\in V. \tag{4.12}$$

In a similar way, if we employ the second inequality of (4.8), requiring that $v=u$ and $q=p+t(s-p)$, with $0<t\leq 1$, we obtain

$$F(s)-F(p)-(\lambda, s-p)\geq 0 \qquad \forall s\in H. \tag{4.13}$$

If we add (4.12) to (4.13), and then add the nonnegative term $(r'/2)|Bw-s|^2$, we obtain

$$F(s)+G(w)+\frac{r'}{2}|Bw-s|^2+(\lambda, Bw-s)-F(p)-G(u)\geq 0.$$

This can be expressed as

$$F(p)+G(u)=\mathcal{L}_{r'}(u, p; \lambda)\leq \mathcal{L}_{r'}(w, s; \lambda) \qquad \forall\{w, s\}\in V\times H,$$

which, together with (4.10), implies that $\{u, p; \lambda\}$ is also a saddle-point of $\mathcal{L}_{r'}$ over $V\times H\times H$. □

4.4. A first algorithm for solving (P). To solve (P) and (Π) we shall determine the saddle-points of $\mathcal{L}_r$ by Uzawa's algorithm that generalizes Algorithm (2.12)–(2.14) of § 2.3.1. Such an algorithm applied to the solution of (4.7) will be referred to as ALG1 and is defined by

ALGORITHM (4.14)–(4.16) (ALG1).

$$\lambda^0\in H, \text{given}; \tag{4.14}$$

then, for $n\geq 0$, with λ^n known, determine u^n, p^n, λ^{n+1} as follows.

$$\begin{cases}\text{Find } \{u^n, p^n\}\in V\times H \text{ such that}\\ \mathcal{L}_r(u^n, p^n; \lambda^n)\leq \mathcal{L}_r(v, q; \lambda^n) \qquad \forall\{v, q\}\in V\times H,\end{cases} \tag{4.15}$$

$$\lambda^{n+1}=\lambda^n+\rho_n(Bu^n-p^n). \tag{4.16}$$

Remark 4.2. The reader may verify that ALG1 can be interpreted as a gradient algorithm applied to the maximization of the dual function

$$h_r(\mu)=\inf_{\{v,q\}\in V\times H}\{\mathcal{L}_r(v, q; \mu)\},$$

since its gradient h'_r is given at μ by

$$h'_r(\mu)=Bu_\mu-p_\mu$$

(where $\mathcal{L}_r(u_\mu, p_\mu; \mu)=h_r(\mu)=\inf_{\{v,q\}\in V\times H}\mathcal{L}_r(v, q; \mu)$).

4.5. Other algorithms for solving (P). The main difficulty in the implementation of ALG1 is clearly the solution at each iteration of the minimization problem (4.15). A natural solution procedure consists of using the block relaxation method given below (where n is fixed)

ALGORITHM (4.17)–(4.19).

$$p^{n,0} = p^{n-1}; \tag{4.17}$$

then, for $k \geq 1$, solve

$$\mathcal{L}_r(u^{n,k}, p^{n,k-1}; \lambda^n) \leq \mathcal{L}_r(v, p^{n,k-1}; \lambda^n) \quad \forall v \in V, \quad u^{n,k} \in V, \tag{4.18}$$

$$\mathcal{L}_r(u^{n,k}, p^{n,k}; \lambda^n) \leq \mathcal{L}_r(u^{n,k}, q; \lambda^n) \quad \forall q \in H, \quad p^{n,k} \in H. \tag{4.19}$$

Algorithm (4.17)–(4.19) is convergent under quite general assumptions on F and G (see, for example, Cea and Glowinski [1973] and Glowinski [1984]). We obtain natural variants of ALG1 by restricting ourselves to a fixed number of block relaxation iterations of Algorithm (4.17)–(4.19). In the limiting case of a single iteration, we obtain the following algorithm (ALG2) for solving (P).

ALGORITHM (4.20)–(4.23) (ALG2).

$$\{p^0, \lambda^1\} \in H \times H \text{ is given}; \tag{4.20}$$

with $\{p^{n-1}, \lambda^n\}$ known, determine successively u^n, p^n, λ^{n+1} as follows.

$$\mathcal{L}_r(u^n, p^{n-1}; \lambda^n) \leq \mathcal{L}_r(v, p^{n-1}; \lambda^n) \quad \forall v \in V, \quad u^n \in V, \tag{4.21}$$

$$\mathcal{L}_r(u^n, p^n; \lambda^n) \leq \mathcal{L}_r(u^n, q; \lambda^n) \quad \forall q \in H, \quad p^n \in H, \tag{4.22}$$

$$\lambda^{n+1} = \lambda^n + \rho_n(Bu^n - p^n). \tag{4.23}$$

A variant of ALG2 can be obtained by exchanging the roles of the variables v and q; depending on the situation, this variant may be more efficient. Actually, it is in general wiser to give a symmetric role to the variables v and q by updating the multiplier λ between steps (4.21) and (4.22). The resulting algorithm (ALG3) is as follows.

ALGORITHM (4.24)–(4.28) (ALG3).

$$\{p^0, \lambda^1\} \in H \times H \text{ is given}; \tag{4.24}$$

with $\{p^{n-1}, \lambda^n\}$ known, determine successively u^n, $\lambda^{n+1/2}$, p^n, λ^{n+1} as follows.

$$\mathcal{L}_r(u^n, p^{n-1}; \lambda^n) \leq \mathcal{L}_r(v, p^{n-1}; \lambda^n) \quad \forall v \in V, \quad u^n \in V, \tag{4.25}$$

$$\lambda^{n+1/2} = \lambda^n + \rho_n(Bu^n - p^{n-1}), \tag{4.26}$$

(4.27) $\mathscr{L}_r(u^n, p^n; \lambda^{n+1/2}) \le \mathscr{L}_r(u^n, q; \lambda^{n+1/2}) \quad \forall q \in H, \quad p^n \in H,$

(4.28) $\lambda^{n+1} = \lambda^{n+1/2} + \rho_n(Bu^n - p^n).$

4.6. Convergence results. We will restrict our discussion to the convergence of ALG2, which is more difficult and therefore more interesting than the convergence of ALG1.

THEOREM 4.2. *Assume that the following hold.*

(i) *The Lagrangian $\mathscr{L}_r$ has a saddle-point $\{u, p; \lambda\}$.*

(ii) *The function $\mathscr{L}_r(\cdot, \cdot; \mu)$ is coercive over $V \times H$ for any fixed μ, is proper on V for any fixed q and μ, and is proper on H for any fixed v and μ.*

(iii) *Either F is uniformly convex on the bounded sets of H with B injective and with* Range (B) *closed, or G is uniformly convex on the bounded sets of V.*

Then, if ρ_n satisfies

$$0 < \rho_n = \rho < \frac{1+\sqrt{5}}{2} r, \tag{4.29}$$

the sequence $\{u^n, p^n; \lambda^n\}$ computed in ALG2 *is well defined and satisfies*

$$\begin{cases} F(p^n) + G(u^n) \to F(p) + G(u) \; (= J(u)), \\ u^n \to u \quad \textit{strongly in } V, \\ p^n \to p \quad \textit{strongly in } H, \\ \lambda^n \to \lambda^* \quad \textit{weakly in } H, \{u, p; \lambda^*\} \textit{ being a saddle-point of } \mathscr{L}_r. \end{cases}$$

Proof. First, observe that, under the above assumptions, problem (4.21) (resp., (4.22)) corresponds to the minimization of a coercive, proper, convex, lower semicontinuous functional over a real Hilbert space and, therefore, from the Weierstrass theorem, admits a solution u^n (resp., p^n). This guarantees that ALG2 is well defined.

Now, since $\{u, p; \lambda\}$ is a solution of the saddle-point problem (4.7), the following extremality relation is satisfied (see (4.12), (4.13)),

$$F(p^n) + G(u^n) + (\lambda, Bu^n - p^n) \ge F(p) + G(u). \tag{4.30}$$

Moreover, by construction, the solutions u^n and p^n of (4.21) and (4.22) satisfy

$$G(v) - G(u^n) + (r(Bu^n - p^{n-1}) + \lambda^n, Bv - Bu^n) \ge 0 \quad \forall v \in V, \tag{4.31}$$

$$F(q) - F(p^n) + (r(Bu^n - p^n) + \lambda^n, p^n - q) \ge 0 \quad \forall q \in H. \tag{4.32}$$

If we require that $v = u$ and $q = p$, and if we denote $u^n - u$ by $\bar{u}^n$ and $p^n - p$ by $\bar{p}^n$, we obtain, by addition of (4.31) and (4.32),

$$\begin{cases} G(u) + F(p) - r|B\bar{u}^n - \bar{p}^n|^2 + r(\bar{p}^{n-1} - \bar{p}^n, B\bar{u}^n) \\ \qquad + (\lambda^n, \bar{p}^n - B\bar{u}^n) \ge F(p^n) + G(u^n). \end{cases} \tag{4.33}$$

Then, if we add (4.33) to (4.30) and denote $\bar{\lambda}^n = \lambda^n - \lambda$, we obtain

$$r|B\bar{u}^n - \bar{p}^n|^2 + r(\bar{p}^n - \bar{p}^{n-1}, B\bar{u}^n) + (\bar{\lambda}^n, B\bar{u}^n - \bar{p}^n) \le 0, \tag{4.34}$$

which, combined with (4.23), yields

$$|\bar{\lambda}^n|^2-|\bar{\lambda}^{n+1}|^2 \geq (2\rho r-\rho^2)|B\bar{u}^n-\bar{p}^n|^2+2\rho r(\bar{p}^n-\bar{p}^{n-1}, B\bar{u}^n). \tag{4.35}$$

On the other hand, if we employ (4.32) at iteration $(n-1)$ instead of n, with $q=p^n$, and if we take $q=p^{n-1}$ in (4.32), we obtain, after addition,

$$r|\bar{p}^n-\bar{p}^{n-1}|^2-(\bar{\lambda}^n-\bar{\lambda}^{n-1}, \bar{p}^n-\bar{p}^{n-1})-r(B\bar{u}^n-B\bar{u}^{n-1}, \bar{p}^n-\bar{p}^{n-1}) \leq 0. \tag{4.36}$$

If we then employ (4.23) at iteration $(n-1)$, (4.36) yields

$$\begin{cases} r|\bar{p}^n-\bar{p}^{n-1}|^2-\rho(B\bar{u}^{n-1}-\bar{p}^{n-1}, \bar{p}^n-\bar{p}^{n-1}) \\ \quad -r(B\bar{u}^n-B\bar{u}^{n-1}, \bar{p}^n-\bar{p}^{n-1}) \leq 0. \end{cases} \tag{4.37}$$

From the identity $(\bar{p}^{n-1}, \bar{p}^n-\bar{p}^{n-1})=\frac{1}{2}(|\bar{p}^n|^2-|\bar{p}^{n-1}|^2-|\bar{p}^n-\bar{p}^{n-1}|^2)$, we then have

$$\begin{cases} r(B\bar{u}^n, \bar{p}^n-\bar{p}^{n-1}) \geq r|\bar{p}^n-\bar{p}^{n-1}|^2+(r-\rho)(B\bar{u}^{n-1}-\bar{p}^{n-1}, \bar{p}^n-\bar{p}^{n-1}) \\ \qquad +\dfrac{r}{2}(|\bar{p}^n|^2-|\bar{p}^{n-1}|^2-|\bar{p}^n-\bar{p}^{n-1}|^2). \end{cases} \tag{4.38}$$

Substituting (4.38) in (4.35), we get

$$\begin{cases} |\bar{\lambda}^n|^2-|\bar{\lambda}^{n+1}|^2 \geq (2\rho r-\rho^2)|B\bar{u}^n-\bar{p}^n|^2+\rho r|\bar{p}^n-\bar{p}^{n-1}|^2+\rho r|\bar{p}^n|^2 \\ -\rho r|\bar{p}^{n-1}|^2+2\rho(r-\rho)(B\bar{u}^{n-1}-\bar{p}^{n-1}, \bar{p}^n-\bar{p}^{n-1}), \end{cases} \tag{4.39}$$

which, from the Cauchy-Schwarz inequality, implies

$$\boxed{\begin{aligned} &(|\bar{\lambda}^n|^2+\rho r|\bar{p}^{n-1}|^2)-(|\bar{\lambda}^{n+1}|^2+\rho r|\bar{p}^n|^2) \\ &\quad \geq \rho(2r-\rho)|B\bar{u}^n-\bar{p}^n|^2+\rho r|\bar{p}^n-\bar{p}^{n-1}|^2 \\ &\quad -\rho|r-\rho|\left(\frac{1}{\alpha}|B\bar{u}^{n-1}-\bar{p}^{n-1}|^2+\alpha|\bar{p}^n-\bar{p}^{n-1}|^2\right) \qquad \forall \alpha>0. \end{aligned}} \tag{4.40}$$

As shown in Fortin and Glowinski [1983, p. 125], such an inequality implies that, for $0<\rho<(1+\sqrt{5})/2$, the sequence $|\bar{\lambda}^n|^2+\rho r|\bar{p}^{n-1}|^2$ is decreasing and also that the sequences $|B\bar{u}^n-\bar{p}^n|^2$ and $|\bar{p}^n-\bar{p}^{n-1}|^2$ converge to zero when n goes to infinity. Therefore, we have

$$\begin{cases} |\lambda^n|^2, |p^n|, |Bu^n| \text{ are bounded}, \\ \lim\limits_{n\to+\infty} |Bu^n-p^n| = \lim\limits_{n\to+\infty} |p^n-p^{n-1}|=0. \end{cases} \tag{4.41}$$

Considering (4.41), and taking the limit in (4.30) and (4.33), we thus get

$$\liminf\,(F(p^n)+G(u^n)) \geq F(p)+G(u) \geq \limsup\,(F(p^n)+G(u^n)),$$

that is,

$$\lim_{n\to+\infty} (F(p^n)+G(u^n))=F(p)+G(u). \tag{4.42}$$

Finally, since $\{u, p; \lambda\}$ is a saddle-point of $\mathscr{L}$ over $V\times H\times H$ (see Theorem 4.1), it follows that

$$-B^t\lambda \in \partial G(u), \tag{4.43}$$

$$\lambda \in \partial F(p), \tag{4.44}$$

and, in particular,

$$G(u^n)+(\lambda, Bu^n)\geq G(u)+(\lambda, Bu), \tag{4.45}$$

$$F(p^n)-(\lambda, p^n)\geq F(p)-(\lambda, p). \tag{4.46}$$

Let us now suppose that F is uniformly convex (as defined in Chap. 2, Remark 2.3) over the bounded sets of H, with B injective and Range (B) closed. Denoting by $\mu^{n+1/2}$ an arbitrary element of $\partial F((p^n+p)/2)$, and employing (4.45) and (4.44), we get

$$\left\{\begin{aligned} &F(p^n)+G(u^n)+(\lambda, Bu^n-p^n)\\ &\quad\geq F\left(\frac{p^n+p}{2}\right)+\left(\mu^{n+1/2}, \frac{p^n-p}{2}\right)+G(u)+(\lambda, Bu-p^n)\\ &\quad\geq F\left(\frac{p^n+p}{2}\right)-F(p)+\left(\lambda, p-\frac{p^n+p}{2}\right)+\left(\mu^{n+1/2}-\lambda, \frac{p^n+p}{2}-p\right)\\ &\qquad\quad +F(p)+G(u)\\ &\quad\geq \psi\left(\frac{|p^n-p|}{2}\right)+F(p)+G(u). \end{aligned}\right. \tag{4.47}$$

The strong convergence of p^n toward p in H is then a direct consequence of (4.47), (4.42), and (4.41). From (4.41), it can be seen that this strong convergence of p^n implies the strong convergence of Bu^n toward Bu and hence the strong convergence of u^n toward u in V because B is injective and has a closed range. Alternatively, if G is uniformly convex, the strong convergence of $\{u^n, p^n\}$ toward $\{u, p\}$ can be proved by exchanging u and p, F and G, and (4.45) and (4.46) in (4.47).

The proof of the theorem is then complete if we observe that the weak convergence of λ^n toward λ^* follows from a direct application of the results of Opial [1967] as shown in Glowinski, Lions, and Tremolières [1981, Appendix 3]. □

Remark 4.3. The convergence properties of ALG1 are similar to those of ALG2, with (4.29) in this case being replaced by

$$0<\alpha_0\leq \rho_n\leq \alpha_1<2r. \tag{4.48}$$

To prove this, one simply has to replace p^{n-1} by p^n in (4.35), which directly yields (4.41) and (4.42) without any further calculation.

Remark 4.4. We can still obtain (4.41) and (4.42) (the convergence of $F(p^n)+G(u^n)$ toward $F(p)+G(u)$), together with weak convergence properties for the sequence $\{u^n, p^n\}$, when no uniform convexity is assumed for F or for G. Indeed, (4.42) was obtained without requiring the uniform convexity of F or of G.

Remark 4.5. Under stronger assumptions ($A=\partial G$ coercive or B^tB bijective, and ∂F^{-1} or ∂G^{-1} coercive), one can prove the linear convergence of ALG1, ALG2, and ALG3. In other words, the sequences $\|u^n-u\|$, $|p^n-p|$, and $|\lambda^n-\lambda|$ that are generated by those algorithms converge to zero at least as fast as θ^n for some θ in $[0, 1)$.

These results, proved by Gabay (see Fortin and Glowinski [1983, Chap. 9, Thms. 4.1, 5.2, 5.3]), are in fact optimal, as will be indicated in § 5. The convergence of ALG3 is proved with $\rho_n=r$, which, numerically, does not appear to be optimal ($\rho_n=\frac{1}{2}r$ is sometime better).

Remark 4.6. The assumptions of Theorem 4.2 are weaker than those used in Fortin and Glowinski [1983, Chap. 3]. They become even weaker if V and H are finite-dimensional spaces. In that case, there always exists a saddle-point for $\mathscr{L}$, and it is not necessary to assume the uniform convexity of F or of G. Indeed, from Remark 4.4, it can be seen that the sequence $\{u^n, p^n\}$ always converges weakly, and therefore strongly, since we are in finite dimension.

In particular, Theorem 4.2 can be applied to the Weber problem where $V=\mathbb{R}^N$, $H=\mathbb{R}^{N\times M}$, $G=0$, and $F(p)=\sum_{i=1}^{M}\alpha_i\|p_i-x_i\|_{\mathbb{R}^N}$ (see Fortin and Glowinski [1982, Chap. 3], [1983, Chap. 3] for more details and numerical results).

4.7. Further comments on the choice of ρ and r. In Remark 4.2, we observed that ALG1 is in fact a gradient algorithm for solving the dual problem

$$\begin{cases}\text{Find } \lambda \in H,\\ h_r(\lambda)\geq h_r(\mu) \qquad \forall \mu\in H,\end{cases}$$

where

$$h_r(\mu)=\inf_{\{v,q\}\in V\times H}\mathscr{L}_r(v, q; \mu).$$

From this observation, it is quite natural to think of using more sophisticated iterative methods such as conjugate-gradient or quasi-Newton methods to improve the speed of convergence of ALG1. In practice, ALG1 or the above methods are difficult to apply, because they require a precise knowledge of the gradient $h_r'(\mu)$, that is, an accurate solution of the minimization problem

$$\mathscr{L}_r(u, p; \mu)\leq \mathscr{L}_r(v, q; \mu) \qquad \forall\{v, q\}\in V\times H, \tag{4.49}$$

which is usually out of reach at a reasonable cost. Therefore, in this case, it is more advisable to use the simpler algorithms ALG2 or ALG3, in sharp contrast with the quadratic situation of § 2.

The complexity of (4.49) is also at the root of the difficulties that arise in the choice of r and $\{\rho_n\}$ in ALG1, ALG2, and ALG3. This will be illustrated in the next chapters, and is summarized briefly in what follows.

(i) For ALG1, with the quasi-optimal choice of $\rho_n = \rho = r$, it is easy to show that convergence is faster the larger the value of r. From a practical point of view, however, the situation is more complex; in fact, the conditioning of problem (4.49) worsens as r increases, so that the speed of convergence of the relaxation Algorithm (4.17)-(4.19) decreases. Experimentally, the combined effect of these factors—namely, an increase of r leading to an acceleration of ALG1 but a slowing down of Algorithm (4.17)-(4.19)—results in an algorithm whose overall speed of convergence (in terms of computation time) depends relatively little on the choice of r in many cases.

(ii) For ALG2, experience indicates that the best choice for ρ_n is $\rho_n = \rho = r$. The choice of r is more problematical than with ALG1, and, in this respect, ALG2 is more sensitive; a good strategy seems to be to choose r of the order of the spectral radius of the Hessian operator of F in H (if such an operator exists).

(iii) For ALG3, the same conclusions hold, with $\rho_n = \rho = \frac{1}{2}r$; however, numerical experiments show that ALG3 is less robust than ALG1 or ALG2, although it is possibly faster. In the following section we shall introduce ALG4, a generalization of ALG3 that has better properties of convergence and stability.

5. Relations between augmented Lagrangian methods and alternating direction methods.

5.1. Motivation. The solution of variational problems such as (P) by methods such as those employed in § 4, that is, by decomposition-coordination methods using augmented Lagrangians, can be interpreted as the numerical integration of associated evolution equations by alternating-direction methods. This analogy appears to be very useful both from theoretical and practical points of view, since it leads to convergence results (Lions and Mercier [1979], Gabay [1983]), and it is at the origin of new algorithms for solving decomposed problems such as (4.7).

5.2. Description of some classical alternating-direction methods. Let H be a real Hilbert space whose scalar product and norm are denoted by $(\cdot,\cdot)$ and $|\cdot|$, respectively. We consider now the following initial value problem in H

$$\text{(5.1)} \qquad \begin{cases} S\dfrac{du}{dt} + A(u) = 0, \\ u(0) = u_0, \end{cases}$$

where S is a symmetric (i.e., $(Sv, w) = (v, Sw)$ for all $v, w \in H$) and H-elliptic (i.e., there exists $\alpha > 0$ such that $(Sv, v) \geq \alpha |v|^2$ for all $v \in H$) isomorphism from H onto itself, where $u_0 \in H$, and where $A\ (=A_1+A_2)$ is an operator from H into itself.

Considering now the numerical integration of (5.1), we introduce a positive time step Δt (>0) and the notation $u^{n+\alpha} \simeq u((n+\alpha)\Delta t)$. If we suppose that A_1 and A_2 are simpler than A, we can take advantage of the decomposition $A=A_1+A_2$ and solve (5.1) by one of the following alternating-direction methods (for other methods, see, for example, Marchouk [1975]).

ALGORITHM (5.2)–(5.4) (Peaceman–Rachford method, Peaceman and Rachford [1955]).

(5.2)
$$u^0 = u_0;$$

then, for $n \geq 0$, with u^n known, solve successively

(5.3)
$$\frac{Su^{n+1/2} - Su^n}{\Delta t/2} + A_1(u^{n+1/2}) + A_2(u^n) = 0,$$

(5.4)
$$\frac{Su^{n+1} - Su^{n+1/2}}{\Delta t/2} + A_1(u^{n+1/2}) + A_2(u^{n+1}) = 0.$$ □

ALGORITHM (5.5)–(5.7) (Douglas–Rachford method, Douglas and Rachford [1956]).

(5.5)
$$u^0 = u_0;$$

then, for $n \geq 0$, with u^n known, solve successively

(5.6)
$$\frac{S\hat{u}^{n+1} - Su^n}{\Delta t} + A_1(\hat{u}^{n+1}) + A_2(u^n) = 0,$$

(5.7)
$$\frac{Su^{n+1} - Su^n}{\Delta t} + A_1(\hat{u}^{n+1}) + A_2(u^{n+1}) = 0.$$ □

ALGORITHM (5.8)–(5.11) (θ-scheme).

(5.8)
$$u^0 = u_0;$$

then, for $n \geq 0$, with u^n known and with $\theta \in (0, \frac{1}{2})$, solve successively

(5.9)
$$\frac{Su^{n+\theta} - Su^n}{\theta \Delta t} + A_1(u^{n+\theta}) + A_2(u^n) = 0,$$

(5.10)
$$\frac{Su^{n+1-\theta} - Su^{n+\theta}}{(1-2\theta)\Delta t} + A_1(u^{n+\theta}) + A_2(u^{n+1-\theta}) = 0,$$

(5.11)
$$\frac{Su^{n+1} - Su^{n+1-\theta}}{\theta \Delta t} + A_1(u^{n+1}) + A_2(u^{n+1-\theta}) = 0.$$ □

If the solution of (5.1) converges to a steady state (a solution of $A(u)=0$), the above algorithms are likely to converge toward this steady solution as n

approaches $+\infty$. Therefore, these algorithms can also be used as iterative methods for solving $A(u)=0$. Of these three methods, the last is the best suited to capture such steady solutions, because it has better asymptotic properties as n approaches $+\infty$ for a fixed Δt (Glowinski [1986]).

5.3. First relations between augmented Lagrangian and alternating-direction methods. To observe such relations, let us consider the particular case of problem (P) where $V=H$ and $B=I$ and where G and F admit as gradients the monotone operators A_1 and A_2, respectively. Problem (P) is then equivalent to

$$A_1(u)+A_2(u)=0. \tag{5.12}$$

Using Algorithm (4.20)-(4.23) (i.e., ALG2) for the solution of (P), we obtain, assuming $\rho=r$,

$$\{p^0, \lambda^1\} \text{ given;} \tag{5.13}$$

then, for $n \geq 1$,

$$r(u^n-p^{n-1})+A_1(u^n)+\lambda^n=0, \tag{5.14}$$

$$r(p^n-u^n)+A_2(p^n)-\lambda^n=0, \tag{5.15}$$

$$\lambda^{n+1}=\lambda^n+r(u^n-p^n). \tag{5.16}$$

By elimination of λ^n, we obtain in turn

$$r(u^n-p^{n-1})+A_1(u^n)+A_2(p^{n-1})=0, \tag{5.17}$$

$$r(p^n-p^{n-1})+A_1(u^n)+A_2(p^n)=0. \tag{5.18}$$

Assuming that $\hat{u}^{n+1}=u^n$, $u^{n+1}=p^n$, $\Delta t=1/r$, and $S=I$, we recognize in (5.17)-(5.18) the Douglas-Rachford method (5.5)-(5.7) applied to the solution of (5.12).

Similarly, if we use Algorithm (4.24)-(4.28) (i.e., ALG3) for the solution of (5.12), we obtain, after elimination of λ^n and $\lambda^{n+1/2}$ and assuming $\rho=r$,

$$p^0 \text{ given;} \tag{5.19}$$

then, for $n \geq 1$,

$$r(u^n-p^{n-1})+A_1(u^n)+A_2(p^{n-1})=0, \tag{5.20}$$

$$r(p^n-u^n)+A_1(u^n)+A_2(p^n)=0. \tag{5.21}$$

Assuming that $u^{n+1}=p^n$, $u^{n+1/2}=u^n$, $\Delta t=2/r$, and $S=I$, (5.19)-(5.21) is in turn equivalent to the Peaceman-Rachford method (5.2)-(5.4) applied to the solution of (5.12).

A problem that naturally follows is finding an augmented Lagrangian formulation of the θ-scheme (5.8)-(5.11). Such a formulation exists, as we shall see in the next section.

5.4. Equivalence theory between augmented Lagrangian and alternating-direction methods for the primal problem (case $B=I$; generalization to $B\neq I$). Let us consider again the problem (P) introduced in § 4.1

$$(P)\qquad \underset{v\in V}{\text{minimize}}\,\{F(Bv)+G(v)\},$$

and its associated saddle-point problem

$$(5.22)\qquad \begin{cases}\text{Find } \{u, p; \lambda\}\in V\times H\times H \text{ such that}\\ \mathcal{L}_r(u, p; \mu)\leq \mathcal{L}_r(u, p; \lambda)\leq \mathcal{L}_r(v, q; \lambda) \qquad \forall\{v, q; \mu\}\in V\times H\times H,\end{cases}$$

where the augmented Lagrangian $\mathcal{L}_r$ is defined by

$$(5.23)\qquad \mathcal{L}_r(v, q; \mu)=F(q)+G(v)+\frac{r}{2}|Bv-q|^2+(\mu, Bv-q).$$

In addition, suppose that, as in § 5.3, F is differentiable on H and that we have

$$V=H \quad\text{and}\quad B=I.$$

If problem (5.22) has a solution, then it follows from Theorem 4.1 that u $(=p)$ is a solution of (P) and satisfies the variational inequalities (4.12) and (4.13), that is, that

$$(5.24)\qquad \begin{cases}G(v)-G(u)+(\lambda, v-u)\geq 0 & \forall v\in H,\\ F(q)-F(u)-(\lambda, q-u)\geq 0 & \forall q\in H.\end{cases}$$

From the definition of $\partial F(u)$ and $\partial G(u)$ (Chap. 2, § 1.3), (5.24) can be expressed as

$$-\lambda\in\partial G(u),\qquad \lambda\in\partial F(u),$$

which is equivalent to

$$(5.25)\qquad \boxed{0\in\partial F(u)+\partial G(u).}$$

We associate with the multivalued "elliptic" equation (5.25) the multivalued initial value problem

$$(5.26)\qquad \begin{cases}0\in\dfrac{du}{dt}+\partial F(u)+\partial G(u),\\ u(0)=u_0.\end{cases}$$

This leads to the following theorem, which can be easily proved by the methods used in § 5.3.

THEOREM 5.1. *Assuming that* $p^0 = u_0$, $\lambda^1 = F'(u_0)$, *and* $\rho_n = \rho = r = 1/\Delta t$, *then* ALG2, *applied to the solution of* (5.22), *is a Douglas-Rachford scheme when applied to the solution of* (5.26). *Similarly, assuming that* $p^0 = u_0$, $\lambda^1 = F'(u_0)$, *and* $\rho_n = \rho = r = 2/\Delta t$, ALG3, *applied to the solution of* (5.22), *is a Peaceman-Rachford scheme when applied to the solution of* (5.26). □

More interesting is the augmented Lagrangian interpretation of the θ-scheme (5.8)-(5.11) when applied to the solution of (5.26). If we suppose that F is differentiable, this algorithm is as follows.

$$u^0 = u_0, \tag{5.27}$$

and, for $n \geq 0$,

$$-\left[\frac{u^{n+\theta} - u^n}{\theta \Delta t} + F'(u^n)\right] \in \partial G(u^{n+\theta}), \tag{5.28}$$

$$-\left[\frac{u^{n+1-\theta} - u^{n+\theta}}{(1-2\theta)\Delta t} + F'(u^{n+1-\theta})\right] \in \partial G(u^{n+\theta}), \tag{5.29}$$

$$-\left[\frac{u^{n+1} - u^{n+1-\theta}}{\theta \Delta t} + F'(u^{n+1-\theta})\right] \in \partial G(u^{n+1}). \tag{5.30}$$

Eliminating formally $\partial G(u^{n+\theta})$ in (5.29) and $F'(u^{n+1-\theta})$ in (5.30), we then obtain

$$u^0 = u_0, \tag{5.31}$$

and, for $n \geq 0$,

$$\lambda^n = F'(u^n), \tag{5.32}$$

$$-\lambda^{n+\theta} = -\left[\lambda^n + \frac{u^{n+\theta} - u^n}{\theta \Delta t}\right] \in \partial G(u^{n+\theta}), \tag{5.33}$$

$$\lambda^{n+1-\theta} = \left[\lambda^{n+\theta} - \frac{u^{n+1-\theta} - u^{n+\theta}}{(1-2\theta)\Delta t}\right] = F'(u^{n+1-\theta}), \tag{5.34}$$

$$-\left[\lambda^{n+1-\theta} + \frac{u^{n+1} - u^{n+1-\theta}}{\theta \Delta t}\right] \in \partial G(u^{n+1}). \tag{5.35}$$

With $\mathcal{L}_r$ still defined by (5.23) and with $r_1 = 1/\theta\Delta t$, $r_2 = 1/(1-2\theta)\Delta t$, $B = I$, and $V = H$, we finally obtain the following algorithm for solving problem (P) from (5.31)-(5.35).

ALGORITHM (5.36)-(5.42).

$$u^0 = u_0, \tag{5.36}$$

and, for $n\geq 0$, with u^n known, determine successively p^n, λ^n, $u^{n+\theta}$, $\lambda^{n+\theta}$, $p^{n+1-\theta}$, $\lambda^{n+1-\theta}$, and u^{n+1} as follows.

$$p^n = Bu^n, \qquad \lambda^n = F'(p^n), \tag{5.37}$$

$$\begin{cases} \mathcal{L}_{r_1}(u^{n+\theta}, p^n; \lambda^n) \leq \mathcal{L}_{r_1}(v, p^n; \lambda^n) & \forall v \in V, \\ u^{n+\theta} \in V, \end{cases} \tag{5.38}$$

$$\lambda^{n+\theta} = \lambda^n + r_1(Bu^{n+\theta} - p^n), \tag{5.39}$$

$$\begin{cases} \mathcal{L}_{r_2}(u^{n+\theta}, p^{n+1-\theta}; \lambda^{n+\theta}) \leq \mathcal{L}_{r_2}(u^{n+\theta}, q; \lambda^{n+\theta}) & \forall q \in H, \\ p^{n+1-\theta} \in H, \end{cases} \tag{5.40}$$

$$\lambda^{n+1-\theta} = \lambda^{n+\theta} + r_2(Bu^{n+\theta} - p^{n+1-\theta}), \tag{5.41}$$

$$\begin{cases} \mathcal{L}_{r_1}(u^{n+1}, p^{n+1-\theta}; \lambda^{n+1-\theta}) \leq \mathcal{L}_{r_1}(v, p^{n+1-\theta}; \lambda^{n+1-\theta}) & \forall v \in V, \\ u^{n+1} \in V. \end{cases} \tag{5.42}$$

Under this form, we observe that Algorithm (5.36)-(5.42) can be easily generalized to the solution of problems (P), where B is different from the identity and where V is different from H. In many applications, such an algorithm appears to be much more efficient than are ALG1, ALG2, or ALG3 for proper values of r_1 and r_2. Unfortunately, general convergence results have not yet been obtained for this algorithm. What is more critical, Algorithm (5.36)-(5.42) cannot be used as written if F is not a differentiable function. In this case, one has to consider a dual formulation of (5.25), as we shall see in the next section.

Remark 5.1. If G is differentiable and if B is one to one, we can easily exchange the roles of the variables v and q in Algorithm (5.36)-(5.42); it suffices to replace (5.37) by

$$u^n = B^{-1}p^n, \qquad \lambda^n = -B^{-t}G'(u^n). \tag{5.43}$$

For more general B, this exchange is impossible.

5.5. Equivalence theory between augmented Lagrangian and alternating-direction methods for the dual problem.

5.5.1. The dual evolution problem. Let us consider again the problem (P) introduced in § 4.1

$$(P) \qquad \underset{v\in V}{\text{minimize}}\,\{F(Bv)+G(v)\}$$

and its associated saddle-point problem (5.22). As soon as $B \neq I$, we encounter difficulties in interpreting the augmented Lagrangian algorithms introduced in § 4 for the solution of (P) as alternating-direction methods operating directly on (P). In particular, as seen in § 5.4, the θ-scheme was difficult to write in a

Lagrangian form in the general case. Following Gabay [1982], [1983], we shall overcome these difficulties by considering a dual formulation of (P).

For that purpose, recall that, if (5.22) has a solution $\{u, p; \lambda\}$, then, from Theorem 4.1, u is a solution of (P) and we have

$$\begin{cases} p = Bu, \\ G(v) - G(u) + (\lambda, Bv - Bu) \geq 0 \qquad \forall v \in V, \\ F(q) - F(p) - (\lambda, q - p) \geq 0 \qquad \forall q \in H. \end{cases} \tag{5.44}$$

It follows from the definition of $\partial G(u)$ and $\partial F(p)$ that (5.44) can be expressed as

$$-B^t\lambda \in \partial G(u), \qquad \lambda \in \partial F(Bu).$$

Equivalently, assuming

$$\partial F^{-1}(\mu) = \{q \in H, \mu \in \partial F(q)\},$$

and with a similar definition for $\partial G^{-1}(B^t\mu)$, we have

$$u \in \partial G^{-1}(-B^t\lambda), \qquad Bu \in \partial F^{-1}(\lambda), \tag{5.45}$$

that is,

$$\boxed{\partial F^{-1}(\lambda) - B\partial G^{-1}(-B^t\lambda) \ni 0.} \tag{P^*}$$

We then associate to (P^*) the multivalued initial value problem

$$\begin{cases} 0 \in \dfrac{d\lambda}{dt} + \partial F^{-1}(\lambda) - B\partial G^{-1}(-B^t\lambda), \\ \lambda(0) = \lambda_0. \end{cases} \tag{5.46}$$

5.5.2. Basic equivalence result. Problem (5.46) can be solved at least formally by the various alternating-direction schemes described in § 5.2. As before, the resulting algorithms can be interpreted as augmented Lagrangian algorithms for solving (P).

THEOREM 5.2. *Assuming that* $\lambda^1 = \lambda_0$, $\rho_n = \rho = r = \Delta t$, *and* $p^0 \in \partial F^{-1}(\lambda_0)$, *then* ALG2 *is a Douglas-Rachford scheme applied to the solution of the dual evolution equation* (5.46). *Similarly, with* $\rho_n = \rho = r = \Delta t/2$, ALG3 *is a Peaceman-Rachford scheme applied to the solution of* (5.46).

Proof. Since the case of ALG3 is similar, we simply prove this result for ALG2. From the definition of $\mathcal{L}_r$, for $\rho_n = r$, (4.21)-(4.23) become

$$\text{(5.47)} \quad \begin{cases} G(v) - G(u^n) + (\lambda^n, B(v-u^n)) + r(Bu^n - p^{n-1}, B(v-u^n)) \geq 0 \\ \forall v \in V, \qquad u^n \in V, \end{cases}$$

$$\text{(5.48)} \quad \begin{cases} F(q) - F(p^n) - (\lambda^n, q-p^n) + r(p^n - Bu^n, q-p^n) \geq 0 \qquad \forall q \in H, \\ p^n \in H, \end{cases}$$

$$\text{(5.49)} \qquad \lambda^{n+1} = \lambda^n + r(Bu^n - p^n).$$

Let us define $\hat{\lambda}^{n+1}$ by

$$\text{(5.50)} \qquad \hat{\lambda}^{n+1} = \lambda^n + r(Bu^n - p^{n-1}).$$

It then follows from (5.47) and from the definition of ∂G that

$$\text{(5.51)} \qquad -B^t\hat{\lambda}^{n+1} \in \partial G(u^n);$$

similarly, from (5.48) and (5.49) we obtain

$$\text{(5.52)} \qquad \lambda^{n+1} \in \partial F(p^n).$$

Using (5.49)-(5.52), we obtain

$$\begin{cases} \dfrac{\hat{\lambda}^{n+1} - \lambda^n}{r} - Bu^n + p^{n-1} = 0, \qquad u^n \in \partial G^{-1}(-B^t\hat{\lambda}^{n+1}), \qquad p^{n-1} \in \partial F^{-1}(\lambda^n), \\ \dfrac{\lambda^{n+1} - \lambda^n}{r} - Bu^n + p^n = 0, \qquad u^n \in \partial G^{-1}(-B^t\hat{\lambda}^{n+1}), \qquad p^n \in \partial F^{-1}(\lambda^{n+1}), \end{cases}$$

which implies

$$\begin{cases} \dfrac{\hat{\lambda}^{n+1} - \lambda^n}{r} - B\partial G^{-1}(-B^t\hat{\lambda}^{n+1}) + \partial F^{-1}(\lambda^n) \ni 0, \\ \dfrac{\lambda^{n+1} - \lambda^n}{r} - B\partial G^{-1}(-B^t\hat{\lambda}^{n+1}) + \partial F^{-1}(\lambda^{n+1}) \ni 0, \end{cases}$$

in which we formally recognize the Douglas–Rachford scheme for the solution of (5.46). □

5.5.3. General interpretation of ALG1, ALG2, and ALG3 as time integrators of the dual evolution problem. The above theorem shows that, for $\rho_n = r$, the dual approach provides an interpretation of ALG2 and ALG3 as alternating-direction methods even if $B \neq I$. Moreover, if $B = I$, we observe that the Douglas–Rachford scheme (5.5)-(5.7) or the Peaceman–Rachford scheme (5.2)-(5.4) applied either to the primal or the dual problem leads to the same Lagrangian algorithm ALG2 or ALG3.

Actually, the interpretation of augmented Lagrangian algorithms as time integrators of the dual evolution problem can be generalized to ALG1, ALG2, and ALG3 for any choice of ρ_n. We have the following theorem.

THEOREM 5.3. *The application of* ALG1 *to the solution of* (P) *corresponds to the time integration of the dual evolution equation* (5.46) *by the following scheme.*

$$0 \in \frac{\lambda^{n+1}-\lambda^{n}}{\Delta_n t}+\partial F^{-1}(\lambda_\rho^{n+1})-B\,\partial G^{-1}(-B^t\lambda_\rho^{n+1}), \quad \textit{with} \tag{5.53}$$

$$\Delta_n t=\rho_n, \qquad \lambda_\rho^{n+1}=\frac{r}{\rho_n}\lambda^{n+1}+\left(1-\frac{r}{\rho_n}\right)\lambda^n.$$

Similarly, ALG2 *corresponds to the scheme*

$$\begin{cases} 0 \in \dfrac{\hat\lambda^{n+1}-\lambda^{n}}{\Delta_n t}+\partial F^{-1}(\lambda_\rho^{n})-B\,\partial G^{-1}(-B^t\hat\lambda_\rho^{n+1}), \\ 0 \in \dfrac{\lambda^{n+1}-\lambda^{n}}{\Delta_n t}+\partial F^{-1}(\lambda_\rho^{n+1})-B\,\partial G^{-1}(-B^t\hat\lambda_\rho^{n+1}), \end{cases} \tag{5.54}$$

with

$$\Delta_n t=\rho_n, \qquad \hat\lambda_\rho^{n+1}=\frac{r}{\rho_n}\hat\lambda^{n+1}+\left(1-\frac{r}{\rho_n}\right)\lambda^n, \qquad \lambda_\rho^{n+1}=\frac{r}{\rho_n}\lambda^{n+1}+\left(1-\frac{r}{\rho_n}\right)\lambda^n.$$

Finally, ALG3 *corresponds to the scheme*

$$\begin{cases} 0 \in \dfrac{\lambda^{n+1/2}-\lambda^{n}}{\Delta_n t/2}+\partial F^{-1}(\lambda_\rho^{n})-B\,\partial G^{-1}(-B^t\lambda_\rho^{n+1/2}), \\ 0 \in \dfrac{\lambda^{n+1}-\lambda^{n+1/2}}{\Delta_n t/2}+\partial F^{-1}(\lambda_\rho^{n+1})-B\,\partial G^{-1}(-B^t\lambda_\rho^{n+1/2}), \end{cases} \tag{5.55}$$

$$\textit{with} \quad \Delta_n t/2=\rho_n, \qquad \lambda_\rho^{n+1/2}=\frac{r}{\rho_n}\lambda^{n+1/2}+\left(1-\frac{r}{\rho_n}\right)\lambda^n,$$

$$\lambda_\rho^{n+1}=\frac{r}{\rho_n}\lambda^{n+1}+\left(1-\frac{r}{\rho_n}\right)\lambda^{n+1/2}.$$

Proof. The proof is identical to that of Theorem 5.2 as soon as we observe that, for ALG1 or ALG2, we have

$$\lambda^n+r(Bu^n-p^n)=\lambda^n+\frac{r}{\rho_n}(\lambda^{n+1}-\lambda^n)=\lambda_\rho^{n+1},$$

and that, for ALG3, we have

$$\lambda^n+r(Bu^n-p^{n-1})=\lambda^n+\frac{r}{\rho_n}(\lambda^{n+1/2}-\lambda^n)=\lambda_\rho^{n+1/2},$$

$$\lambda^{n+1/2}+r(Bu^n-p^n)=\lambda^{n+1/2}+\frac{r}{\rho_n}(\lambda^{n+1}-\lambda^{n+1/2})=\lambda_\rho^{n+1}. \qquad \square$$

Observe finally that ALG1, ALG2, and ALG3 correspond, respectively, to a backward Euler scheme, a Douglas–Rachford scheme, and a Peaceman–Rachford scheme, with the replacement of λ^{n+1} by a linear combination of λ^n and λ^{n+1} in each operator. In particular, for $\rho_n = r$ (resp., $\rho_n = 2r$), (5.53) (resp., ALG1) is a genuine backward Euler scheme (resp., a Crank–Nicholson scheme).

5.5.4. Augmented Lagrangian interpretation of the θ-scheme. We have seen that the application of the θ-scheme (5.8)–(5.11) to the primal problem (5.26) leads to the Lagrangian Algorithm (5.36)–(5.42). In view of the nice interpretation of the alternating-directions schemes that we obtained from using them to solve the dual evolution problem (5.46), it is natural to apply the θ-scheme to this dual problem as well. We shall see that this leads to a new general Lagrangian algorithm for solving (P). Indeed, the θ-scheme (5.8)–(5.11), when applied, with $S = I$, to the solution of the dual evolution equation (5.46) formally yields

$$(5.56)\quad \begin{cases} \dfrac{\lambda^{n+\theta}-\lambda^n}{\theta\Delta t} - B\,\partial G^{-1}(-B^t\lambda^{n+\theta}) + \partial F^{-1}(\lambda^n) \ni 0, \\ \dfrac{\lambda^{n+1-\theta}-\lambda^{n+\theta}}{(1-2\theta)\Delta t} - B\,\partial G^{-1}(-B^t\lambda^{n+\theta}) + \partial F^{-1}(\lambda^{n+1-\theta}) \ni 0, \\ \dfrac{\lambda^{n+1}-\lambda^{n+1-\theta}}{\theta\Delta t} - B\,\partial G^{-1}(-B^t\lambda^{n+1}) + \partial F^{-1}(\lambda^{n+1-\theta}) \ni 0. \end{cases}$$

To overcome the ambiguity associated with the sum of two multivalued operators in (5.56), we can express (5.56) under the form

$$(5.57)\quad p^n \in \partial F^{-1}(\lambda^n),$$

$$\begin{cases} (5.58)\quad u^{n+\theta} \in \partial G^{-1}(-B^t\lambda^{n+\theta}), \\ (5.59)\quad \dfrac{\lambda^{n+\theta}-\lambda^n}{\theta\Delta t} - Bu^{n+\theta} + p^n = 0, \end{cases}$$

$$\begin{cases} (5.60)\quad p^{n+1-\theta} \in \partial F^{-1}(\lambda^{n+1-\theta}), \\ (5.61)\quad \dfrac{\lambda^{n+1-\theta}-\lambda^{n+\theta}}{(1-2\theta)\Delta t} - Bu^{n+\theta} + p^{n+1-\theta} = 0, \end{cases}$$

$$\begin{cases} (5.62)\quad u^{n+1} \in \partial G^{-1}(-B^t\lambda^{n+1}), \\ (5.63)\quad \dfrac{\lambda^{n+1}-\lambda^{n+1-\theta}}{\theta\Delta t} - Bu^{n+1} + p^{n+1-\theta} = 0. \end{cases}$$

By elimination of $\lambda^{n+\theta}$ between (5.58) and (5.59), and from the definition of ∂G^{-1}, (5.58) yields

$$(5.64)\quad -B^t[\lambda^n + \theta\Delta t(Bu^{n+\theta} - p^n)] \in \partial G(u^{n+\theta}).$$

Similarly, (5.60) and (5.61) and (5.62) and (5.63) imply

$$(5.65)\qquad \lambda^{n+\theta}+(1-2\theta)\Delta t(Bu^{n+\theta}-p^{n+1-\theta})\in\partial F(p^{n+1-\theta}),$$

and

$$(5.66)\qquad -B^t[\lambda^{n+1-\theta}+\theta\Delta t(Bu^{n+1}-p^{n+1-\theta})]\in\partial G(u^{n+1}),$$

respectively. Expressing (5.57), (5.64), (5.59), (5.65), (5.61), (5.66), and (5.63) in augmented Lagrangian form, we obtain the following algorithm (ALG4) for the solution of (P).

ALGORITHM (5.67)-(5.74) (ALG4).

$$(5.67)\qquad \lambda^0\in H, \text{ given};$$

then, with λ^n known, determine successively p^n, $u^{n+\theta}$, $\lambda^{n+\theta}$, $p^{n+1-\theta}$, $\lambda^{n+1-\theta}$, u^{n+1}, and λ^{n+1} as follows.

$$(5.68)\qquad \mathcal{L}_0(v,p^n;\lambda^n)\le\mathcal{L}_0(v,q;\lambda^n)\qquad \forall\{v,q\}\in V\times H,\quad p^n\in H;$$

$$(5.69)\qquad \mathcal{L}_{\theta\Delta t}(u^{n+\theta},p^n;\lambda^n)\le\mathcal{L}_{\theta\Delta t}(v,p^n;\lambda^n)\qquad \forall v\in V,\quad u^{n+\theta}\in V;$$

$$(5.70)\qquad \lambda^{n+\theta}=\lambda^n+\theta\Delta t(Bu^{n+\theta}-p^n);$$

$$(5.71)\qquad \begin{cases}\mathcal{L}_{(1-2\theta)\Delta t}(u^{n+\theta},p^{n+1-\theta};\lambda^{n+\theta})\le\mathcal{L}_{(1-2\theta)\Delta t}(u^{n+\theta},q;\lambda^{n+\theta})\\ \forall q\in H,\quad p^{n+1-\theta}\in H;\end{cases}$$

$$(5.72)\qquad \lambda^{n+1-\theta}=\lambda^{n+\theta}+(1-2\theta)\Delta t(Bu^{n+\theta}-p^{n+1-\theta});$$

$$(5.73)\qquad \mathcal{L}_{\theta\Delta t}(u^{n+1},p^{n+1-\theta};\lambda^{n+1-\theta})\le\mathcal{L}_{\theta\Delta t}(v,p^{n+1-\theta};\lambda^{n+1-\theta})\qquad \forall v\in V,\quad u^{n+1}\in V;$$

$$(5.74)\qquad \lambda^{n+1}=\lambda^{n+1-\theta}+\theta\Delta t(Bu^{n+1}-p^{n+1-\theta}).$$

Remark 5.2. Observe that it is very easy to exchange the roles of the variables v and q in ALG4. In this case, (5.68) becomes

$$(5.75)\qquad \mathcal{L}_0(u^n,q;\lambda^n)\le\mathcal{L}_0(v,q;\lambda^n)\qquad \forall\{v,q\}\in V\times H,\quad u^n\in V.$$

Remark 5.3. For ALG4 to be well defined, it is necessary that problem (5.68) has a solution, that is, that the set $\partial F^{-1}(\lambda^n)$ is not empty for any λ^n in H. Alternatively, if we have exchanged the roles of the variables v and q, it follows from (5.75) that ALG4 is well defined in that case only if $\partial G^{-1}(-B^t\lambda^n)$ is not empty for any λ^n in H.

Remark 5.4. No convergence result has yet been proven for ALG4. Numerically, depending on the values of θ and Δt, this algorithm can be much faster than ALG2 or ALG3. Its convergence will be discussed with more detail in the next section.

5.6. Analysis of the algorithm on a model problem.

5.6.1. Formulation of the model problem. Let us again consider problem

$$(P) \qquad \underset{v \in V}{\text{minimize}} \{F(Bv) + G(v)\},$$

where we take spaces V and H both to be finite-dimensional, and F and G to be defined by

$$F(q) = \frac{1}{2}|q|^2, \qquad G(v) = \frac{\varepsilon}{2}\|v\|^2 - ((\ell, v)),$$

with ε a positive number and ℓ a given element of V.

The solution of problem (P) is clearly

$$u = (\varepsilon I + B^t B)^{-1}\ell.$$

The saddle-point problem associated with (P) is still (5.22), with $\mathcal{L}_r$ defined by

$$\mathcal{L}_r(v, q; \mu) = \frac{1}{2}|q|^2 + \frac{\varepsilon}{2}\|v\|^2 - ((\ell, v)) + \frac{r}{2}|Bv - q|^2 + (\mu, Bv - q). \tag{5.76}$$

In addition, from the definition of F and G, the dual problem (P^*) is simply

$$\lambda + \frac{1}{\varepsilon} BB^t\lambda - \frac{1}{\varepsilon} B\ell = 0 \tag{5.77}$$

and the associated dual evolution problem (5.46) reduces to the initial-value problem

$$\begin{cases} \dfrac{d\lambda}{dt} + \lambda + \dfrac{1}{\varepsilon} BB^t\lambda - \dfrac{1}{\varepsilon} B\ell = 0, \\ \lambda(0) = \lambda_0. \end{cases} \tag{5.78}$$

5.6.2. Convergence properties of ALG1. As seen in §5.5, solving (P) through the use of ALG1 in fact corresponds to the time integration of (5.78) by the following scheme if $\rho_n = \rho = \Delta t$.

$$\begin{cases} \dfrac{\lambda^{n+1} - \lambda^n}{\rho} + \left[\dfrac{r}{\rho}\lambda^{n+1} + \left(1 - \dfrac{r}{\rho}\right)\lambda^n\right] + \dfrac{1}{\varepsilon} BB^t\left[\dfrac{r}{\rho}\lambda^{n+1} + \left(1 - \dfrac{r}{\rho}\right)\lambda^n\right] \\ -\dfrac{1}{\varepsilon} B\ell = 0. \end{cases} \tag{5.79}$$

The expansion of (5.79) over an eigenvector basis of BB^t gives us (with obvious notation)

$$\bar{\lambda}_i^{n+1} = K_i \bar{\lambda}_i^n, \tag{5.80}$$

with $\bar{\lambda}_i^n$ being the ith component of the residual $\lambda^n - \lambda$ (λ: solution of (5.77)) and with

$$K_i = \frac{1-(\rho-r)(1+\alpha_i/\varepsilon)}{1+r(1+\alpha_i/\varepsilon)}, \tag{5.81}$$

where α_i is the ith eigenvalue of BB^t.

From (5.81), we can easily deduce the following properties of ALG1.

(i) $0<|K_i|<1$ if and only if, for all i,

$$0<\frac{\rho}{r}<2\left[1+\frac{1}{r(1+\alpha_i/\varepsilon)}\right],$$

implying the stability of scheme (5.79) and the linear convergence of λ^n toward λ if $0<\rho/r<2$.

(ii) For $\rho=r=\Delta t$, we have $\lim_{\rho\to+\infty} K_i=0$, implying the stiff-$A$ stability of scheme (5.79).

(iii) If $\rho=r$ and for $\alpha_i=0$, i.e., for those components of $\bar{\lambda}^n$ belonging to $\ker(B^t)$, the convergence of $\bar{\lambda}_i^n$ to 0 is linear with an asymptotic constant $1/(1+r)$. As for the other components of $\bar{\lambda}^n$, the convergence is still linear, but, when ε goes to zero, the corresponding asymptotic constant also converges to zero. Actually, in the limit case $\varepsilon=0$, these other components converge to zero in one iteration, implying the convergence of u^n also in one iteration, as already mentioned in Fortin and Glowinski [1983, pp. 120–121].

5.6.3. Convergence properties of ALG2 for $\rho=r$. From Theorem 5.2, for $\rho_n=\rho=r=\Delta t$, solving (P) through the use of ALG2 corresponds in fact to the time integration of (5.78) by a Douglas–Rachford scheme defined by

$$\begin{cases} \dfrac{\hat{\lambda}^{n+1}-\lambda^n}{r}+\lambda^n+\dfrac{1}{\varepsilon}BB^t\hat{\lambda}^{n+1}-\dfrac{1}{\varepsilon}B\ell=0, \\ \dfrac{\lambda^{n+1}-\lambda^n}{r}+\lambda^{n+1}+\dfrac{1}{\varepsilon}BB^t\hat{\lambda}^{n+1}-\dfrac{1}{\varepsilon}B\ell=0. \end{cases} \tag{5.82}$$

In the case of $\bar{\lambda}^n=\lambda^n-\lambda$, eliminating $\hat{\lambda}^{n+1}$, (5.82) becomes

$$(1+r)\bar{\lambda}^{n+1}=\left(I-\frac{r}{\varepsilon}BB^t\left[I+\frac{r}{\varepsilon}BB^t\right]^{-1}(1-r)\right)\bar{\lambda}^n,$$

or, equivalently, after an expansion over an eigenvector basis of BB^t and under the notations of § 5.6.2,

$$\begin{cases} \bar{\lambda}_i^{n+1}=K_i\bar{\lambda}_i^n, \\ K_i=\dfrac{(1+\alpha_i r^2/\varepsilon)}{(1+r)(1+\alpha_i r/\varepsilon)}. \end{cases} \tag{5.83}$$

From (5.83), we observe that

(i) $0<K_i<1$, for all i and $r>0$, implying the unconditional stability of scheme (5.82) and the linear convergence of λ^n to λ.

(ii) If $\alpha_i \neq 0$, then $\lim_{r\to+\infty} K_i = 1$, and thus scheme (5.82) is not stiff-A-stable.

(iii) For $\alpha_i = 0$, i.e., for those components of $\bar{\lambda}^n$ belonging to $\ker(B^t)$, the convergence of $\bar{\lambda}_i^n$ to 0 is linear with an asymptotic constant $1/(1+r)$. As for the other components of $\bar{\lambda}^n$, the convergence is also linear with an asymptotic constant converging to $r/(1+r)$ when ε goes to zero. Overall, the convergence of $\bar{\lambda}^n$ to 0 is therefore linear, but the asymptotic constant cannot be better than $\max\{1/(1+r), r/(1+r)\}$, that is, at best 1/2.

(iv) For $r=1$ and $\varepsilon=0$, we have, nevertheless, the convergence of $\hat{\lambda}^n$ to zero in one iteration, implying the convergence of u^n also in one iteration.

5.6.4. Convergence properties of ALG3 with $\rho = r$. It follows from Theorem 5.2, for $\rho_n = \rho = r = \Delta t/2$, that solving (P) using ALG3 corresponds to the time integration of (5.78) by a Peaceman–Rachford scheme defined by

$$\text{(5.84)} \qquad \begin{cases} \dfrac{\lambda^{n+1/2} - \lambda^n}{r} + \lambda^n + \dfrac{1}{\varepsilon} BB^t \lambda^{n+1/2} - \dfrac{1}{\varepsilon} B\ell = 0, \\[2ex] \dfrac{\lambda^{n+1} - \lambda^{n+1/2}}{r} + \lambda^{n+1} + \dfrac{1}{\varepsilon} BB^t \lambda^{n+1/2} - \dfrac{1}{\varepsilon} B\ell = 0. \end{cases}$$

Eliminating $\lambda^{n+1/2}$, we obtain from (5.84)

$$(1+r)\bar{\lambda}^{n+1} = \left(I - \frac{r}{\varepsilon} BB^t\right)\left(I + \frac{r}{\varepsilon} BB^t\right)^{-1} (1-r)\bar{\lambda}^n,$$

or, equivalently, after expansion over an eigenvector basis of BB^t,

$$\text{(5.85)} \qquad \begin{cases} \bar{\lambda}_i^{n+1} = K_i \bar{\lambda}_i^n, \\[1ex] K_i = \dfrac{(1 - \alpha_i r/\varepsilon)(1-r)}{(1+\alpha_i r/\varepsilon)(1+r)}. \end{cases}$$

From (5.85), we can deduce the following.

(i) $0 < |K_i| < 1$, for all i, $r > 0$, implying the unconditional stability of scheme (5.84) and the linear convergence of λ^n to λ.

(ii) If $\alpha_i \neq 0$, then $\lim_{r\to+\infty} |K_i| = 1$, and thus scheme (5.84) is also not stiff-A-stable.

(iii) For $r = 1$, there is convergence of all the components of $\bar{\lambda}^n$ to zero in one iteration. This implies, in turn, the convergence of u^n and p^n in one iteration. For r close to 1, convergence does not occur in one iteration, but it is still very fast.

5.6.5. Convergence properties of ALG3 with $\rho \neq r$. In the above section, we studied ALG3 with $\rho_n = \rho = r$; however, choosing $\rho_n = \rho < r$ can improve the convergence properties of ALG3. To illustrate this point, let us consider the solution of (P) with ALG3, choosing $\rho_n = \rho \neq r$. As seen in Theorem 5.3, this corresponds to the time integration of (5.78) by the scheme

$$(5.86)\qquad \begin{cases} \dfrac{\lambda^{n+1/2}-\lambda^n}{\rho}+\lambda_\rho^n+\dfrac{1}{\varepsilon}BB^t\lambda_\rho^{n+1/2}-\dfrac{1}{\varepsilon}B\ell=0, \\ \dfrac{\lambda^{n+1}-\lambda^{n+1/2}}{\rho}+\lambda_\rho^{n+1}+\dfrac{1}{\varepsilon}BB^t\lambda_\rho^{n+1/2}-\dfrac{1}{\varepsilon}B\ell=0 \end{cases}$$

with $\lambda_\rho^{n+1/2}$ and λ_ρ^n as defined in Theorem 5.3. After projection of (5.86) onto $\ker(B^t)$ and elimination, we obtain

$$(5.87)\qquad \bar{\lambda}_2^{n+1}=\left[1-\frac{2\rho}{(r+1)}\right]\bar{\lambda}_2^n,$$

where $\bar{\lambda}_2^n$ denotes the projection of the residual $\lambda^n-\lambda$ onto $\ker(B^t)$. It follows from (5.87) that $\bar{\lambda}_2^n$ converges to 0 if and only if

$$(5.88)\qquad 0<\rho<r+1.$$

Thus, for large r, the choice $\rho_n=\rho=r$ appears almost to be a limiting value for convergence. On the other hand, $\bar{\lambda}_2^n$ converges to 0 in one iteration if $\rho_n=\rho=(r+1)/2$, the optimal value of which gets close to $r/2$ when r is large.

Of course, this super-convergence of $\bar{\lambda}_2^n$ for $\rho_n=\rho=(r+1)/2$ is tempered by the linear convergence of the projection of λ^n onto $\mathrm{Im}(B)$. The optimal choice for ρ_n is in fact problem-dependent and must be determined by numerical tests. But, in any case, (5.87) strongly advocates the possible use of $\rho_n \neq r$ in ALG3.

5.6.6. Convergence properties of ALG4. In its derivation, ALG4 was constructed as a time integrator of (5.46) by the θ-scheme (5.8)-(5.11). This θ-scheme, when applied to the initial-value problem (5.78), reduces to

$$(5.89)\qquad \begin{cases} \dfrac{\lambda^{n+\theta}-\lambda^n}{\theta\Delta t}+\lambda^n+\dfrac{1}{\varepsilon}BB^t\lambda^{n+\theta}-\dfrac{1}{\varepsilon}B\ell=0, \\ \dfrac{\lambda^{n+1-\theta}-\lambda^{n+\theta}}{(1-2\theta)\Delta t}+\lambda^{n+1-\theta}+\dfrac{1}{\varepsilon}BB^t\lambda^{n+\theta}-\dfrac{1}{\varepsilon}B\ell=0, \\ \dfrac{\lambda^{n+1}-\lambda^{n+1-\theta}}{\theta\Delta t}+\lambda^{n+1-\theta}+\dfrac{1}{\varepsilon}BB^t\lambda^{n+1}-\dfrac{1}{\varepsilon}B\ell=0. \end{cases}$$

After elimination of $\lambda^{n+\theta}$ and $\lambda^{n+1-\theta}$, we obtain from (5.89)

$$\bar{\lambda}^{n+1}=\left[I+\theta\Delta t\frac{BB^t}{\varepsilon}\right]^{-1}(1-\theta\Delta t)(1+(1-2\theta)\Delta t)^{-1}\left[I-(1-2\theta)\Delta t\frac{BB^t}{\varepsilon}\right]$$
$$\cdot\left[I+\theta\Delta t\frac{BB^t}{\varepsilon}\right]^{-1}(1-\theta\Delta t)\bar{\lambda}^n,$$

or, equivalently, after expansion over an eigenvector basis of BB^t

(5.90)
$$\begin{cases} \bar{\lambda}_i^{n+1} = K_i \bar{\lambda}_i^n, \\ K_i = \dfrac{(1-\theta\Delta t)^2(1-(1-2\theta)\Delta t\alpha_i/\varepsilon)}{(1+(1-2\theta)\Delta t)(1+\theta\Delta t\alpha_i/\varepsilon)^2}. \end{cases}$$

From (5.90), we can deduce the following.

(i) For $\alpha_i = 0$ (i.e., for those components of $\bar{\lambda}^n$ in $\ker(B^t)$), we have

$$K_i = K_0 = \frac{(1-\theta\Delta t)^2}{1+(1-2\theta)\Delta t},$$

which is less than 1 if and only if

(5.91)
$$\boxed{\theta^2\Delta t < 1.}$$

Therefore, the scheme is only conditionally stable on $\ker(B^t)$; it is stable and the components of $\bar{\lambda}^n$ in $\ker(B^t)$ converge linearly to 0 if and only if (5.91) is satisfied.

(ii) For $\alpha_i \neq 0$ and ε going to 0, we have

$$K_i = K_0 \frac{(2\theta-1)}{\alpha_i\theta^2\Delta t}\varepsilon + 0(\varepsilon^2),$$

and, therefore, if ε is sufficiently small, the components of $\bar{\lambda}^n$ in $\operatorname{Im}(B)$ converge linearly toward 0 with an asymptotic constant of order ε. This convergence is then very fast, even for large Δt (stiff-A stability of the scheme on $\operatorname{Im}(B)$ for ε small); at the limit $\varepsilon = 0$, this convergence occurs in one iteration, implying a similar convergence for u^n.

(iii) In the general case, for $\alpha_i \neq 0$, we have

$$K_i = K_0\beta_i = K_0 \frac{[1-(1-2\theta)\alpha_i\Delta t/\varepsilon]}{(1+\theta\alpha_i\Delta t/\varepsilon)^2},$$

with K_0 defined as in (i). Therefore, if condition (5.91) is satisfied, a sufficient condition to obtain $|K_i| < 1$ for all i, is to obtain $|\beta_i| < 1$ for all i, which occurs if

(5.92)
$$\begin{cases} \theta > \dfrac{2-\sqrt{2}}{4} = 0.146\cdots, \quad \text{or} \\ \alpha_i \dfrac{\Delta t}{\varepsilon} > \dfrac{(1-4\theta)+\sqrt{(1-4\theta)^2-8\theta^2}}{2\theta^2}, \quad \text{or} \\ \alpha_i \dfrac{\Delta t}{\varepsilon} < \dfrac{(1-4\theta)-\sqrt{(1-4\theta)^2-8\theta^2}}{2\theta^2}. \end{cases}$$

In summary, we obtain linear convergence of λ^n toward λ when, for each i, (5.91) and one of the conditions of (5.92) is satisfied.

(iv) For $\theta\Delta t=1$, the convergence of λ^n toward λ occurs in one iteration.

Remark 5.5. By inverting the roles of F and G in ALG4, we obtain

$$\begin{cases}\bar{\lambda}_i^{n+1}=K_i\bar{\lambda}_i^n,\\ K_i=\dfrac{(1-\theta\alpha_i\Delta t/\varepsilon)^2(1-(1-2\theta)\Delta t)}{(1+\theta\Delta t)^2(1+(1-2\theta)\alpha_i\Delta t/\varepsilon)}.\end{cases}$$

Then, ALG4 converges linearly on ker (B^t) if and only if one of the conditions of (5.92) is satisfied, with α_i/ε replaced by 1. It converges linearly on Im (B) if, in addition, $\theta^2\alpha_i\Delta t<\varepsilon$ for any i, and diverges when ε goes to zero. For $\varepsilon\neq 0$, ALG4 converges in one iteration for $(1-2\theta)\Delta t=1$.

6. Application to the solution of linear and nonlinear eigenvalue problems.

6.1. Generalities and synopsis. The main goal of this section is to show that the concepts introduced in §§ 4 and 5 also apply to eigenvalue calculations. Since the resulting methods belong to the class of so-called inverse power methods, they bring very little to bear on linear eigenvalue problems (for which the basic references are Parlett [1980] and Chatelin [1983]), but are nicely suited to solve nonlinear eigenvalue problems such as the Hartree equation in quantum physics.

In § 6.2, we will discuss the calculation of the smallest eigenvalue of a symmetric matrix. Then, in § 6.3, we will consider the calculation of the ground-state solution of the Hartree equation for the helium atom.

6.2. The linear eigenvalue problem

6.2.1. Formulation of the problem. Let $\mathbf{A}$ and $\mathbf{S}$ be two symmetric $N\times N$, real matrices with $\mathbf{S}$ positive-definite. We denote by Λ_i the eigenvalues of $\mathbf{S}^{-1}\mathbf{A}$, and we suppose that $\Lambda_1\leq\Lambda_2\leq\cdots\leq\Lambda_N$. We will concentrate on the calculation of the smallest eigenvalue of $\mathbf{S}^{-1}\mathbf{A}$ (i.e., Λ_1) and of an associated eigenvector. It is well known that Λ_1 satisfies

$$\Lambda_1=\min_{v\in\Sigma}(\mathbf{A}\mathbf{v},\mathbf{v})=(\mathbf{A}\mathbf{u},\mathbf{u}),\tag{6.1}$$

where $\mathbf{u}$ is an eigenvector of norm one associated to Λ_1 and where

$$\Sigma=\{\mathbf{v}\,|\,\mathbf{v}\in\mathbb{R}^N,(\mathbf{S}\mathbf{v},\mathbf{v})=1\},\tag{6.2}$$

$$(\mathbf{v},\mathbf{w})=\sum_{i=1}^N v_iw_i\qquad\forall\mathbf{v}=\{v_i\}_{i=1}^N,\quad \mathbf{w}=\{w_i\}_{i=1}^N.\tag{6.3}$$

Let us introduce now the function I_Σ defined by

$$I_\Sigma(\mathbf{v})=\begin{cases}0 & \text{if } \mathbf{v}\in\Sigma,\\ +\infty & \text{if } \mathbf{v}\notin\Sigma;\end{cases}\tag{6.4}$$

I_Σ is the indicator function of Σ. We clearly have

$$\frac{\Lambda_1}{2}=\frac{1}{2}(\mathbf{Au},\mathbf{u})+I_\Sigma(\mathbf{u})\le\frac{1}{2}(\mathbf{Av},\mathbf{v})+I_\Sigma(\mathbf{v})\qquad\forall\mathbf{v}\in\mathbb{R}^N. \tag{6.5}$$

Observe that I_Σ is neither convex nor differentiable.

6.2.2. Augmented Lagrangian formulation. Defining V, H, B, F, and G by

$$V=H=\mathbb{R}^N,\qquad B=\mathbf{Id},$$

$$F(\mathbf{q})=I_\Sigma(\mathbf{q}),\qquad G(\mathbf{v})=\tfrac{1}{2}(\mathbf{Av},\mathbf{v}),$$

respectively, we observe that (6.5) is formally a problem (P) as defined in § 4.1. Therefore, following the approach of § 4, we associate with (6.5) the augmented Lagrangian $\mathcal{L}_r$ defined by

$$\mathcal{L}_r(\mathbf{v},\mathbf{q};\boldsymbol{\mu})=I_\Sigma(\mathbf{q})+G(\mathbf{v})+\frac{r}{2}(\mathbf{S}(\mathbf{v}-\mathbf{q}),\mathbf{v}-\mathbf{q})+(\mathbf{S}\boldsymbol{\mu},\mathbf{v}-\mathbf{q}) \tag{6.6}$$

and reduce the solution of (6.5) to finding the local saddle-points of $\mathcal{L}_r$ over $(\mathbb{R}^N\times\mathbb{R}^N)\times\mathbb{R}^N$. This last problem can then be numerically solved by one of the algorithms ALG2, ALG3, or ALG4 of §§ 4 and 5. We do not consider ALG1 in this case because its implementation for Problem (6.5) leads to rather costly computations.

6.2.3. Practical formulations of the augmented Lagrangian algorithms. The application of ALG2 to the calculation of local saddle-points of the augmented Lagrangian $\mathcal{L}_r$ defined by (6.6) leads to the following algorithm.

ALGORITHM (6.7)–(6.10).

$$\{\mathbf{p}^0,\boldsymbol{\lambda}^1\}\in\mathbb{R}^N\times\mathbb{R}^N \text{ is given;} \tag{6.7}$$

with $\{\mathbf{p}^{n-1},\boldsymbol{\lambda}^n\}$ known, determine successively $\mathbf{u}^n$, $\mathbf{p}^n$, $\boldsymbol{\lambda}^{n+1}$ as follows.

$$\mathcal{L}_r(\mathbf{u}^n,\mathbf{p}^{n-1};\boldsymbol{\lambda}^n)\le\mathcal{L}_r(\mathbf{v},\mathbf{p}^{n-1};\boldsymbol{\lambda}^n)\qquad\forall\mathbf{v}\Leftrightarrow(r\mathbf{S}+\mathbf{A})\mathbf{u}^n=\mathbf{S}(r\mathbf{p}^{n-1}-\boldsymbol{\lambda}^n), \tag{6.8}$$

$$\mathcal{L}_r(\mathbf{u}^n,\mathbf{p}^n;\boldsymbol{\lambda}^n)\le\mathcal{L}_r(\mathbf{u}^n,\mathbf{q};\boldsymbol{\lambda}^n)\qquad\forall\mathbf{q}\Leftrightarrow\mathbf{p}^n=\frac{\boldsymbol{\lambda}^n+r\mathbf{u}^n}{(\mathbf{S}(\boldsymbol{\lambda}^n+r\mathbf{u}^n),\boldsymbol{\lambda}^n+r\mathbf{u}^n)^{1/2}}, \tag{6.9}$$

$$\boldsymbol{\lambda}^{n+1}=\boldsymbol{\lambda}^n+\rho(\mathbf{u}^n-\mathbf{p}^n). \tag{6.10}$$

Similarly, the application of ALG3 to the same problem leads to the following algorithm.

ALGORITHM (6.11)–(6.15).

$$\{\mathbf{p}^0,\boldsymbol{\lambda}^1\}\in\mathbb{R}^N\times\mathbb{R}^N \text{ is given;} \tag{6.11}$$

with $\{\mathbf{p}^{n-1}, \boldsymbol{\lambda}^n\}$ known, determine successively $\mathbf{u}^n$, $\boldsymbol{\lambda}^{n+1/2}$, $\mathbf{p}^n$, and $\boldsymbol{\lambda}^{n+1}$ as follows.

$$(6.12)\qquad (r\mathbf{S}+\mathbf{A})\mathbf{u}^n = \mathbf{S}(r\mathbf{p}^{n-1}-\boldsymbol{\lambda}^n),$$

$$(6.13)\qquad \boldsymbol{\lambda}^{n+1/2} = \boldsymbol{\lambda}^n + \rho(\mathbf{u}^n - \mathbf{p}^{n-1}),$$

$$(6.14)\qquad \mathbf{p}^n = \frac{\boldsymbol{\lambda}^{n+1/2} + r\mathbf{u}^n}{(\mathbf{S}(\boldsymbol{\lambda}^{n+1/2}+r\mathbf{u}^n), \boldsymbol{\lambda}^{n+1/2}+r\mathbf{u}^n)^{1/2}},$$

$$(6.15)\qquad \boldsymbol{\lambda}^{n+1} = \boldsymbol{\lambda}^{n+1/2} + \rho(\mathbf{u}^n - \mathbf{p}^n).$$

Finally, applying Algorithm (5.36)-(5.42) and taking Remark 5.1 into account, we obtain yet another algorithm for our eigenvalue problem.

ALGORITHM (6.16)-(6.22).

$$(6.16)\qquad \mathbf{p}^0 \in \mathbb{R}^N \text{ given},$$

and, for $n \geq 0$, with $\mathbf{p}^n$ known, determine successively $\mathbf{u}^n$, $\boldsymbol{\lambda}^n$, $\mathbf{p}^{n+\theta}$, $\boldsymbol{\lambda}^{n+\theta}$, $\mathbf{u}^{n+1-\theta}$, $\boldsymbol{\lambda}^{n+1-\theta}$, and $\mathbf{p}^{n+1}$ as follows.

$$(6.17)\qquad \mathbf{u}^n = \mathbf{p}^n, \qquad \mathbf{S}\boldsymbol{\lambda}^n = -\mathbf{A}\mathbf{u}^n,$$

$$(6.18)\qquad \mathbf{p}^{n+\theta} = \frac{\boldsymbol{\lambda}^n + r_1\mathbf{u}^n}{(\mathbf{S}(\boldsymbol{\lambda}^n + r_1\mathbf{u}^n), \boldsymbol{\lambda}^n + r_1\mathbf{u}^n)^{1/2}},$$

$$(6.19)\qquad \boldsymbol{\lambda}^{n+\theta} = \boldsymbol{\lambda}^n + r_1(\mathbf{u}^n - \mathbf{p}^{n+\theta}),$$

$$(6.20)\qquad (\mathbf{A}+r_2\mathbf{S})\mathbf{u}^{n+1-\theta} = \mathbf{S}(r_2\mathbf{p}^{n+\theta} - \boldsymbol{\lambda}^{n+\theta}),$$

$$(6.21)\qquad \boldsymbol{\lambda}^{n+1-\theta} = \boldsymbol{\lambda}^{n+\theta} + r_2(\mathbf{u}^{n+1-\theta} - \mathbf{p}^{n+\theta}),$$

$$(6.22)\qquad \mathbf{p}^{n+1} = \frac{\boldsymbol{\lambda}^{n+1-\theta} + r_1\mathbf{u}^{n+1-\theta}}{(\mathbf{S}(\boldsymbol{\lambda}^{n+1-\theta} + r_1\mathbf{u}^{n+1-\theta}), \boldsymbol{\lambda}^{n+1-\theta} + r_1\mathbf{u}^{n+1-\theta})^{1/2}}.$$

In Algorithm (6.16)-(6.22), parameters r_1 and r_2 are defined by $r_1 = 1/\theta\Delta t$ and $r_2 = 1/(1-2\theta)\Delta t$, with $r = 1/\Delta t$.

In all the above algorithms, we observe that at each iteration we must solve a linear system associated with a matrix of type $\alpha\mathbf{S}+\mathbf{A}$ as well as perform one or two vector normalizations. Therefore, these algorithms do appear as shifted inverse power methods for computing the smallest eigenvalue of a symmetric matrix.

6.2.4. Interpretation as time integrators. Suppose that I_Σ is differentiable (which is definitively not the case) and denote by ∂I_Σ its gradient; if $\mathbf{u}$ were a minimizer in (6.5), i.e., an eigenvector of norm one associated to Λ_1, then $\mathbf{u}$ would satisfy

$$(6.23)\qquad \mathbf{A}\mathbf{u} + \partial I_\Sigma(\mathbf{u}) = 0.$$

It is then quite natural to associate with the "nonlinear equation" (6.23) the following initial-value problem

$$\mathbf{S}\frac{d\mathbf{u}}{dt}+\mathbf{A}\mathbf{u}+\partial I_\Sigma(\mathbf{u})=\mathbf{0}, \qquad \mathbf{u}(0)=\mathbf{p}_0, \tag{6.24}$$

and to look for the steady-state solutions of (6.24), i.e., solutions that go to the limit of $\mathbf{u}(t)$ as t goes to infinity, if such a limit exists.

If we now apply to (6.24) one of the alternating-direction schemes of § 5.2 (that is, the Douglas–Rachford scheme (5.2)–(5.4), the Peaceman–Rachford scheme (5.5)–(5.7), or the θ-scheme (5.8)–(5.11)), we will recover exactly the algorithms of § 6.2.3, as already observed in § 5.3. In other words, the algorithms described in § 6.2.3 can be interpreted as standard time integrators of the evolution "equation" (6.24).

6.2.5. Numerical results. Let us consider the particular case of Problem (6.5) with $N=40$, $\mathbf{S}=\mathbf{Id}$, and $\mathbf{A}$ being tridiagonal, having diagonal coefficients equal to 2 and subdiagonal coefficients equal to -1. Such a finite-dimensional eigenvalue problem is obtained when using finite differences to discretize the eigenvalue problem

$$-\frac{d^2u(x)}{dx^2}=\lambda u(x) \quad \text{on } (0,1), \qquad u(0)=u(1)=0.$$

The results presented in Table 6.1 have been obtained by Bourgat [1984]; they give, for each algorithm of § 6.2.3, the value of r that leads to the best possible

TABLE 6.1.
Convergence results for $-u''=\lambda u$, $u(0)=u(1)=0$.

Algorithm	Value of ρ or θ	Optimal value of r	Number of iterations
Algorithm (6.7)–(6.10) (ALG2)	$\rho=r$	88	15
Algorithm (6.11)–(6.15) (ALG3)	$\rho=r$	216	19
	$\rho=r/2$	120	15
Algorithm (6.16)–(6.22) (ALG4)	$\theta=1/3$	800	83
	$\theta=1-\sqrt{2}/2$	600	65
	$\theta=0.25$	440	51
	$\theta=0.2$	336	41
	$\theta=0.15$	192	27
	$\theta=0.1$	96	17
	$\theta=0.05$	33.60	10
	$\theta=0.025$	16	8
	$\theta=0.01$	5.12	7
	$\theta=0.001$	0.56	6

convergence and, for each of these values of r, the number of iterations that are necessary to obtain $(u^n - u^{n-1}, u^n - u^{n-1}) \leq 10^{-10}$, with $u_i^0 = 1/\sqrt{N}$ and $\lambda_i^0 = 0$.

6.3. Solution procedure for the Hartree equation.

6.3.1. Formulation of the problem. Expressed in atomic units and supposing that the two electrons are on the same spatial orbital, the Hartree equation of the helium atom is

$$-\frac{1}{2}\Delta u(\mathbf{x}) - \frac{2}{|\mathbf{x}|}u(\mathbf{x}) + \left(\int_{\mathbb{R}^3} \frac{u^2(\mathbf{y})}{|\mathbf{x}-\mathbf{y}|}\,d\mathbf{y}\right)u(\mathbf{x}) = \Lambda u(\mathbf{x}), \tag{6.25}$$

where $u \in H^1(\mathbb{R}^3)$ and $\Lambda \in \mathbb{R}$.

Equation (6.25) was introduced by Hartree [1928], [1957] as an approximation of the Schrödinger equation of the helium atom and assumes that the wave function ψ of the helium atom can be decomposed into the form

$$\psi(\mathbf{x}, \mathbf{y}) = u(\mathbf{x})u(\mathbf{y}).$$

For physical reasons, we are interested in the normalized solutions of (6.25), that is, solutions that satisfy

$$\int_{\mathbb{R}^3} u^2(\mathbf{x})\,d\mathbf{x} = 1. \tag{6.26}$$

Among these solutions, we are particularly interested by the ground-state solutions, that is, those solutions of (6.25), (6.26) that correspond to the smallest possible energy Λ. It is then a classical result that these ground-state solutions are in fact those that minimize G over $\Sigma \cap H^1(\mathbb{R}^3)$, with

$$\boxed{\Sigma = \{v \in L^2(\mathbb{R}^3),\ \|v\|_{L^2(\mathbb{R}^3)} = 1\},} \tag{6.27}$$

the function G being defined by

$$\boxed{G(v) = \frac{1}{4}\int_{\mathbb{R}^3} |\nabla v|^2\,d\mathbf{x} - \int_{\mathbb{R}^3} \frac{v^2}{|\mathbf{x}|}\,d\mathbf{x} + \frac{1}{4}\int_{\mathbb{R}^3\times\mathbb{R}^3} \frac{v^2(\mathbf{x})v^2(\mathbf{y})}{|\mathbf{x}-\mathbf{y}|}\,d\mathbf{x}\,d\mathbf{y}.} \tag{6.28}$$

Introducing, as in (6.4), the indicator function I_Σ of Σ, our final problem is thus the following minimization problem.

$$\boxed{\text{Find } u \in H^1(\mathbb{R}^3) \text{ such that } G(u) + I_\Sigma(u) \leq G(v) + I_\Sigma(v) \quad \forall v \in H^1(\mathbb{R}^3).} \tag{6.29}$$

Problem (6.29) then appears as a nonlinear eigenvalue problem; because we are looking for the smallest eigenvalue, we will solve this problem using algorithms directly inspired by those of § 6.2.3. (For further references and numerical results, see De Loura [1986] and Gogny and Lions [1986].)

Remark 6.1. Denoting by ϕ the potential

$$\phi(\mathbf{x})=\int_{\mathbb{R}^3}\frac{u^2(\mathbf{y})}{|\mathbf{x}-\mathbf{y}|}\,d\mathbf{y}$$

that appears in (6.28), it is well known from potential theory that ϕ satisfies the Poisson equation

$$-\Delta\phi=4\pi u^2. \tag{6.30}$$

6.3.2. Augmented Lagrangian algorithms for the solution of the Hartree problem (6.29). Defining the function G as in (6.28), and V, H, B, and F by

$$V=H^1(\mathbb{R}^3),\qquad H=L^2(\mathbb{R}^3),$$
$$B=\text{injection from } H^1(\mathbb{R}^3) \text{ into } L^2(\mathbb{R}^3),$$
$$F(q)=I_\Sigma(q),$$

respectively, we observe that (6.29) is again formally a problem (P) in the sense of § 4.1. Then as in § 6.2.2, we associate with (6.29) the augmented Lagrangian $\mathcal{L}_r$ defined by

$$\mathcal{L}_r(v,q;\mu)=I_\Sigma(q)+G(v)+\frac{r}{2}\int_{\mathbb{R}^3}(v-q)^2\,d\mathbf{x}+\int_{\mathbb{R}^3}\mu(v-q)\,d\mathbf{x}, \tag{6.31}$$

so that we can reduce the solution of (6.29) to finding the local saddle-points of $\mathcal{L}_r$ over $(H^1(\mathbb{R}^3)\times L^2(\mathbb{R}^3))\times L^2(\mathbb{R}^3)$. The above saddle-point problem can be solved by ALG3 of § 4.5. Making explicit the necessary conditions associated to the two minimization steps (4.25) and (4.27) of ALG3, we obtain the following algorithm for solving (6.29) (with good convergence properties for $r/2\le\rho\le r$).

ALGORITHM (6.32)–(6.36).

$$\{p^0,\lambda^1\}\in L^2(\mathbb{R}^3)\times L^2(\mathbb{R}^3) \text{ given;} \tag{6.32}$$

with $\{p^{n-1},\lambda^n\}$ known, determine successively u^n, $\lambda^{n+1/2}$, p^n, and λ^{n+1} by

$$\left\{\begin{aligned} & r\int_{\mathbb{R}^3}u^n v\,d\mathbf{x}+\frac{1}{2}\int_{\mathbb{R}^3}\nabla u^n\cdot\nabla v\,d\mathbf{x}-2\int_{\mathbb{R}^3}u^n v\frac{d\mathbf{x}}{|\mathbf{x}|}\\ &\qquad +\int_{\mathbb{R}^3}\left[\int_{\mathbb{R}^3}\frac{[u^n(\mathbf{y})]^2}{|\mathbf{x}-\mathbf{y}|}\,d\mathbf{y}\right]u^n(\mathbf{x})v(\mathbf{x})\,d\mathbf{x}\\ &\qquad =\int_{\mathbb{R}^3}(rp^{n-1}-\lambda^n)v\,d\mathbf{x}\qquad \forall v\in H^1(\mathbb{R}^3),\quad u^n\in H^1(\mathbb{R}^3); \end{aligned}\right. \tag{6.33}$$

$$\lambda^{n+1/2}=\lambda^n+\rho(u^n-p^{n-1}); \tag{6.34}$$

$$p^n=\frac{\lambda^{n+1/2}+ru^n}{\|\lambda^{n+1/2}+ru^n\|_{L^2(\mathbb{R}^3)}}; \tag{6.35}$$

$$\lambda^{n+1}=\lambda^{n+1/2}+\rho(u^n-p^n). \tag{6.36}$$

□

The only nontrivial step in the above algorithm is the solution of the nonlinear variational problem (6.33). To solve (6.33), one can use, for example, a conjugate-gradient algorithm preconditioned by the linear elliptic operator $(rI-\frac{1}{2}\Delta)$ (see Glowinski [1984, Chap. 7], De Loura [1986]).

Actually, according to Remark 6.1, we can simplify Algorithm (6.32)-(6.36) by replacing in (6.33) the quantity

$$\int_{\mathbb{R}^3} \frac{[u^n(\mathbf{y})]^2}{|\mathbf{x}-\mathbf{y}|}\,d\mathbf{y}$$

by the solution $\phi^{n-1}(\mathbf{x})$ of the linear Poisson problem

$$-\Delta\phi^{n-1}=4\pi(p^{n-1})^2.$$

Then, the nonlinear variational problem (6.33) can be replaced by the following linear one.

$$(6.37)\quad \begin{cases} r\displaystyle\int_{\mathbb{R}^3} u^n v\,d\mathbf{x}+\frac{1}{2}\int_{\mathbb{R}^3}\nabla u^n\cdot\nabla v\,d\mathbf{x}-2\int_{\mathbb{R}^3} u^n v\frac{d\mathbf{x}}{|\mathbf{x}|} \\ +\displaystyle\int_{\mathbb{R}^3}\phi^{n-1}u^n v\,d\mathbf{x}=\int_{\mathbb{R}^3}(rp^{n-1}-\lambda^n)v\,d\mathbf{x}\quad \forall v\in H^1(\mathbb{R}^3),\quad u^n\in H^1(\mathbb{R}^3). \end{cases}$$

This last problem can be solved by the conjugate-gradient Algorithm (3.48)-(3.55), with $H^1(\mathbb{R}^3)$ equipped with the scalar product

$$(u,v)=r\int_{\mathbb{R}^3} uv\,d\mathbf{x}+\frac{1}{2}\int_{\mathbb{R}^3}\nabla u\cdot\nabla v\,d\mathbf{x}.$$

Remark 6.2. In practice, we shall replace $\mathbb{R}^3$ by an open ball of radius R centered at the origin. Then, we shall take advantage of the spherical symmetry of the ground-state solution, reducing (6.29) to a one-dimensional problem set on the interval $(0, R)$. This one-dimensional problem will be finally approximated by either finite differences or finite elements, taking as a boundary condition $u(R)=0$. More details can be found in De Loura [1986], where it is shown, in particular, that $R=10$ provides already excellent numerical results.

7. Liquid crystals theory and further comments.

7.1. Formulation of the problem. To conclude this chapter, and to show the efficiency of the augmented Lagrangian and alternating-direction methods that were introduced in the previous sections, we shall apply these methods to the numerical solution of a problem originating in the mathematical theory of liquid crystals. For this purpose, we first introduce additional notation.

Let Ω be a bounded domain of $\mathbb{R}^3$; we denote by Γ the boundary of Ω, and we suppose that Γ is sufficiently smooth (for example, Lipschitz-continuous). We now define the space

$$\mathbf{H}^1(\Omega)=(H^1(\Omega))^3;$$

the function

$$(7.1)\quad J(\mathbf{v})=\frac{1}{2}\int_\Omega |\nabla \mathbf{v}|^2\,d\mathbf{x}\quad \left(=\frac{1}{2}\sum_{i=1}^{3}\int_\Omega |\nabla v_i|^2\,d\mathbf{x}\right),\qquad \forall \mathbf{v}=\{v_i\}_{i=1}^3\in \mathbf{H}^1(\Omega).$$

Finally, the set

$$(7.2)\qquad E=\{\mathbf{v}\,|\,\mathbf{v}\in \mathbf{H}^1(\Omega),\ \mathbf{v}=\mathbf{g} \text{ on } \Gamma,\ |\mathbf{v}(\mathbf{x})|=1 \text{ a.e.}\},$$

where $|\mathbf{v}|=(\sum_{i=1}^3 v_i^2)^{1/2}$. We suppose that $\mathbf{g}$ is such that $E\neq\emptyset$.

Remark 7.1. Consider $\mathbf{a}\in\mathbb{R}^3$ and define $\boldsymbol{\phi}_a$ as the restriction to Ω of the function

$$\mathbf{x}\to\frac{\mathbf{x}-\mathbf{a}}{|\mathbf{x}-\mathbf{a}|}.$$

We clearly have $|\boldsymbol{\phi}_a(\mathbf{x})|=1$ a.e.; furthermore, we can easily prove that $\boldsymbol{\phi}_a\in \mathbf{H}^1(\Omega)$ (even if $\mathbf{a}\in\bar{\Omega}$).

With all of the preceding notation, we consider now the following variational problem

$$(7.3)\qquad \boxed{\begin{cases}\text{Find } \mathbf{u}\in E, \text{ such that}\\ J(\mathbf{u})\le J(\mathbf{v})\qquad \forall \mathbf{v}\in E.\end{cases}}$$

Using the fact that E is weakly closed in $\mathbf{H}^1(\Omega)$, it can easily be proved from the Weierstrass theorem that problem (7.3) has at least a solution; further mathematical properties of (7.3) are discussed in Hart and Kinderlehrer [1986], Hart, Kinderlehrer, and Lin [1986], and Brezis, Coron, and Lieb [1986]. Problem (7.3) is associated with the mathematical modeling of quite interesting physical phenomena (as discussed in Brezis, Coron, and Lieb [1986, § 1]), some of which occur in the physics of liquid crystals (see De Gennes [1974], Ericksen [1976], and Chandrasekhar [1977] for further information on the physics of liquid crystals).

7.2. Numerical solution of problem (7.3). At first glance, problem (7.3) seems to be a nontrivial problem of the calculus of variations. In fact, the solution of (7.3) is quite easy to achieve by the augmented Lagrangian and alternating-direction methods of §§ 4 and 5. Indeed, this follows from the fact that problem (7.3) is equivalent to

$$(7.4)\qquad \begin{cases}\text{Find } \mathbf{u}\in \mathbf{H}_g^1 \text{ such that}\\ J(\mathbf{u})+I_\Sigma(\mathbf{u})\le J(\mathbf{v})+I_\Sigma(\mathbf{v})\qquad \forall \mathbf{v}\in \mathbf{H}_g^1,\end{cases}$$

where, with $\mathbf{L}^2(\Omega)=(L^2(\Omega))^3$,

$$\mathbf{H}_g^1=\{\mathbf{v}\,|\,\mathbf{v}\in \mathbf{H}^1(\Omega),\ \mathbf{v}=\mathbf{g} \text{ on } \Gamma\},$$

$$\Sigma=\{\mathbf{v}\,|\,\mathbf{v}\in \mathbf{L}^2(\Omega),\ |\mathbf{v}(\mathbf{x})|=1 \text{ a.e.}\},$$

and where $I_\Sigma : \mathbf{L}^2(\Omega) \to \mathbb{R} \cup \{+\infty\}$ is defined by

$$I_\Sigma(\mathbf{v}) = \begin{cases} 0 & \text{if } \mathbf{v} \in \Sigma, \\ +\infty & \text{if } \mathbf{v} \notin \Sigma. \end{cases}$$

Defining V, H, G, F, and B by

$$V = \mathbf{H}^1_g, \qquad H = \mathbf{L}^2(\Omega),$$
$$G = J, \qquad F = I_\Sigma,$$
$$B = \text{injection from } \mathbf{H}^1(\Omega) \text{ into } \mathbf{L}^2(\Omega),$$

respectively, it is clear that problem (7.4) is a problem (P) in the sense of § 4.1. Then, as in §§ 6.2.2 and 6.3.2, we associate with problem (7.4) the augmented Lagrangian $\mathscr{L}_r$ defined by

$$\mathscr{L}_r(\mathbf{v}, \mathbf{q}; \boldsymbol{\mu}) = J(\mathbf{v}) + I_\Sigma(\mathbf{q}) + \frac{r}{2}\int_\Omega |\mathbf{v}-\mathbf{q}|^2 \, d\mathbf{x} + \int_\Omega \boldsymbol{\mu} \cdot (\mathbf{v}-\mathbf{q}) \, d\mathbf{x}, \tag{7.5}$$

(where $\boldsymbol{\mu} \cdot \mathbf{q} = \sum_{i=1}^3 \mu_i q_i$ for all $\boldsymbol{\mu}$, $\mathbf{q}$), and reduce the solution of (7.4) to finding the (possibly local) saddle-points of $\mathscr{L}_r$ over $(\mathbf{H}^1_g \times \mathbf{L}^2(\Omega)) \times \mathbf{L}^2(\Omega)$. The above saddle-point problem can in turn be solved by the algorithms of §§ 4 and 5. Taking, for example, ALG2, we obtain the following quite simple algorithm for solving problem (7.3), (7.4).

ALGORITHM (7.6)–(7.9).

$$\{\mathbf{p}^0, \boldsymbol{\lambda}^1\} \in \mathbf{L}^2(\Omega) \times \mathbf{L}^2(\Omega) \text{ given;} \tag{7.6}$$

with $\{\mathbf{p}^{n-1}, \boldsymbol{\lambda}^n\}$ known, compute successively $\mathbf{u}^n$, $\mathbf{p}^n$, and $\boldsymbol{\lambda}^{n+1}$ as follows.

$$\begin{cases} r\mathbf{u}^n - \Delta \mathbf{u}^n = r\mathbf{p}^{n-1} - \boldsymbol{\lambda}^n & \text{in } \Omega, \\ \mathbf{u}^n = \mathbf{g} \quad \text{on } \Gamma, \end{cases} \tag{7.7}$$

$$\mathbf{p}^n = \frac{r\mathbf{u}^n + \boldsymbol{\lambda}^n}{|r\mathbf{u}^n + \boldsymbol{\lambda}^n|}, \tag{7.8}$$

$$\boldsymbol{\lambda}^{n+1} = \boldsymbol{\lambda}^n + \rho(\mathbf{u}^n - \mathbf{p}^n). \tag{7.9}$$

The remarkably simple relation (7.8) follows from the fact that, over Σ, the nonconstant part of the functional

$$\mathbf{q} \to \mathscr{L}_r(\mathbf{u}^n, \mathbf{q}; \boldsymbol{\lambda}^n)$$

reduces to the linear functional

$$\mathbf{q} \to -\int_\Omega (r\mathbf{u}^n + \boldsymbol{\lambda}^n) \cdot \mathbf{q} \, d\mathbf{x},$$

whose minimizer (over Σ) is precisely given by (7.8), assuming that $r\mathbf{u}^n + \boldsymbol{\lambda}^n \neq \mathbf{0}$.

Here again, a good choice for ρ appears to be $\rho = r$. Actually, we can employ the other algorithms of §§ 4 and 5 (namely, ALG1, ALG3, and ALG4) to solve problem (7.3), (7.4), using either their augmented Lagrangian or alternating-direction formulations. Indeed, in this simple case where operator B is the injection from $\mathbf{H}^1(\Omega)$ into $\mathbf{L}^2(\Omega)$, it is quite easy to obtain the alternating-direction algorithms by direct consideration of the following nonlinear parabolic "equation" associated with problem (7.3), (7.4).

$$\text{(7.10)}\qquad \begin{cases} \dfrac{\partial \mathbf{u}}{\partial t} - \Delta \mathbf{u} + \partial I_\Sigma(\mathbf{u}) = 0, \\ \mathbf{u} = \mathbf{g} \quad \text{on } \Gamma, \\ \mathbf{u}(0) = \mathbf{u}_0, \end{cases}$$

where ∂I_Σ is the "subgradient" of I_Σ at $\mathbf{u}$. Concentrating on the θ-scheme introduced in § 5.2, since it is the most complicated scheme, we obtain the following algorithm, where $\Delta t > 0$ and $\theta \in (0, \frac{1}{2})$.

ALGORITHM (7.11)-(7.14).

$$\text{(7.11)}\qquad \mathbf{u}^0 = \mathbf{u}_0, \text{ given in } \mathbf{H}^1_g;$$

for $n \geq 0$, $\mathbf{u}^n$ being known, we compute $\mathbf{u}^{n+\theta}$, $\mathbf{u}^{n+1-\theta}$, $\mathbf{u}^{n+1}$ as follows.

$$\text{(7.12)}\qquad \frac{\mathbf{u}^{n+\theta} - \mathbf{u}^n}{\theta\, \Delta t} - \Delta \mathbf{u}^n + \partial I_\Sigma(\mathbf{u}^{n+\theta}) = \mathbf{0},$$

$$\text{(7.13)}\qquad \begin{cases} \dfrac{\mathbf{u}^{n+1-\theta} - \mathbf{u}^{n+\theta}}{(1-2\theta)\, \Delta t} - \Delta \mathbf{u}^{n+1-\theta} + \partial I_\Sigma(\mathbf{u}^{n+\theta}) = \mathbf{0}, \\ \mathbf{u}^{n+1-\theta} = \mathbf{g} \quad \text{on } \Gamma, \end{cases}$$

$$\text{(7.14)}\qquad \frac{\mathbf{u}^{n+1} - \mathbf{u}^{n+1-\theta}}{\theta\, \Delta t} - \Delta \mathbf{u}^{n+1-\theta} + \partial I_\Sigma(\mathbf{u}^{n+1}) = \mathbf{0}.$$

The augmented Lagrangian formulation of this algorithm is left to the reader. When using Algorithm (7.11)-(7.14) for practical calculations, one has to further define the two multivalued equations (7.12) and (7.14). The interpretation given to (7.12) will be

$$\text{(7.15)}\qquad \begin{cases} \mathbf{u}^{n+\theta} \in \Sigma, \qquad \mathbf{u}^{n+\theta} \text{ minimizes over } \Sigma \text{ the function} \\ \mathbf{v} \to \dfrac{1}{2} \displaystyle\int_\Omega |\mathbf{v}|^2 \, d\mathbf{x} - \int_\Omega (\mathbf{u}^n + \theta\, \Delta t\, \Delta \mathbf{u}^n) \cdot \mathbf{v} \, d\mathbf{x}; \end{cases}$$

the solution of which clearly is given by

$$\text{(7.16)}\qquad \mathbf{u}^{n+\theta} = \frac{\mathbf{u}^n + \theta\, \Delta t\, \Delta \mathbf{u}^n}{|\mathbf{u}^n + \theta\, \Delta t\, \Delta \mathbf{u}^n|}.$$

Similarly, the solution of (7.14) is given by

$$\mathbf{u}^{n+1}=\frac{\mathbf{u}^{n+1-\theta}+\theta\,\Delta t\,\Delta\mathbf{u}^{n+1-\theta}}{|\mathbf{u}^{n+1-\theta}+\theta\,\Delta t\,\Delta\mathbf{u}^{n+1-\theta}|}. \tag{7.17}$$

Once $\mathbf{u}^{n+\theta}$ is known, we obtain $\partial I_\Sigma(\mathbf{u}^{n+\theta})$ from (7.12) and use the result to compute $\mathbf{u}^{n+1-\theta}$ in (7.13) via the solution of a Dirichlet problem for the elliptic operator

$$\mathbf{v}\to\mathbf{v}-(1-2\theta)\,\Delta t\,\Delta\mathbf{v}.$$

From these observations, it can be seen that the only costly step of Algorithm (7.11)-(7.14) is the solution of the Dirichlet problem (7.13).

7.3. Numerical experiments. The numerical methods described in § 7.2 have been applied to the solution of various test problems (see also Cohen, Hart, Kinderlehrer, Lin, and Luskin [1987] for related numerical experiments). These test problems have in common Ω, defined here by

$$\Omega=[(0,1)]^3. \tag{7.18}$$

In this section, we shall consider only three cases for $\mathbf{g}$, namely,

(i) $\mathbf{g}=\mathbf{g}_1$, where $\mathbf{g}_1=\boldsymbol{\phi}_a$, with $\boldsymbol{\phi}_a$ defined by

$$\boldsymbol{\phi}_a(\mathbf{x})=\frac{\mathbf{x}-\mathbf{a}}{|\mathbf{x}-\mathbf{a}|},\qquad \mathbf{a}=\left\{\frac{1}{2},\frac{1}{2},\frac{1}{2}\right\}. \tag{7.19}$$

(ii) $\mathbf{g}=\mathbf{g}_2$, with $\mathbf{g}_2$ defined by

$$\mathbf{g}_2=\boldsymbol{\nu}, \tag{7.20}$$

where $\boldsymbol{\nu}$ is the unit outward normal vector of Γ.

(iii) $\mathbf{g}=\mathbf{g}_3$, with $\mathbf{g}_3$ defined by

$$\left\{\begin{aligned}
&\mathbf{g}_3(x_1,x_2,1)=\{0,0,1\}\quad \text{if } 0<x_1<1,\quad 0<x_2<1,\\
&\mathbf{g}_3(x_1,x_2,0)=\{0,0,-1\}\quad \text{if } 0<x_1<1,\quad 0<x_2<1,\\
&\mathbf{g}_3(1,x_2,x_3)=\begin{cases}\{1,0,0\} & \text{if } 0<x_2<\tfrac12,\quad 0<x_3<1,\\ \{-1,0,0\} & \text{if } \tfrac12<x_2<1,\quad 0<x_3<1,\end{cases}\\
&\mathbf{g}_3(x_1,1,x_3)=\begin{cases}\{0,-1,0\} & \text{if } 0<x_1<\tfrac12,\quad 0<x_3<1,\\ \{0,1,0\} & \text{if } \tfrac12<x_1<1,\quad 0<x_3<1,\end{cases}\\
&\mathbf{g}_3(0,x_2,x_3)=\begin{cases}\{1,0,0\} & \text{if } 0<x_2<\tfrac12,\quad 0<x_3<1,\\ \{-1,0,0\} & \text{if } \tfrac12<x_2<1,\quad 0<x_3<1,\end{cases}\\
&\mathbf{g}_3(x_1,0,x_3)=\begin{cases}\{0,-1,0\} & \text{if } 0<x_1<\tfrac12,\quad 0<x_3<1,\\ \{0,1,0\} & \text{if } \tfrac12<x_1<1,\quad 0<x_3<1.\end{cases}
\end{aligned}\right. \tag{7.21}$$

Since the functions $\mathbf{g}_2$ and $\mathbf{g}_3$ are not the traces of functions of $\mathbf{H}^1(\Omega)$, the corresponding problem (7.3) has no solution. However, since the associated discrete problems do have solutions, we determined it would be of interest to see how the numerical algorithms discussed previously would behave in such pathological situations. In the case of (7.3), with $\mathbf{g}=\mathbf{g}_1$, it follows from Brezis, Coron, and Lieb [1986] that this problem has a unique solution precisely given by

$$\mathbf{u}_1 = \boldsymbol{\phi}_{a|\Omega}. \tag{7.22}$$

Because of the simplicity of Ω, it is quite convenient to approximate problem (7.3) by a finite-difference method such as the following.

Let N be a positive integer. We define a space-discretization step h by $h=1/(N+1)$, and then define the discrete set

$$\{M_{ijk}\}_{0\le i,j,k\le N+1} \quad \text{with } M_{ijk}=\{ih, jh, kh\}.$$

With $\mathbf{v}_h=\{\{v_{ijk}^{\ell}\}_{\ell=1}^{3}\}_{0\le i,j,k\le N+1}$, we approximate $J(\mathbf{v})$ by

$$\begin{aligned} J_h(\mathbf{v}_h)=\frac{h^3}{4}\sum_{\ell=1}^{3}\sum_{1\le i,j,k\le N}\Bigg\{&\left|\frac{v_{i+1jk}^{\ell}-v_{ijk}^{\ell}}{h}\right|^2 \\ &+\left|\frac{v_{i-1jk}^{\ell}-v_{ijk}^{\ell}}{h}\right|^2+\left|\frac{v_{ij+1k}^{\ell}-v_{ijk}^{\ell}}{h}\right|^2+\left|\frac{v_{ij-1k}^{\ell}-v_{ijk}^{\ell}}{h}\right|^2 \\ &+\left|\frac{v_{ijk+1}^{\ell}-v_{ijk}^{\ell}}{h}\right|^2+\left|\frac{v_{ijk-1}^{\ell}-v_{ijk}^{\ell}}{h}\right|^2\Bigg\}, \end{aligned} \tag{7.23}$$

and then approximate E by

$$E_h=\left\{\mathbf{v}_h \,\middle|\, \sum_{\ell=1}^{3}|v_{ijk}^{\ell}|^2=1,\ 1\le i,j,k\le N;\ \mathbf{v}_{ijk}=\mathbf{g}_{ijk}\ \forall M_{ijk}\in\Gamma\right\}. \tag{7.24}$$

In (7.24), if $\mathbf{g}$ is continuous at M_{ijk}, $\mathbf{g}_{ijk}=\mathbf{g}(M_{ijk})$; if $\mathbf{g}$ is not continuous at M_{ijk}, $\mathbf{g}_{ijk}$ is obtained from $\mathbf{g}$ by a simple averaging technique.

Finally, problem (7.3) is approximated by

$$\begin{cases} \text{Find } \mathbf{u}_h\in E_h \text{ such that} \\ J_h(\mathbf{u}_h)\le J_h(\mathbf{v}_h) \quad \forall \mathbf{v}_h\in E_h. \end{cases} \tag{7.25}$$

Applying the algorithms discussed in § 7.2 (Algorithms (7.6)–(7.9), (7.11)–(7.14)) to the finite-dimensional problem (7.25) is quite easy, since (7.25) has the same structure and decomposition properties as (7.3).

The numerical results that follow were obtained by J. F. Bourgat and C. H. Li. All calculations were initialized by $\mathbf{u}_h^0$, the finite-difference approximation of the solution $\mathbf{u}^0$ of the Dirichlet problem

$$-\Delta\mathbf{u}^0=\mathbf{0} \quad \text{in } \Omega, \qquad \mathbf{u}^0=\mathbf{g} \quad \text{on } \Gamma. \tag{7.26}$$

As convergence criteria, we used (with obvious notation):

$$(7.27)\qquad \left(h^3 \sum_{1\le i,j,k\le N} |\mathbf{u}_{ijk}^{n+1}-\mathbf{u}_{ijk}^{n}|^2\right)^{1/2} \Big/ \left(h^3 \sum_{1\le i,j,k\le N} |\mathbf{u}_{ijk}^{n+1}|^2\right)^{1/2} \le 10^{-4}.$$

Numerical results for $\mathbf{g}=\mathbf{g}_1$. This example is quite interesting, since the exact solution $\mathbf{u}=\mathbf{u}_1$ of (7.3) is known and is given by (cf. (7.22))

$$\mathbf{u}_1(\mathbf{x}) = \frac{\mathbf{x}-\mathbf{a}}{|\mathbf{x}-\mathbf{a}|} \quad \text{a.e. on } \Omega.$$

We can therefore estimate the $L^2(\Omega)$-norm of the approximation error; we chose as an estimator of the L^2-error the quantity ρ_h, defined by

$$(7.28)\qquad \rho_h = \left(h^3 \sum_{1\le i,j,k\le N} |\mathbf{u}_1(M_{ijk})-\mathbf{u}_{ijk}|^2\right)^{1/2}$$

(we took here $\mathbf{u}_1(\frac{1}{2},\frac{1}{2},\frac{1}{2}) = \{\frac{1}{\sqrt{3}},\frac{1}{\sqrt{3}},\frac{1}{\sqrt{3}}\}$). The numerical results associated with $\mathbf{g}=\mathbf{g}_1$ are summarized in the following tables.

The results summarized in Table 7.1 were obtained using the point Gauss–Seidel method to solve the discrete linear elliptic problems encountered at each full step of the above algorithms; as an initializer we used the value provided by the previous time step, and as a stopping criterion one similar to (7.27) but with $\varepsilon = 10^{-6}$.

In Table 7.2, we show, at each time step, the number of Gauss–Seidel iterations necessary in order to obtain convergence for the test problem $\mathbf{g}=\mathbf{g}_1$, under the conditions described above. If, instead of the Gauss–Seidel method, we used an over-relaxation one, performances of the Peaceman–Rachford and Douglas–Rachford schemes were practically unchanged. However, the performance of the θ-scheme was dramatically improved (particularly for the first time step), and, for the test problem $\mathbf{g}=\mathbf{g}_1$ (with the same values of θ, h,

TABLE 7.1
Comparison of numerical results using the Douglas–Rachford [1956], *Peaceman–Rachford* [1955], *and θ-schemes.*

Scheme	ρ_h	θ	h	Δt	No. of time steps to reach steady state	VAX 11/780 CPU time
θ (5.8)–(5.11)	$0.39\ 10^{-1}$	0.1	1/20	1/200	8	11 mn 18 s
Douglas–Rachford (5.5)–(5.7)	$0.36\ 10^{-1}$	—	1/20	1/2000	10	10 mn 42 s
Peaceman–Rachford (5.2)–(5.4)	$0.43\ 10^{-1}$	—	1/20	1/1000	67	34 mn 21 s

TABLE 7.2
Variations of the number of Gauss–Seidel iterations with the time step using the Douglas–Rachford [1956], *Peaceman–Rachford* [1955], *and θ-schemes.*

θ scheme (5.8)–(5.11)		Douglas–Rachford scheme (5.5)–(5.7)		Peaceman–Rachford scheme (5.2)–(5.4)	
Time step	Gauss–Seidel iterations	Time step	Gauss–Seidel iterations	Time step	Gauss–Seidel iterations
				1–16	7
1	26	1	4	17–30	6
2	4	2	3	31–44	5
3	2	3–5	2	45–55	4
4–8	1	6–10	1	56–61	3
				62–64	2
				65–67	1

and Δt) we obtained convergence in 5 time steps (instead of 8), the computational time being reduced to 3 mn 44 s (instead of 11 mn 18 s, cf. Table 7.1). In fact, the L^2-error was also substantially reduced, since we then had $\rho_h = 0.23\times 10^{-1}$ (instead of 0.39×10^{-1}).

Table 7.3 shows the variations of the L^2-error ρ_h as a function of h and Δt for the various schemes discussed above (results for the θ-scheme were obtained using the over-relaxation instead of the Gauss–Seidel method; the case $h = 1/40$ was computed on the CRAY-XMPS201 and took approximately 12 s).

The results shown in Table 7.3 suggest that the L^2-approximation error is $O(h)$ at best; the analysis of such an error is a most interesting problem in itself.

TABLE 7.3
Variations of the L^2-error ρ_h as a function of h and Δt, using the Douglas–Rachford [1956], *Peaceman–Rachford* [1955], *and θ schemes.*

Scheme	h	Δt	θ	ρ_h
θ scheme	1/10	1/100	1/10	0.74×10^{-1}
(5.8)–(5.11)	1/20	1/200	1/10	0.23×10^{-1}
	1/40	1/2000	1/5	0.12×10^{-1}
Douglas–Rachford	1/10	1/1000	—	0.63×10^{-1}
(5.5)–(5.7)	1/20	1/2000	—	0.36×10^{-1}
Peaceman–Rachford	1/10	1/1000	—	0.72×10^{-1}
(5.2)–(5.4)	1/20	1/1000	—	0.43×10^{-1}
	1/20	1/2000	—	0.43×10^{-1}

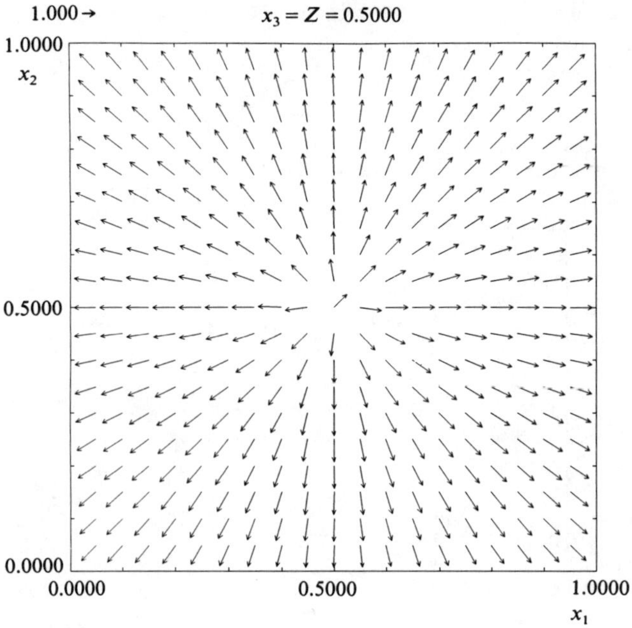

FIG. 7.1. $\mathbf{g} = \mathbf{g}_1$.

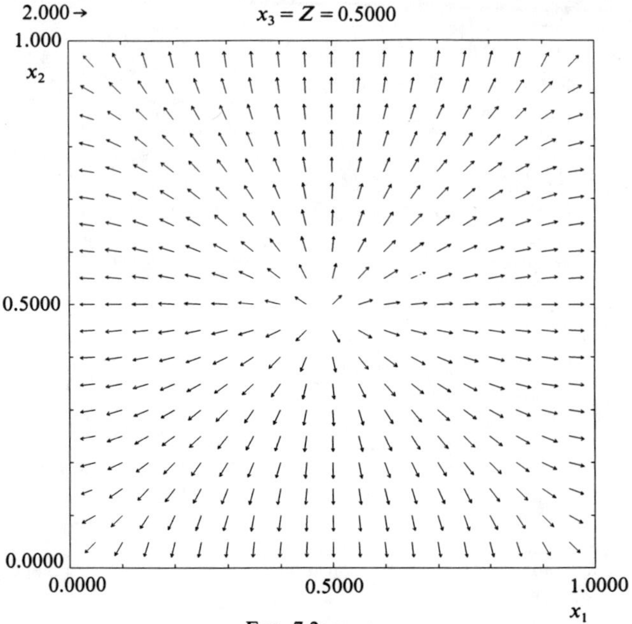

FIG. 7.2. $\mathbf{g} = \mathbf{g}_2$.

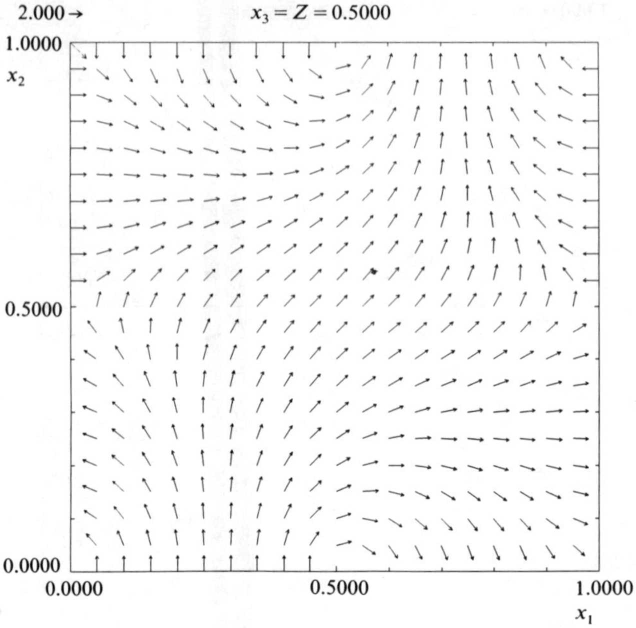

FIG. 7.3. $\mathbf{g} = \mathbf{g}_3$; $\mathbf{u}_h^0$ *computed by Gauss–Seidel.*

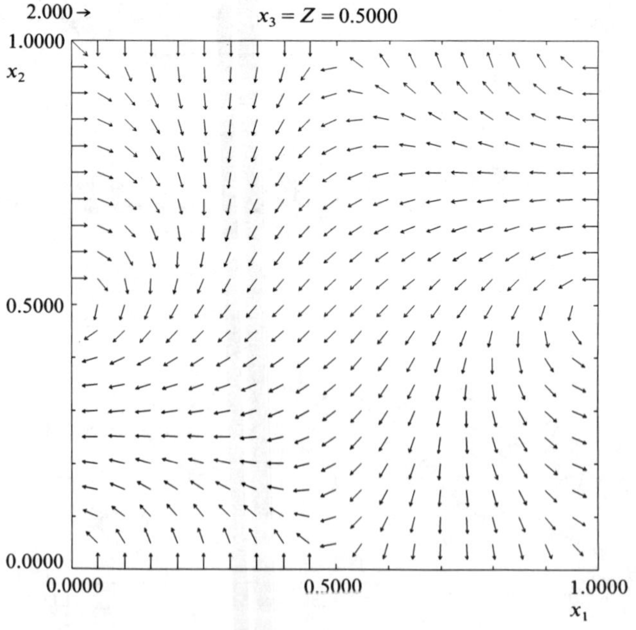

FIG. 7.4. $\mathbf{g} = \mathbf{g}_3$; $\mathbf{u}_h^0$ *computed by over-relaxation with* $\omega = 1.8$.

Figure 7.1 shows the computed solution $\mathbf{u}_{1h}$ associated with the grid points belonging to the plane $x_3 = \frac{1}{2}$; we have shown, in fact, the orthogonal projection of the solution on that plane (multiplied by an appropriate scaling factor). We observe that, outside of a small neighborhood of $\mathbf{a} = \{\frac{1}{2}, \frac{1}{2}, \frac{1}{2}\}$, the solution is radial and of constant norm (in fact, we observe a uniform convergence outside of the above neighborhood).

Numerical results for $\mathbf{g} = \mathbf{g}_2$. We have $\mathbf{g}_2 = \boldsymbol{\nu}$; therefore, problem (7.3) has no solution. Nonetheless, the corresponding discrete problem (7.25) has a solution. Using a θ-scheme, with $\theta = \frac{1}{10}$, $h = \frac{1}{20}$, and $\Delta t = \frac{1}{200}$, a discrete solution was obtained in 5 time steps (3 mn 44 s on the VAX 11/780); the discrete elliptic problem was solved using an over-relaxation method.

The computed solution is shown in Fig. 7.2, using the same principles that underlie Fig. 7.1.

Numerical results for $\mathbf{g} = \mathbf{g}_3$. Recall that the function $\mathbf{g}_3$ was defined in (7.21). Again, the corresponding problem (7.3) has no solution. Nevertheless, using a θ-scheme with $\theta = \frac{1}{20}$, $h = \frac{1}{20}$, and $\Delta t = \frac{1}{200}$, we obtained a discrete solution in 43 time steps (26 mn 17 s on the VAX 11/780). Indeed, in this case we observed a quite interesting phenomenon. The steady-state solution obtained by the θ-scheme using as an initializer $\mathbf{u}_h^0$ computed from (7.26) by the Gauss–Seidel method was different than that obtained using $\mathbf{u}_h^0$ computed by the over-relaxation method (with $\omega = 1.8$). The two solutions give the same value to J_h and, in fact, are obtained one from another by a rotation of $\pi/4$. These two solutions have been visualized (again, in the plane $x_3 = \frac{1}{2}$) on Figs. 7.3 and 7.4, and a vortex-like pattern can be observed.

7.4. Further comments. Augmented Lagrangian methods, once popular, no longer seem to be well regarded by the mathematical programming community. It is our opinion, however, that combined with natural decomposition principles and using the special structure that most practical nonlinear problems present, they can lead to very efficient algorithms for the solution of difficult problems of great practical interest.

This point of view is supported by the fact that, as seen in this chapter, classical methods of numerical analysis such as alternating-direction methods for time-dependent problems or power methods for eigenvalue calculations can be interpreted as augmented Lagrangian algorithms. To our knowledge, the relation between alternating-direction and augmented Lagrangian methods was first observed by Chan and Glowinski [1978] and then systematically analyzed by Gabay [1979], [1982], [1983].

The efficiency of augmented Lagrangian methods will be further illustrated in the following chapters, where, in specific mechanical situations, we will exhibit special structures that make the use of such methods both possible and efficient.

Chapter 4

Viscoplasticity and Elastoviscoplasticity in Small Strains

1. Introduction.

1.1. Description of the next chapters. In the next chapters, we apply the general numerical techniques described in Chapter 3 (mainly those of §§ 4 and 5) to specific classes of mechanical problems. These include elastoviscoplasticity (Chap. 4), limit load analysis (Chap. 5), flows of viscoplastic fluids (Chap. 6), and finite elasticity (Chaps. 7, 8). These techniques require that the original mechanical problems, whose variational formulations are given in Chapter 2, must first be decomposed. Herein, the decomposition strategy will always be the same.

(i) The primal variable $\mathbf{v}$ will be either the velocity or the displacement field.

(ii) The dual variable $\boldsymbol{\lambda}$ will measure stresses.

(iii) The relation $-B^T\boldsymbol{\lambda} \in \partial G(\mathbf{v})$ will express the virtual work theorem.

(iv) The relation $\boldsymbol{\lambda} \in \partial F(B\mathbf{v})$ will express the nonlinear part of the constitutive equations and will be satisfied pointwise.

1.2. Elastoviscoplasticity. We consider in this chapter the problem of computing the quasi-static viscous flows of elastoviscoplastic materials in small strains subjected to given distributions of external loads. The constitutive law that models the behavior of the considered materials (Chap. 1, (4.9)), together with the reference configuration of the body that it composes, is given. The unknowns in this model are the velocities and the stresses inside the body resulting from the application of external loads.

The materials involved in such problems include steel, concrete, bituminous cements, polymers at high temperatures, frozen soils, and different types of muds. When subjected to external loads, these materials flow viscously in a nonreversible and mostly incompressible pattern and develop stresses of both viscous and elastic origin.

From the numerical point of view, problems arise, first, from the difficulty of approximating incompressible velocity fields and, second, from the poor conditioning and the possible lack of differentiability of the involved functions. Both difficulties can be surmounted using the augmented Lagrangian techniques presented in Chapter 3.

1.3. Synopsis. In this chapter we apply the numerical techniques of Chapter 3 to problems in elastoviscoplasticity. First, we will review the variational formulation of the mechanical problems under consideration (§ 2), and then introduce finite-element approximations of the above formulation (§§ 3, 4). An algorithm will then be proposed for the numerical solution of these approximate problems based on elimination of the velocity field and on time integration of the resulting stress evolution problem by alternating-direction techniques (§ 5). As was the case in Chapter 3, § 5, this algorithm turns out to be equivalent to that generated by use of the augmented Lagrangian methods of Chapter 3, § 4 to numerically solve the associated stationary problem. Indeed, the stress evolution problem is precisely the dual evolution problem that the techniques of Chapter 3, § 5 would derive from this stationary problem. This problem differs from that of Chapter 3 in that neither the primal stationary problem nor the dual evolution problem is a mathematical artifact but both are problems of physical interest, since they determine the final velocity and the stress history inside the body, respectively. Next, the nonlinear local problems appearing in the decomposition are studied in detail (§ 6), and finally, numerical examples (§ 7) are presented.

1.4. Viscosplasticity. From the mechanical point of view (Chap. 1, § 4), viscoplasticity can be obtained from elastoviscoplasticity by cancelling the elasticity term $\mathbf{A}^{-1}\dot{\boldsymbol{\sigma}}$ in the constitutive law and inverting this law. Because this suppresses all evolution terms in the mechanical equations, the resulting viscoplastic problem must therefore be equivalent to stationary elastoviscoplasticity.

Indeed, we will see in § 4 that the finite-element approximation of the viscoplastic problems studied in Chapter 2 is identical to the finite-element formulation of the stationary elastoviscoplastic problems. From the numerical point of view, viscoplasticity is then completely equivalent to stationary elastoviscoplasticity and will be treated as such here.

2. Mixed variational formulations of elastoviscoplasticity.

2.1. The three-dimensional formulation. As in Chapter 2, § 4, we consider a continuous body made of an elastoviscoplastic material that occupies a domain $\Omega \subset \mathbb{R}^3$ in its reference configuration, that is fixed on the part Γ_1 of the boundary Γ of Ω, and that is subjected to given body forces $\mathbf{f}$ and surface tractions $\mathbf{g}$ applied on the part $\Gamma_2 = \Gamma - \Gamma_1$ of its boundary. The problem then consists in determining, in the time interval $[0, t_1]$ and for given initial values $\mathbf{u}_0$ and $\boldsymbol{\sigma}_0$, the history $\mathbf{u}(\cdot, t)$ and $\boldsymbol{\sigma}(\cdot, t)$ of the displacement and of the second Piola–Kirchhoff stress tensor inside this body (see Fig. 2.1 in Chap. 2).

In Chapter 2 we derived the following well-posed variational formulation for this problem, for a given temperature field, assuming small strains and neglecting inertia effects.

$$
(2.1)\quad \begin{cases}
\text{Find } \{\mathbf{u}, \boldsymbol{\sigma}\}: [0, t_1] \to V^{m^*} \times \Sigma^m \text{ such that} \\[4pt]
\displaystyle\int_\Omega \boldsymbol{\sigma} \cdot \mathbf{D}(\mathbf{w})\, d\mathbf{x} = \int_\Omega \mathbf{f} \cdot \mathbf{w}\, d\mathbf{x} + \int_{\Gamma_2} \mathbf{g} \cdot \mathbf{w}\, da \\
\hfill \forall \mathbf{w} \in V^{m^*} \quad \text{a.e. in } (0, t_1), \\[4pt]
\displaystyle\int_\Omega (\mathbf{D}(\dot{\mathbf{u}}) - \mathbf{A}^{-1}\dot{\boldsymbol{\sigma}}) \cdot (\boldsymbol{\tau} - \boldsymbol{\sigma})\, d\mathbf{x} \le \int_\Omega \frac{\lambda}{q} \{|j_c(\mathbf{x}, \boldsymbol{\tau})|^q - |j_c(\mathbf{x}, \boldsymbol{\sigma})|^q\}\, d\mathbf{x} \\
\hfill \forall \boldsymbol{\tau} \in \Sigma^m \quad \text{a.e. in } (0, t_1), \\[4pt]
\mathbf{u}(\mathbf{x}, 0) = \mathbf{u}_0(\mathbf{x}), \qquad \boldsymbol{\sigma}(\mathbf{x}, 0) = \boldsymbol{\sigma}_0(\mathbf{x}),
\end{cases}
$$

under the notation

$$
(2.2)\qquad \mathbf{D}(\mathbf{w}) = \tfrac{1}{2}(\nabla \mathbf{w} + \nabla \mathbf{w}^T),
$$

$$
(2.3)\quad \begin{cases}
\Sigma^s = \{\boldsymbol{\tau}: \Omega \to \mathbb{R}^9_{\text{sym}}, \operatorname{tr}(\boldsymbol{\tau}) \in L^2(\Omega), \boldsymbol{\tau}_D = (\boldsymbol{\tau} - \frac{1}{3}\operatorname{tr}(\boldsymbol{\tau})\mathbf{Id}) \in (L^s(\Omega))^9\} \\
\hfill \text{(plastically incompressible case)}, \\
\Sigma^s = \{\boldsymbol{\tau}: \Omega \to \mathbb{R}^9_{\text{sym}}, \boldsymbol{\tau} \in (L^s(\Omega))^9\} \qquad \text{(plastically compressible case)},
\end{cases}
$$

$$
(2.4)\qquad V^s = \{\mathbf{w}, \mathbf{D}(\mathbf{w}) \in \Sigma^s, \mathbf{w} = \mathbf{0} \text{ on } \Gamma_1\},
$$

$$
(2.5)\qquad m = \max(2, q), \qquad m^* = \frac{m}{m-1},
$$

$$
(2.6)\qquad j_c(\mathbf{x}, \boldsymbol{\tau}) = \inf\{\mu > 0, \boldsymbol{\tau} \in \mu \mathbf{C}(\mathbf{x})\}.
$$

In the foregoing equations, an overdot denotes, as usual, partial differentiation with respect to time. $\mathbf{D}(\dot{\mathbf{u}}) = \dot{\mathbf{E}}(\mathbf{u})$ is the time derivative of the linearized strain tensor, $\mathbf{A}$ is the fourth-order linear elasticity tensor, $\mathbf{C}(\mathbf{x})$ is the closed convex set of locally admissible elastic stresses and is included in the space $\mathbb{R}^9_{\text{sym}}$ of symmetric second-order tensors operating on $\mathbb{R}^3$, and λ and q are material constants. Both $\mathbf{f}$, $\mathbf{g}$, $\mathbf{A}$, and λ may depend on time, for example, through a change of temperature or because the material is aging, and on the material coordinates $\mathbf{x}$ of the body.

Recall, in addition, that plastically incompressible bodies are characterized by functions $j_c(\mathbf{x}, \cdot)$ which depend only on the deviatoric part $\boldsymbol{\tau}_D = (\boldsymbol{\tau} - \frac{1}{3}\operatorname{tr}(\boldsymbol{\tau})\mathbf{Id})$ of the stress tensors. A typical example is given by Maxwell–Norton materials, for which we have

$$j_c(\mathbf{x}, \boldsymbol{\tau}) = (k(\mathbf{x})\sqrt{2})^{-1}|\boldsymbol{\tau}_D|, \tag{2.7}$$

and for which the variational inequalities in (2.1) correspond, respectively, to the virtual work theorem and to the constitutive law

$$\dot{\mathbf{E}}(\mathbf{u}) = \mathbf{A}^{-1}\dot{\boldsymbol{\sigma}} + \lambda(k(\mathbf{x})\sqrt{2})^{-q}|\boldsymbol{\tau}_D|^{q-2}\boldsymbol{\tau}_D.$$

Remark 2.1. In the remainder of this chapter, in order to be more general, we will allow j_c to be any convex continuous function defined over $\mathbb{R}^9_{\text{sym}}$ instead of considering only functions j_c given by (2.6).

2.2. Plane strains formulation. In plane strains, it is supposed, in addition, that the considered body undergoes no motion along the x_3 direction. This assumption is very realistic for bodies that are thick and invariant along x_3, that are subjected to adequate boundary conditions, and that are loaded in the plane (x_1, x_2) uniformly in x_3.

Cancelling the x_3 component of the displacement $\mathbf{u}$ and of the test functions $\mathbf{w}$ in (2.1), we obtain a variational formulation of this plane strains problem. While it is still given by (2.1), Ω now represents the section of the body in the plane (x_1, x_2) of its reference configuration; the functions $\mathbf{w}$ of V^s are defined on Ω and have values in $\mathbb{R}^2$; $\mathbf{u}(x_1, x_2)$ is the in-plane displacement of any particle $\mathbf{x} = \{x_1, x_2, x_3\}$ of the body; and the components of $\mathbf{D}(\mathbf{w})$ are given by

$$\begin{cases} (\mathbf{D}(\mathbf{w}))_{ij} = \dfrac{1}{2}\left(\dfrac{\partial w_j}{\partial x_i} + \dfrac{\partial w_i}{\partial x_j}\right) & \text{for } 1 \leq i, \quad j \leq 2, \\ (\mathbf{D}(\mathbf{w}))_{i3} = (\mathbf{D}(\mathbf{w}))_{3i} = 0 & \text{for } 1 \leq i \leq 3. \end{cases} \tag{2.8}$$

2.3. Plane stresses formulation. In plane stresses, the body is supposed to be very thin along x_3 and loaded in its plane so that, in a first approximation, all stresses along x_3 are 0. Cancelling the x_3 components of the stress field $\boldsymbol{\sigma}$ and of the test functions $\boldsymbol{\tau}$ in (2.1), we obtain a variational formulation of this plane stresses problem. After elimination of the x_3 components of $\mathbf{u}$ and $\mathbf{w}$, which are no longer relevant, this formulation is identical to (2.1). However, Ω now represents the middle-plane section of the body in its reference configuration; the functions $\mathbf{w}$ and $\boldsymbol{\tau}$ of V^s and Σ^s are defined on Ω and have values, respectively, in $\mathbb{R}^2$ and $(\mathbb{R}^2 \times \{0\}) \otimes (\mathbb{R}^2 \times \{0\})$; $\mathbf{u}(x_1, x_2)$ and $\boldsymbol{\sigma}(x_1, x_2)$ represent, respectively, the in-plane displacement and the stress tensor at the particle $\mathbf{x} = \{x_1, x_2, 0\}$; and the components of $\mathbf{D}(\mathbf{w})$ are given by (2.8).

2.4. Final formulation. In summary, the general variational formulation of quasi-static elastoviscoplastic problems valid both for three-dimensional, plane strains, and plane stresses situations is as follows.

$$(2.9)\quad \begin{cases} \text{Find } \{\mathbf{u}, \boldsymbol{\sigma}\} : [0, t_1] \to V^{m^*} \times \Sigma^m \text{ such that} \\ \displaystyle\int_\Omega \boldsymbol{\sigma} \cdot \mathbf{D}(\mathbf{w})\, d\mathbf{x} = \int_\Omega \mathbf{f} \cdot \mathbf{w}\, d\mathbf{x} + \int_{\Gamma_2} \mathbf{g} \cdot \mathbf{w}\, da \\ \qquad \forall \mathbf{w} \in V^{m^*} \quad \text{a.e. in } (0, t_1), \\ \displaystyle\int_\Omega (\mathbf{D}(\dot{\mathbf{u}}) - \mathbf{A}^{-1}\dot{\boldsymbol{\sigma}}) \cdot (\boldsymbol{\tau} - \boldsymbol{\sigma})\, d\mathbf{x} \le \int_\Omega \frac{\lambda}{q} \{|j_c(\mathbf{x}, \boldsymbol{\tau})|^q - |j_c(\mathbf{x}, \boldsymbol{\sigma})|^q\}\, d\mathbf{x} \\ \qquad \forall \boldsymbol{\tau} \in \Sigma^m \quad \text{a.e. in } (0, t_1), \\ \mathbf{u}(\mathbf{x}, 0) = \mathbf{u}_0(\mathbf{x}), \qquad \boldsymbol{\sigma}(\mathbf{x}, 0) = \boldsymbol{\sigma}_0(\mathbf{x}), \end{cases}$$

with

$$(2.10)\quad \begin{cases} N = 3 & \text{(three-dimensional case)}, \\ \quad = 2 & \text{(plane strains, plane stresses)}, \end{cases}$$

$$(2.11)\quad \begin{cases} \Omega \subset \mathbb{R}^N = \text{interior of body } (N = 3) \text{ or middle-plane} \\ \qquad \text{section } (N = 2) \text{ in its reference configuration,} \end{cases}$$

$$(2.12)\quad \begin{cases} (\mathbf{D}(\mathbf{w}))_{ij} = \dfrac{1}{2}\left(\dfrac{\partial w_j}{\partial x_i} + \dfrac{\partial w_i}{\partial x_j}\right) \quad \text{for } 1 \le i, \quad j \le N, \\ \qquad\qquad = 0 \quad \text{for } i > N, \quad j > N, \end{cases}$$

$$(2.13)\quad \begin{cases} S^9 = \mathbb{R}^9_{\text{sym}} & \text{(three-dimensional or plane strains case)}, \\ S^9 = \{\boldsymbol{\tau} \in \mathbb{R}^9_{\text{sym}}, \tau_{i3} = \tau_{3i} = 0\ \forall i = 1, 3\} & \text{(plane stresses case)}, \end{cases}$$

$$(2.14)\quad \begin{cases} \Sigma^s = \{\boldsymbol{\tau} : \Omega \to S^9, \operatorname{tr}(\boldsymbol{\tau}) \in L^2(\Omega), \boldsymbol{\tau}_D \in (L^s(\Omega))^9\} \\ \qquad\qquad\qquad\qquad \text{(plastically incompressible case)}, \\ \Sigma^s = \{\boldsymbol{\tau} : \Omega \to S^9, \boldsymbol{\tau} \in (L^s(\Omega))^9\} \qquad \text{(plastically compressible case)}, \end{cases}$$

$$(2.15)\quad \begin{cases} V^s = \{\mathbf{w} : \Omega \to \mathbb{R}^N, \mathbf{D}(\mathbf{w}) \in \Sigma^s, \mathbf{w} = \mathbf{0} \text{ on } \Gamma_1\}, \\ m = \max(2, q), \qquad m^* = \dfrac{m}{m-1}. \end{cases}$$

Remark 2.2. With minor modifications, the existence Theorem 4.1 of Chapter 2 applies to problem (2.9), which is therefore well posed.

3. Finite-element formulations of elastoviscoplasticity.

3.1. The discrete spaces. The numerical solution of the variational system (2.9) begins with its approximation by a system of finite-dimensional nonlinear equations. This is classically done in the finite-element method by replacing the spaces V^{m^*} and Σ^m of unknown velocities and stresses with finite-element spaces V_h and Σ_h. Here, we will simply use Lagrange simplicial elements of order 1 (Ciarlet [1978]) and construct V_h and Σ_h as follows.

Let Ω be a polygonal (resp., polyhedral) domain of $\mathbb{R}^2$ (resp., $\mathbb{R}^3$). We first decompose Ω into a finite number N_h of triangles (resp., tetrahedrons) Ω_ℓ such that

$$\Omega = \bigcup_{\ell=1,N_h} \Omega_\ell,$$

the diameter of any Ω_ℓ is bounded by h, any Ω_ℓ contains a ball of radius αh with α given once and for all, and two different elements Ω_ℓ have either nothing, a vertex, an edge, or a face in common. Such a decomposition is called a regular triangulation $\mathcal{T}_h$ of Ω (see Figs. 3.1 and 3.2). With $\mathcal{T}_h$ given, Σ_h and V_h are defined, respectively, by

$$\Sigma_h = \{\boldsymbol{\tau}_h : \Omega \to S^9,\ \boldsymbol{\tau}_{h|\Omega_\ell} = \text{const}\ \forall \ell = 1, \cdots, N_h\}, \tag{3.1}$$

$$V_h = \{\mathbf{w}_h \in \mathbf{C}^0(\bar{\Omega}),\ \mathbf{w}_h = \mathbf{0} \text{ on } \Gamma_1,\ \mathbf{w}_{h|\Omega_\ell} \in \mathbf{P}_1(\Omega_\ell)\ \forall \ell = 1, N_h\}, \tag{3.2}$$

$\mathbf{P}_1(\Omega_\ell)$ denoting the space of first-degree polynomials defined over Ω_ℓ that have values in $\mathbb{R}^N$. In other words, Σ_h is a space of piecewise constant functions, and V_h is a space of continuous piecewise linear functions.

When the maximal diameter h of the triangulation $\mathcal{T}_h$ goes to zero, the spaces (Σ_h) and (V_h) defined by (3.1)–(3.2) form converging sequences of finite-dimensional approximations of Σ^m and V^{m^*}, and we have (Ciarlet [1978])

$$\begin{cases} \dim \Sigma_h < +\infty, \qquad \forall 1 < s < +\infty \quad \Sigma_h \subset \Sigma^s, \\ \forall \boldsymbol{\tau} \in \Sigma^s \quad \lim\limits_{h \to 0} \left\{ \inf\limits_{\boldsymbol{\tau}_h \in \Sigma_h} \|\boldsymbol{\tau} - \boldsymbol{\tau}_h\|_{\Sigma^s} \right\} = 0; \end{cases} \tag{3.3}$$

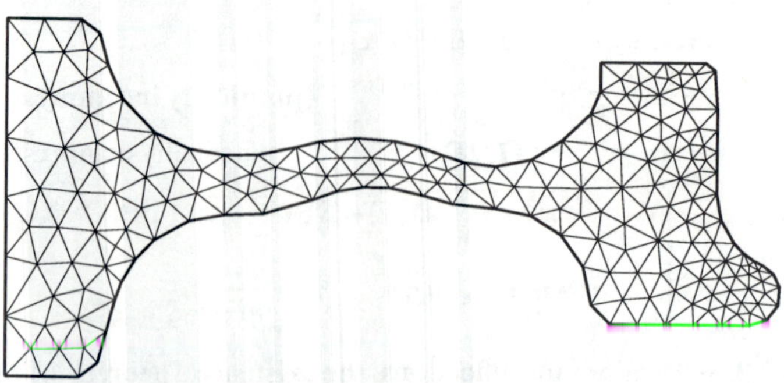

FIG. 3.1. *Decomposition of a two-dimensional domain into triangles.*

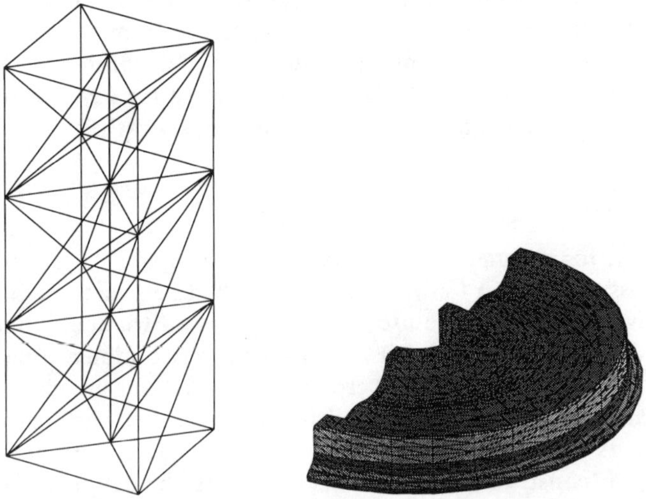

FIG. 3.2. *Decomposition of three-dimensional domains into tetrahedrons.*

$$(3.4)\qquad \begin{cases} \dim V_h < +\infty, \qquad \forall 1 < s < +\infty \quad V_h \subset V^s, \\ \forall \mathbf{w} \in V^s \quad \lim\limits_{h \to 0} \left\{ \inf\limits_{\mathbf{w}_h \in V_h} \|\mathbf{w} - \mathbf{w}_h\|_{V^s} \right\} = 0. \end{cases}$$

3.2. The discrete variational system. The discrete variational system is simply obtained by replacing Σ^m and V^{m^*} with Σ_h and V_h in the variational formulation (2.9), which gives the following.

$$(3.5)\qquad \begin{cases} \text{Find } \{\mathbf{u}_h, \boldsymbol{\sigma}_h\} : [0, t_1] \to V_h \times \Sigma_h \text{ such that} \\ \displaystyle\int_\Omega \boldsymbol{\sigma}_h \cdot \mathbf{D}(\mathbf{w}_h)\, d\mathbf{x} = \int_\Omega \mathbf{f} \cdot \mathbf{w}_h\, d\mathbf{x} + \int_{\Gamma_2} \mathbf{g} \cdot \mathbf{w}_h\, da \\ \qquad\qquad\qquad\qquad \forall \mathbf{w}_h \in V_h \quad \text{a.e. in } (0, t_1), \\ \displaystyle\int_\Omega (\mathbf{D}(\dot{\mathbf{u}}_h) - \mathbf{A}^{-1}\dot{\boldsymbol{\sigma}}_h) \cdot (\boldsymbol{\tau}_h - \boldsymbol{\sigma}_h)\, d\mathbf{x} \le \int_\Omega \frac{\lambda}{q} \{|j_c(\mathbf{x}, \boldsymbol{\tau}_h)|^q \\ \qquad\qquad - |j_c(\mathbf{x}, \boldsymbol{\sigma}_h)|^q\}\, d\mathbf{x} \qquad \forall \boldsymbol{\tau}_h \in \Sigma_h \quad \text{a.e. in } (0, t_1), \\ \mathbf{u}_h(\mathbf{x}, 0) = \mathbf{u}_{0h}(\mathbf{x}), \qquad \boldsymbol{\sigma}_h(\mathbf{x}, 0) = \boldsymbol{\sigma}_{0h}(\mathbf{x}). \end{cases}$$

In (3.5), we will suppose that the restrictions of λ, $j_c(\mathbf{x}, \boldsymbol{\tau}_h)$, and $\mathbf{A}$ on Ω_ℓ are constant over Ω_ℓ.

3.3. Modifications for the plastically incompressible case. In the case of plastically incompressible materials in three dimensions or in plane strains (j_c

depending only on the deviatoric part of $\boldsymbol{\tau}$), and at the stationary limit where $\dot{\mathbf{u}}$ and $\boldsymbol{\sigma}$ no longer depend on time, it is easy to see from (2.9) and (3.5) that the divergence of both the continuous solution $\dot{\mathbf{u}}$ and the finite-element solution $\dot{\mathbf{u}}_h$ must be equal to zero. In this situation, if we want $\dot{\mathbf{u}}_h$ to approximate $\dot{\mathbf{u}}$ correctly, the finite-element space V_h must obviously satisfy

$$\forall \mathbf{w} \text{ in } V^{m^*} \text{ with } \operatorname{div} \mathbf{w} = 0, \lim_{h\to 0} \left\{ \inf_{\substack{\mathbf{w}_h \in V_h \\ \operatorname{div} \mathbf{w}_h = 0}} \|\mathbf{w} - \mathbf{w}_h\|_{V^{m^*}} \right\} = 0. \tag{3.6}$$

Unfortunately, for a general triangulation $\mathcal{T}_h$, the space V_h defined by (3.2) does not satisfy condition (3.6). This difficulty can be overcome in two ways.

The first way is through the use of special crossed triangulations $\mathcal{T}_h$ of Ω. This is the simplest method and it is valid for the two-dimensional case only (i.e., $\Omega \subset \mathbb{R}^2$). More precisely, these crossed triangulations are obtained by first dividing Ω into a finite number of quadrilaterals and then cutting each quadrilateral along its diagonals to obtain four triangles per quadrilateral (see Fig. 3.3). Once $\mathcal{T}_h$ is constructed in this special manner, the definitions (3.1), (3.2), and (3.5) of the finite-element spaces and of the discrete variational system are kept unchanged. This is the method that will be adopted in this chapter.

The second method (Le Tallec and Ravachol [1988]), more general and usually more accurate, uses more elaborate finite elements to construct V_h and Σ_h and changes the definition of j_c into

$$j_c(\boldsymbol{\tau}) = \sup_{\mathbf{D}\in\Sigma_{dh}} \left\{ \int_\Omega (\mathbf{D} \cdot \boldsymbol{\tau} - j_q^*(\mathbf{D}))\, d\mathbf{x} \right\},$$

where j_q^* denotes the dual (conjugate function) of $\lambda |j_c(\cdot)|^q/q$ on S^9, and where we have that

$$\Sigma_{dh} = \left\{ \mathbf{D} \in \Sigma_h, \int_\Omega q_h \operatorname{tr}(\mathbf{D}_h)\, d\mathbf{x} = 0\ \forall q_h \in P_h \right\},$$

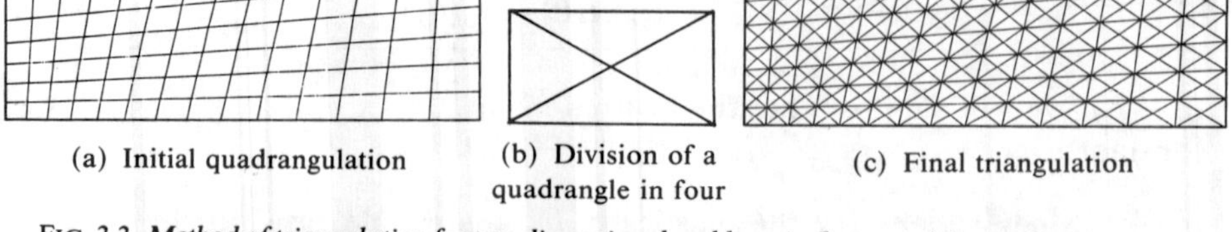

(a) Initial quadrangulation (b) Division of a quadrangle in four (c) Final triangulation

FIG. 3.3. *Method of triangulation for two-dimensional problems in the case of plastically incompressible materials.* (a) *Divide* Ω *into a finite number of quadrilaterals.* (b) *Divide each quadrilateral along diagonals to obtain* (c).

with P_h denoting an auxiliary pressure space. This approach is rather technical and, for simplicity, will not be described here. In simple terms, it amounts to the imposition of the kinematic constraint tr $(\mathbf{E}^{an})=0$ weakly in P_h instead of everywhere.

3.4. Subdifferential calculus. To treat (3.5) numerically, it will be useful to formulate it either as a stress evolution problem to be solved by alternating-direction methods, or as a stationary problem associated with (3.5) to be solved by augmented Lagrangian techniques. For that purpose, we must first compute different subgradients. For $\mathbf{D}$ and $\boldsymbol{\tau}$ in Σ_h, let us introduce

$$F_t(\mathbf{D})=\int_\Omega \sup_{\boldsymbol{\tau}\in S^9}\left\{\boldsymbol{\tau}\cdot\mathbf{D}-\frac{\lambda}{q}(\mathbf{x},t)|j_c(\mathbf{x},\boldsymbol{\tau})|^q\right\}d\mathbf{x}, \tag{3.7}$$

$$\psi_t(\boldsymbol{\tau})=\int_\Omega \frac{\lambda}{q}(\mathbf{x},t)|j_c(\mathbf{x},\boldsymbol{\tau})|^q\,d\mathbf{x}, \tag{3.8}$$

$$\begin{cases} S_h(t)=\left\{\boldsymbol{\tau}_h\in\Sigma_h,\ \int_\Omega \boldsymbol{\tau}_h\cdot\mathbf{D}(\mathbf{w}_h)\,d\mathbf{x}\right.\\ \qquad\left.=\int_\Omega \mathbf{f}(\mathbf{x},t)\cdot\mathbf{w}_h\,d\mathbf{x}+\int_{\Gamma_2}\mathbf{g}(\mathbf{x},t)\cdot\mathbf{w}_h\,da\ \forall\mathbf{w}_h\in V_h\right\},\end{cases} \tag{3.9}$$

$$\begin{cases} I_t^h(\boldsymbol{\tau})=0 & \text{if } \boldsymbol{\tau}\in S_h(t),\\ \qquad\ =+\infty & \text{if not.}\end{cases} \tag{3.10}$$

Then, let us endow Σ_h with the L^2 scalar product

$$\langle\boldsymbol{\tau},\mathbf{D}\rangle=\int_\Omega \boldsymbol{\tau}\cdot\mathbf{D}\,d\mathbf{x}\quad \forall\boldsymbol{\tau},\mathbf{D}\in\Sigma_h. \tag{3.11}$$

We have the following lemma.

LEMMA 3.1. *The functions F_t and ψ_t are dual (conjugate) from each other, and, therefore, $\boldsymbol{\tau}\in\partial F_t(\mathbf{D})$ if and only if $\mathbf{D}\in\partial\psi_t(\boldsymbol{\tau})$.*

Proof. We first recall that the dual (conjugate function) F^* and the subdifferential ∂F of a real, convex, lower semicontinuous function F defined on Σ_h are given, respectively, by

$$F^*(\boldsymbol{\tau})=\sup_{\mathbf{H}\in\Sigma_h}\{\langle\boldsymbol{\tau},\mathbf{H}\rangle-F(\mathbf{H})\},$$

$$\partial F(\mathbf{D})=\{\boldsymbol{\tau}\in\Sigma_h,\langle\boldsymbol{\tau},\mathbf{H}\rangle\le F(\mathbf{D}+\mathbf{H})-F(\mathbf{D})\ \forall\mathbf{H}\in\Sigma_h\}.$$

From these definitions, we can easily verify (Ekeland and Temam [1976]) that $\boldsymbol{\tau}\in\partial F(\mathbf{D})$ if and only if we have

$$F(\mathbf{D})+F^*(\boldsymbol{\tau})=\langle\boldsymbol{\tau},\mathbf{D}\rangle,$$

that is, substituting F^* for F, if and only if $\mathbf{D} \in \partial F^*(\boldsymbol{\tau})$. Therefore, the whole lemma will be proved if we can verify the identity $F_t(\mathbf{D}) = \psi_t^*(\mathbf{D})$.

Since Σ_h is made of piecewise constant functions whose values on each element Ω_ℓ are independent from one another, a direct calculation yields

$$\begin{aligned}
\psi_t^*(\mathbf{D}) &= \sup_{\boldsymbol{\tau}\in\Sigma_h} \left\{ \int_\Omega \mathbf{D}\cdot\boldsymbol{\tau}\, d\mathbf{x} - \int_\Omega \frac{\lambda}{q} |j_c(\mathbf{x},\boldsymbol{\tau})|^q\, d\mathbf{x} \right\} \\
&= \sup_{\boldsymbol{\tau}\in\Sigma_h} \left\{ \sum_{\ell=1}^{N_h} \int_{\Omega_\ell} \left\{ \mathbf{D}\cdot\boldsymbol{\tau} - \frac{\lambda}{q} |j_c(\mathbf{x},\boldsymbol{\tau})|^q \right\} d\mathbf{x} \right\} \\
&= \sup_{\boldsymbol{\tau}\in\Sigma_h} \sum_{\ell=1}^{N_h} \operatorname{meas}(\Omega_\ell) \left\{ \mathbf{D}\cdot\boldsymbol{\tau} - \frac{\lambda}{q} |j_c(\boldsymbol{\tau})|^q \right\} \\
&= \sum_{\ell=1}^{N_h} \operatorname{meas}(\Omega_\ell) \sup_{\boldsymbol{\tau}\in S^9} \left\{ \mathbf{D}\cdot\boldsymbol{\tau} - \frac{\lambda}{q} |j_c(\boldsymbol{\tau})|^q \right\} \\
&= \int_\Omega \sup_{\boldsymbol{\tau}\in S^9} \left\{ \mathbf{D}\cdot\boldsymbol{\tau} - \frac{\lambda}{q} |j_c(\mathbf{x},\boldsymbol{\tau})|^q \right\} d\mathbf{x} = F_t(\mathbf{D}). \qquad \square
\end{aligned}$$

Remark 3.1. If we are not considering plane stresses and if j_c is given by (2.6), then

$$F_t(\mathbf{D}) = \int_\Omega \sup_{\boldsymbol{\tau}\in\mathbb{R}^9_{\text{sym}}} \left\{ \mathbf{D}\cdot\boldsymbol{\tau} - \frac{\lambda}{q} |\inf_{\mu>0} (\mu, \boldsymbol{\tau} \in \mu \mathbf{C}(\mathbf{x}))|^q \right\} d\mathbf{x},$$

from which, through the change of variable $\boldsymbol{\tau} = \boldsymbol{\tau}/\mu_{\min}$, we deduce

$$F_t(\mathbf{D}) = \int_\Omega \sup_{\boldsymbol{\tau}\in\mathbf{C}(\mathbf{x})} \sup_{\mu>0} \left\{ \mu\mathbf{D}\cdot\boldsymbol{\tau} - \frac{\lambda}{q}\mu^q \right\} d\mathbf{x}.$$

Since the function $g(\mu) = -\lambda\mu^q/q + \mu\mathbf{D}\cdot\boldsymbol{\tau}$ has, for maximal value, either 0 if $\mathbf{D}\cdot\boldsymbol{\tau}$ is negative or $\lambda^{1-s}/s(\mathbf{D}\cdot\boldsymbol{\tau})^s$ with $s = q/(q-1)$ if it is not, we finally obtain, for j_c given by (2.6),

$$F_t(\mathbf{D}) = \int_\Omega \frac{\lambda^{1-s}}{s} |\sup_{\boldsymbol{\tau}\in\mathbf{C}(\mathbf{x})} (\boldsymbol{\tau}\cdot\mathbf{D})|^s\, d\mathbf{x}.$$

Remark 3.2. For the plastically incompressible case in three dimensions or in plane strains ($S^9 = R^9_{\text{sym}}, j_c(\mathbf{x},\boldsymbol{\tau}) = j_c(\mathbf{x},\boldsymbol{\tau}_D)$), we can easily deduce the following from the definition of F_t in (3.7).

(i) If $\operatorname{tr}(\mathbf{D}) = 0$ a.e. in Ω, then $F_t(\mathbf{D})$ is finite, and $\partial F_t(\mathbf{D})$ is not empty and is invariant by translation along the tensor **Id**;

(ii) if $\operatorname{tr}(\mathbf{D}) \neq 0$ a.e. in Ω, then $F_t(\mathbf{D})$ takes an infinite value and $\partial F_t(\mathbf{D})$ is empty.

LEMMA 3.2. *Under the notation of* (3.9)-(3.11), *we also have that*

$$\text{(3.12)}\qquad \boxed{\begin{aligned} \partial I_t^h(\boldsymbol{\tau}_0) &= \mathbf{D}(V_h) = \{\mathbf{D}(\mathbf{w}_h), \mathbf{w}_h \in V_h\} \quad \textit{if } \boldsymbol{\tau}_0 \in S_h(t), \\ &= \textit{empty set} \quad \textit{if not.} \end{aligned}}$$

Proof. Let us denote by $\mathbf{D}(V_h)^\perp$ the orthogonal of $\mathbf{D}(V_h)$ in Σ_h, that is

$$\mathbf{D}(V_h)^\perp = \{\boldsymbol{\tau} \in \Sigma_h, \langle \boldsymbol{\tau}, \mathbf{D} \rangle = 0 \; \forall \mathbf{D} \in \mathbf{D}(V_h)\},$$

and let $\boldsymbol{\tau}_0$ be given in $S_h(t)$. By definition of $S_h(t)$, we have that

$$\text{(3.13)}\qquad \begin{cases} S_h(t) = \{\boldsymbol{\tau} \in \Sigma_h, \langle \boldsymbol{\tau}, \mathbf{D}(\mathbf{w}_h) \rangle = \langle \boldsymbol{\tau}_0, \mathbf{D}(\mathbf{w}_h) \rangle \; \forall \mathbf{w}_h \in V_h\} \\ \qquad = \boldsymbol{\tau}_0 + \mathbf{D}(V_h)^\perp. \end{cases}$$

Then, using the definitions of I_t^h and its subgradient, it follows that

$$\begin{aligned} \partial I_t^h(\boldsymbol{\tau}_0) &= \{\mathbf{H} \in \Sigma_h, \langle \mathbf{H}, (\boldsymbol{\tau} - \boldsymbol{\tau}_0) \rangle \le I_t^h(\boldsymbol{\tau}) - I_t^h(\boldsymbol{\tau}_0) \; \forall \boldsymbol{\tau} \in \Sigma_h\} \\ &= \{\mathbf{H} \in \Sigma_h, \langle \mathbf{H}, \boldsymbol{\tau} - \boldsymbol{\tau}_0 \rangle \le 0 \; \forall \boldsymbol{\tau} \in S_h(t)\} \\ &= \{\mathbf{H} \in \Sigma_h, \langle \mathbf{H}, \boldsymbol{\tau} - \boldsymbol{\tau}_0 \rangle \le 0 \; \forall \boldsymbol{\tau} \in \{\boldsymbol{\tau}_0 + \mathbf{D}(V_h)^\perp\}\} \\ &= \{\mathbf{H} \in \Sigma_h, \langle \mathbf{H}, \boldsymbol{\tau} \rangle \le 0 \; \forall \boldsymbol{\tau} \in \mathbf{D}(V_h)^\perp\} \\ &= (\mathbf{D}(V_h)^\perp)^\perp = \mathbf{D}(V_h). \end{aligned}$$

On the other hand, for $\boldsymbol{\tau}_0$ given in $\Sigma_h - S_h(t)$, $I_t^h(\boldsymbol{\tau}_0)$ takes the value $+\infty$, and no element $\mathbf{H}$ of Σ_h will ever satisfy

$$\langle \mathbf{H}, \boldsymbol{\tau} - \boldsymbol{\tau}_0 \rangle \le -\infty \qquad \forall \boldsymbol{\tau} \in S_h(t),$$

which then means that $\partial I_t^h(\boldsymbol{\tau}_0)$ is empty. □

3.5. The discrete stress evolution problem. The displacement field $\mathbf{u}_h$ can be eliminated from the discrete variational system (3.5) of elastoviscoplasticity in the same way as for the continuous problem (Chap. 2, Thm. 4.1).

THEOREM 3.3. *Under the notation of* (3.7)-(3.11), *the discrete variational system* (3.5) *is equivalent to the following stress evolution problem.*

$$\text{(3.14)}\qquad \boxed{\begin{aligned} &\textit{Find } \boldsymbol{\sigma}_h : [0, t_1] \to \Sigma_h \textit{ with } \boldsymbol{\sigma}_h(\mathbf{x}, 0) = \boldsymbol{\sigma}_{0h}(\mathbf{x}) \textit{ and} \\ &-\mathbf{A}^{-1}\dot{\boldsymbol{\sigma}}_h \in \{\partial\psi_t(\boldsymbol{\sigma}_h) + \partial I_t^h(\boldsymbol{\sigma}_h)\} \quad \textit{in } \Sigma_h \quad \textit{a.e. in } (0, t_1). \end{aligned}}$$

Moreover, under the assumptions of Theorem 4.1 *of Chapter* 2, *problem* (3.14) *has a unique solution* $\boldsymbol{\sigma}_h$ *in* $W^{1,2}(0, T; \Sigma_h)$.

Proof. Step 1. Let $\{\dot{\mathbf{u}}_h, \boldsymbol{\sigma}_h\}$ be a solution of (3.5). Under the notation of (3.7)-(3.11), (3.5) can be expressed as

$$\text{(3.15)}\qquad \boldsymbol{\sigma}_h \in S_h(t), \qquad (\mathbf{D}(\dot{\mathbf{u}}_h) - \mathbf{A}^{-1}\dot{\boldsymbol{\sigma}}_h) \in \partial\psi_t(\boldsymbol{\sigma}_h);$$

that is, employing Lemma 3.2,

$$\partial I_t^h(\boldsymbol{\sigma}_h) = \mathbf{D}(V_h), \qquad (-\mathbf{A}^{-1}\dot{\boldsymbol{\sigma}}_h + \mathbf{D}(\dot{\mathbf{u}}_h)) \in \partial\psi_t(\boldsymbol{\sigma}_h). \tag{3.16}$$

It can be seen from the first relation of (3.16) that $-\mathbf{D}(\dot{\mathbf{u}}_h)$ belongs to $\partial I_t^h(\boldsymbol{\sigma}_h)$, and, thus, (3.16) implies

$$-\mathbf{A}^{-1}\dot{\boldsymbol{\sigma}}_h \in \{\partial\psi_t(\boldsymbol{\sigma}_h) + \partial I_t^h(\boldsymbol{\sigma}_h)\}, \tag{3.17}$$

which means that $\boldsymbol{\sigma}_h$ is a solution of (3.14).

Step 2. Conversely, let $\boldsymbol{\sigma}_h$ be a solution of (3.14). Then, by definition of the sum of two sets, there exists $\mathbf{D}$ in $\partial I_t^h(\boldsymbol{\sigma}_h)$ such that

$$(-\mathbf{A}^{-1}\dot{\boldsymbol{\sigma}}_h - \mathbf{D}) \in \partial\psi_t(\boldsymbol{\sigma}_h), \qquad \mathbf{D} \in \partial I_t^h(\boldsymbol{\sigma}_h). \tag{3.18}$$

In particular, $\partial I_t^h(\boldsymbol{\sigma}_h)$ is not empty. On the basis of Lemma 3.2, this implies that $\boldsymbol{\sigma}_h$ belongs to $S_h(t)$ and that $\partial I_t^h(\boldsymbol{\sigma}_h)$ is equal to the space $\mathbf{D}(V_h)$. Therefore, there exists $\dot{\mathbf{u}}_h$ in V_h such that

$$\mathbf{D} = -\mathbf{D}(\dot{\mathbf{u}}_h). \tag{3.19}$$

If we substitute (3.19) into (3.18) (and considering that $\boldsymbol{\sigma}_h$ belongs to $S_h(t)$), we obtain (3.15), and, thus, $\{\dot{\mathbf{u}}_h, \boldsymbol{\sigma}_h\}$ is a solution of the variational system (3.5).

Step 3. To sketch the proof of the existence and uniqueness of the solution $\boldsymbol{\sigma}_h$ of the stress evolution problem (3.14), we now introduce the solution $\mathbf{v}_h$ of the well-posed linear elliptic problem,

$$\begin{cases} \text{find } \mathbf{v}_h \in W^{1,2}(0, t_1; V_h) \text{ such that} \\ \displaystyle\int_\Omega \mathbf{A}\mathbf{D}(\mathbf{v}_h) \cdot \mathbf{D}(\mathbf{w}_h)\, d\mathbf{x} = \int_\Omega \mathbf{f} \cdot \mathbf{w}_h\, d\mathbf{x} + \int_{\Gamma_2} \mathbf{g} \cdot \mathbf{w}_h\, d\mathbf{a} \qquad \forall \mathbf{w}_h \in V_h \quad \text{a.e. in } (0, t_1), \end{cases}$$

together with the auxiliary unknown

$$\boldsymbol{\sigma}_h^1 = \boldsymbol{\sigma}_h - \mathbf{A}\mathbf{D}(\mathbf{v}_h). \tag{3.20}$$

Considering the family of real, convex, lower semicontinuous functions ϕ_t defined on Σ_h by

$$\phi_t(\boldsymbol{\tau}) = \psi_t(\boldsymbol{\tau} + \mathbf{A}\mathbf{D}(\mathbf{v}_h)) + I_t^h(\boldsymbol{\tau} + \mathbf{A}\mathbf{D}(\mathbf{v}_h)),$$

we can express (3.14) under an equivalent form as

$$\begin{cases} \text{find } \boldsymbol{\sigma}_h^1 \in W^{1,2}(0, t_1; \Sigma_h) \text{ such that} \\ \dot{\boldsymbol{\sigma}}_h^1 + \partial_{\mathbf{A}}\phi_t(\boldsymbol{\sigma}_h^1) \ni \mathbf{A}\dot{\mathbf{D}}(\mathbf{v}_h), \\ \boldsymbol{\sigma}_h^1(\mathbf{x}, 0) = \boldsymbol{\sigma}_{0h}(\mathbf{x}) - \mathbf{A}\mathbf{D}(\mathbf{v}_h)(\mathbf{x}, 0), \end{cases} \tag{3.21}$$

provided that we compute $\partial_{\mathbf{A}}\phi_t$ in Σ_h by endowing Σ_h with the time-dependent scalar product

$$\langle \boldsymbol{\tau}, \mathbf{H}\rangle_t = \int_\Omega \mathbf{A}^{-1}(\mathbf{x}, t)\boldsymbol{\tau} \cdot \mathbf{H}\, d\mathbf{x}.$$

In other words, in (3.21), $\partial_{\mathrm{A}}\phi_t$ is defined by

$$\partial_{\mathrm{A}}\phi_t(\boldsymbol{\tau}) = \{\mathbf{D} \in \Sigma_h, \langle \mathbf{D}, \mathbf{H}\rangle_t \leq \phi_t(\boldsymbol{\tau}+\mathbf{H}) - \phi_t(\boldsymbol{\tau})\ \forall \mathbf{H} \in \Sigma_h\}.$$

Problems like (3.21) have been studied by Damlamian [1974], who proved the existence and uniquenesss of the solution of (3.21) under quite general regularity conditions on ϕ_t and $\langle\cdot,\cdot\rangle_t$. In our case, it can be seen that the regularity conditions are easily satisfied from the assumptions of Theorem 4.1 of Chapter 2, as verified by Blanchard and Le Tallec [1986] in their existence proof. Therefore, (3.21) and, hence, (3.14) does have a unique solution. □

Remark 3.3. The stress evolution problem (3.14) of elastoviscoplasticity is formally of the form

$$S\dot{\boldsymbol{\sigma}}_h + A_1\boldsymbol{\sigma}_h + A_2\boldsymbol{\sigma}_h = 0,$$

with $S = \mathbf{A}^{-1}$, $A_1 = \partial\psi_t$, and $A_2 = \partial I_t^h$. It can be solved numerically using the alternating-direction methods of Chapter 3, § 5. This will be the purpose of § 5 of this chapter.

3.6. Augmented Lagrangian formulation of the discrete stationary elastoviscoplastic problem. Let us go back to the discrete variational formulation (3.5) of our elastoviscoplastic problem, assuming now that the external forces $\mathbf{f}$ and $\mathbf{g}$, the elasticity tensor $\mathbf{A}$, and the function λj_c^q are independent of time. We are now interested in the limits, as time goes to infinity, of the solutions $\{\dot{\mathbf{u}}_h, \boldsymbol{\sigma}_h\}$ of Problem (3.5). If such limits exist, they are time-independent (stationary) solutions of (3.5) and can be obtained as saddle-points of an augmented Lagrangian defined over $V_h \times \Sigma_h \times \Sigma_h$.

Indeed, let r be positive arbitrary, and let us define the augmented Lagrangian $\mathcal{L}_r: V_h \times \Sigma_h \times \Sigma_h \to \bar{\mathbb{R}}$ by

$$\mathcal{L}_r(\mathbf{w}_h, \mathbf{H}_h; \boldsymbol{\mu}_h) = F(\mathbf{H}_h) + G(\mathbf{w}_h) + \frac{r}{2}|B\mathbf{w}_h - \mathbf{H}_h|^2 + (\boldsymbol{\mu}_h, B\mathbf{w}_h - \mathbf{H}_h), \tag{3.22}$$

where F is as defined in (3.7),

$$G(\mathbf{w}_h) = -\int_\Omega \mathbf{f}\cdot\mathbf{w}_h\,d\mathbf{x} - \int_{\Gamma_2} \mathbf{g}\cdot\mathbf{w}_h\,da, \tag{3.23}$$

$$B\mathbf{w}_h = \mathbf{D}(\mathbf{w}_h), \tag{3.24}$$

$$(\boldsymbol{\mu}_h, \mathbf{H}_h) = \int_\Omega \mathbf{A}\boldsymbol{\mu}_h \cdot \mathbf{H}_h\,d\mathbf{x}, \tag{3.25}$$

$$|\mathbf{H}_h|^2 = (\mathbf{H}_h, \mathbf{H}_h)^{1/2}. \tag{3.26}$$

As in Chapter 3, § 4.3, we associate with $\mathcal{L}_r$ the following saddle-point problem.

$$(3.27)\quad \begin{cases} \text{Find } \{\dot{\mathbf{u}}_h, \mathbf{D}_h ; \boldsymbol{\lambda}_h\} \in V_h \times \Sigma_h \times \Sigma_h \text{ such that} \\ \mathscr{L}_r(\dot{\mathbf{u}}_h, \mathbf{D}_h ; \boldsymbol{\mu}_h) \le \mathscr{L}_r(\dot{\mathbf{u}}_h, \mathbf{D}_h ; \boldsymbol{\lambda}_h) \le \mathscr{L}_r(\mathbf{w}_h, \mathbf{H}_h ; \boldsymbol{\lambda}_h) \\ \forall \{\mathbf{w}_h, \mathbf{H}_h ; \boldsymbol{\mu}_h\} \in V_h \times \Sigma_h \times \Sigma_h. \end{cases}$$

We then have the following theorem.

THEOREM 3.4. *To each stationary solution* $\{\dot{\mathbf{u}}_h, \boldsymbol{\sigma}_h\}$ *of the variational system* (3.5) *of elastoviscoplasticity corresponds a saddle-point* $\{\dot{\mathbf{u}}_h, \mathbf{D}(\dot{\mathbf{u}}_h); A^{-1}\boldsymbol{\sigma}_h\}$ *of* (3.27) *and conversely.*

Proof. Let $\{\dot{\mathbf{u}}_h, \boldsymbol{\sigma}_h\}$ be a stationary (invariant in time) solution of (3.5). From the second relation of (3.5), we obtain

$$\mathbf{D}(\dot{\mathbf{u}}_h) \in \partial\psi(\boldsymbol{\sigma}_h) \quad \text{in } \Sigma_h.$$

Thus, from Lemma 3.1, we can see that $\boldsymbol{\sigma}_h \in \partial F(\mathbf{D}(\dot{\mathbf{u}}_h))$, that is

$$(3.28)\quad \int_\Omega \boldsymbol{\sigma}_h \cdot (\mathbf{H}_h - \mathbf{D}(\dot{\mathbf{u}}_h))\, d\mathbf{x} \le F(\mathbf{H}_h) - F(\mathbf{D}(\dot{\mathbf{u}}_h)) \quad \forall \mathbf{H}_h \in \Sigma_h.$$

Adding (3.28) to the first relation of (3.5) yields

$$(3.29)\quad \begin{cases} F(\mathbf{D}(\dot{\mathbf{u}}_h)) + G(\dot{\mathbf{u}}_h) \le F(\mathbf{H}_h) + G(\mathbf{w}_h) + \displaystyle\int_\Omega \boldsymbol{\sigma}_h \cdot (\mathbf{D}(\dot{\mathbf{u}}_h) - \mathbf{H}_h)\, d\mathbf{x} \\ \qquad + \displaystyle\int_\Omega \boldsymbol{\sigma}_h \cdot (\mathbf{D}(\mathbf{w}_h) - \mathbf{D}(\dot{\mathbf{u}}_h))\, d\mathbf{x} \quad \forall \{\mathbf{w}_h, \mathbf{H}_h\} \in V_h \times \Sigma_h. \end{cases}$$

If we add the positive term $\frac{r}{2}|B\mathbf{w}_h - \mathbf{H}_h|^2$ to the right-hand side of (3.29) and use the definition (3.22) of $\mathscr{L}_r$, (3.29) can be expressed as

$$\mathscr{L}_r(\dot{\mathbf{u}}_h, \mathbf{D}(\dot{\mathbf{u}}_h); \mathbf{A}^{-1}\boldsymbol{\sigma}_h) \le \mathscr{L}_r(\mathbf{w}_h, \mathbf{H}_h ; \mathbf{A}^{-1}\boldsymbol{\sigma}_h) \quad \forall \{\mathbf{w}_h, \mathbf{H}_h\} \in V_h \times \Sigma_h.$$

On the other hand, by definition of B, we have that

$$\mathscr{L}_r(\dot{\mathbf{u}}_h, \mathbf{D}(\dot{\mathbf{u}}_h), \boldsymbol{\mu}_h) = \mathscr{L}_r(\dot{\mathbf{u}}_h, \mathbf{D}(\dot{\mathbf{u}}_h), \mathbf{A}^{-1}\boldsymbol{\sigma}_h) \quad \forall \boldsymbol{\mu}_h \in \Sigma_h,$$

and therefore, finally, that $\{\dot{\mathbf{u}}_h, \mathbf{D}(\dot{\mathbf{u}}_h); \mathbf{A}^{-1}\boldsymbol{\sigma}_h\}$ is a saddle-point of (3.27).

Conversely, let $\{\dot{\mathbf{u}}_h, \mathbf{D}_h ; \boldsymbol{\lambda}_h\}$ be a saddle-point of (3.27). By applying Theorem 4.1 of Chapter 3, we see that for such a saddle-point, equations (4.9), (4.12), and (4.13) of Chapter 3 hold; in the present context, the equations are

$$(3.30)\quad \mathbf{D}_h = B\dot{\mathbf{u}}_h = \mathbf{D}(\dot{\mathbf{u}}_h),$$

$$(3.31)\quad G(\mathbf{w}_h) - G(\dot{\mathbf{u}}_h) + (\boldsymbol{\lambda}_h, \mathbf{D}(\mathbf{w}_h - \dot{\mathbf{u}}_h)) \ge 0 \quad \forall \mathbf{w}_h \in V_h,$$

$$(3.32)\quad F(\mathbf{H}_h) - F(\mathbf{D}_h) - (\boldsymbol{\lambda}_h, \mathbf{H}_h - \mathbf{D}_h) \ge 0 \quad \forall \mathbf{H}_h \in \Sigma_h.$$

From the linearity of G and the definition of the scalar product $(\cdot, \cdot)$, (3.31) yields

$$\int_\Omega \mathbf{A}\boldsymbol{\lambda}_h \cdot \mathbf{D}(\mathbf{w}_h) = -\mathbf{G}(\mathbf{w}_h) \quad \forall \mathbf{w}_h \in V_h. \tag{3.33}$$

On the other hand, from the definition of ∂F and Lemma 3.1, (3.32) can be expressed as

$$\mathbf{D}(\dot{\mathbf{u}}_h) = \mathbf{D}_h \in \partial\psi(\mathbf{A}\boldsymbol{\lambda}_h). \tag{3.34}$$

From (3.33)–(3.34), $\{\dot{\mathbf{u}}_h, \mathbf{A}\boldsymbol{\lambda}_h\}$ can be seen to be a stationary solution of (3.5), and our proof is complete. □

4. Quasi-static viscoplasticity.

4.1. Variational formulation. As in Chapter 2, § 2, we now consider the problem of computing the velocity field inside a viscoplastic solid when the solid flows in a quasi-static way under the action of given body forces **f**, given surface tractions **g** applied on Γ_2 and imposed zero velocity $\mathbf{v}_1 = \mathbf{0}$ on $\Gamma_1 = \Gamma - \Gamma_2$. In Chapter 2, § 2, assuming small strains, we derived the following well-posed variational formulation for this problem.

$$\begin{cases} \text{Find } \mathbf{v} \text{ in } X \text{ such that} \\ J(\mathbf{v}) \le J(\mathbf{w}) \quad \forall \mathbf{w} \in X. \end{cases} \tag{4.1}$$

In (4.1), the space X and the function J from X into $\mathbb{R}$ are defined, respectively, by

$$J(\mathbf{w}) = \int_\Omega \mathcal{D}_1(\mathbf{x}, \mathbf{D}(\mathbf{w}))\, d\mathbf{x} - \int_\Omega \mathbf{f} \cdot \mathbf{w}\, d\mathbf{x} - \int_{\Gamma_2} \mathbf{g} \cdot \mathbf{w}\, da, \tag{4.2}$$

$$X = \{\mathbf{w} \in \mathbf{W}^{1,s}(\Omega), \operatorname{div} \mathbf{w} = 0, \mathbf{w} = \mathbf{0} \text{ on } \Gamma_1\}. \tag{4.3}$$

The internal dissipation potential $\mathcal{D}_1(\mathbf{x}, \mathbf{D})$ is a known function of $\mathbf{x}$ and $\mathbf{D}$ and is measurable in $\mathbf{x}$, convex in $\mathbf{D}$, and such that

$$C_1|\mathbf{D}|^s \le \mathcal{D}_1(\mathbf{x}, \mathbf{D}) \le C_2 + C_3|\mathbf{D}|^s \tag{4.4}$$

almost everywhere in Ω and for any zero trace tensor $\mathbf{D}$ of $\mathbb{R}^9_{\text{sym}}$. In particular, for Norton, Tresca, or Bingham viscoplastic materials, we have

$$\mathcal{D}_1(\mathbf{x}, \mathbf{D}) = \frac{\lambda^{1-s}}{s}(k(\mathbf{x})\sqrt{2})^s|\mathbf{D}|^s \qquad \text{(Norton)},$$

$$\mathcal{D}_1(\mathbf{x}, \mathbf{D}) = \frac{\lambda^{1-s}}{s}(\sigma_0(\mathbf{x}))^s \sup_{1\le i\le 3} \{|D_i|^s\} \qquad \text{(Tresca)},$$

$$\mathcal{D}_1(\mathbf{x}, \mathbf{D}) = \sqrt{2}g(\mathbf{x})|\mathbf{D}| + \mu(\mathbf{x})|\mathbf{D}|^2 \qquad \text{(Bingham)},$$

which correspond to the constitutive laws

$$\boldsymbol{\sigma} = \lambda^{1-s}(k\sqrt{2})^s|\mathbf{D}(\mathbf{v})|^{s-2}\mathbf{D}(\mathbf{v}) - p\mathbf{Id}, \qquad \operatorname{div} \mathbf{v} = 0 \qquad \text{(Norton)},$$

$$(\boldsymbol{\sigma} + p\mathbf{Id}) \in \partial\left\{\frac{\lambda^{1-s}}{s}\sigma_0^s \sup\{|D_i|^s\}\right\}, \qquad \operatorname{div} \mathbf{v} = 0 \qquad \text{(Tresca)},$$

$$\boldsymbol{\sigma} = (\sqrt{2}g|\mathbf{D}(\mathbf{v})|^{-1} + 2\mu)\mathbf{D}(\mathbf{v}) - p\mathbf{Id}, \qquad \operatorname{div} \mathbf{v} = 0 \qquad \text{(Bingham)}.$$

Remark 4.1. As defined, problem (4.1) models the three-dimensional situation but, using the corrections introduced in § 2.2, it will apply to plane flows as well.

Remark 4.2. From Chapter 2, § 3, Problem (4.1) can be seen to correspond also to the problem of computing the stationary velocity $\mathbf{v}$ (referred to Ω) of an incompressible *viscoplastic fluid* flowing viscously inside a given domain Ω. In this case, Ω corresponds to the present configuration of the body and not to a fixed reference configuration of the considered material.

4.2. Finite-element formulations. The discrete variational formulation of the viscoplastic problem (4.1) is obtained simply by replacing the space X with a finite-dimensional approximation X_h given by

$$X_h = \{\mathbf{w}_h \in V_h, \operatorname{div} \mathbf{w}_h = 0\}, \tag{4.5}$$

with V_h as defined in (3.2). We thus obtain the following formulation.

$$\begin{cases} \text{Find } \mathbf{v}_h \in X_h \text{ such that} \\ J(\mathbf{v}_h) \le J(\mathbf{w}_h) \quad \forall \mathbf{w}_h \in X_h. \end{cases} \tag{4.6}$$

If we define B and G as in (3.23)-(3.24), and if we introduce the function

$$\begin{cases} F: \Sigma_h \to \bar{\mathbb{R}}, \\ F(\mathbf{H}_h) = \displaystyle\int_\Omega \mathcal{D}_1(\mathbf{x}, \mathbf{H}_h)\, d\mathbf{x} \quad \text{if } \operatorname{tr}(\mathbf{H}_h) = 0, \\ \qquad\quad = +\infty \quad \text{if not}, \end{cases} \tag{4.7}$$

then the discrete variational formulation of (4.1) finally becomes

$$\boxed{F(B\mathbf{v}_h) + G(\mathbf{v}_h) \le F(B\mathbf{w}_h) + G(\mathbf{w}_h) \quad \forall \mathbf{w}_h \in V_h, \quad \mathbf{v}_h \in V_h.} \tag{4.8}$$

Under the notation of Chapter 3, this is a problem of type (P). Thus, if r denotes an arbitrary positive number, and $(\cdot, \cdot)$ and $|\cdot|$ are defined by (3.25) and (3.26), respectively, (4.8) can be associated with the augmented Lagrangian

$$\mathcal{L}_r(\mathbf{w}_h, \mathbf{H}_h; \boldsymbol{\mu}_h) = F(\mathbf{H}_h) + G(\mathbf{w}_h) + \frac{r}{2}|B\mathbf{w}_h - \mathbf{H}_h|^2 + (\boldsymbol{\mu}_h, B\mathbf{w}_h - \mathbf{H}_h) \tag{4.9}$$

and with the following saddle-point problem.

$$(4.10)\qquad \begin{cases}\text{Find } \{\mathbf{v}_h, \mathbf{D}_h ; \boldsymbol{\lambda}_h\} \in V_h \times \Sigma_h \times \Sigma_h \text{ such that} \\ \mathscr{L}_r(\mathbf{v}_h, \mathbf{D}_h ; \boldsymbol{\mu}_h) \le \mathscr{L}_r(\mathbf{v}_h, \mathbf{D}_h ; \boldsymbol{\lambda}_h) \le \mathscr{L}_r(\mathbf{w}_h, \mathbf{H}_h ; \boldsymbol{\lambda}_h) \\ \forall \{\mathbf{w}_h, \mathbf{H}_h ; \boldsymbol{\mu}_h\} \in V_h \times \Sigma_h \times \Sigma_h.\end{cases}$$

4.3. Relation to stationary elastoviscoplasticity. Let us again consider the elastoviscoplastic problem of § 3 with $\mathbf{A}$ arbitrary, $\mathbf{f}$ and $\mathbf{g}$ as given in (4.2), and j_c defined from $\mathscr{D}_1$ through the relation

$$(4.11)\qquad j_c(\mathbf{x}, \boldsymbol{\tau}) = \left\{ q \sup_{\mathbf{H} \in \mathbb{R}^9_{\text{sym}}, \text{tr}(\mathbf{H})=0} (\boldsymbol{\tau} \cdot \mathbf{H} - \mathscr{D}_1(\mathbf{x}, \mathbf{H})) \right\}^{1/q}, \qquad \frac{1}{q} + \frac{1}{s} = 1.$$

For example, for Norton, Tresca, or Bingham materials, we have

$$j_c(\mathbf{x}, \boldsymbol{\tau}) = (k(\mathbf{x})\sqrt{2})^{-1} |\boldsymbol{\tau}_D| \qquad \text{(Norton)},$$

$$j_c(\mathbf{x}, \boldsymbol{\tau}) = (\sigma_0(\mathbf{x}))^{-1} \sup_{i,j} |\tau_i - \tau_j| \qquad \text{(Tresca)},$$

$$j_c(\mathbf{x}, \boldsymbol{\tau}) = (\mu(\mathbf{x}))^{-1/2} \max\{0, |\boldsymbol{\tau}_D| - 2g(\mathbf{x})\} \qquad \text{(Bingham)}.$$

THEOREM 4.1. *Under the above notation, the augmented Lagrangian problem* (4.10) *is identical to the augmented Lagrangian formulation* (3.27) *of stationary elastoviscoplasticity. Problem* (4.10) *is also equivalent to the original viscoplastic problem* (4.8) *and has at least a solution.*

Proof. Step 1. By construction, Problems (4.10) and (3.27) are identical within the definitions of F given by (3.7) in (3.27) and by (4.7) in (4.10). As in Lemma 3.1, it is straightforward to observe that the function ψ defined by (3.8), with j_c as given in (4.11), is the dual of the function F given in (4.7). Therefore, both definitions of F correspond to the same dual function ψ and are thus equivalent.

Moreover, from Theorem 4.1 of Chapter 3, any saddle-point of $\mathscr{L}_r$ corresponds to a solution of (P), and, thus, any saddle-point $\{\mathbf{v}_h, \mathbf{D}_h ; \boldsymbol{\lambda}_h\}$ of (4.10) corresponds to a solution $\mathbf{v}_h$ of (4.8).

Step 2. Conversely, let $\mathbf{v}_h = \dot{\mathbf{u}}_h$ be a solution of (4.8). As in Theorem 2.2 of Chapter 2, let us introduce the spaces

$$P_h = \{q_h, \exists \boldsymbol{\sigma}_h \in \Sigma_h \text{ with } q_h = \text{tr}(\boldsymbol{\sigma}_h)\} = \text{Im (trace)},$$

$$Y_h = \{\mathbf{D}_h \in \Sigma_h, \text{tr}(\mathbf{D}_h) = 0\} = \ker(\text{trace}),$$

together with the convex function

$$\Phi(\mathbf{w}_h, \mathbf{H}_h) = F(\mathbf{D}(\dot{\mathbf{u}}_h + \mathbf{w}_h) - \mathbf{H}_h) - F(\mathbf{D}(\dot{\mathbf{u}}_h)) + G(\mathbf{w}_h + \dot{\mathbf{u}}_h).$$

By construction, Φ is finite on $X_h \times Y_h$ and, hence, continuous. Moreover, it can be seen from (4.8) that $\Phi(\mathbf{w}_h, \mathbf{0})$ is bounded below on X_h by $\Phi(\mathbf{0}, \mathbf{0})$. Then, from Theorem 1.10 of Chapter 2 about convex analysis, there exists $\boldsymbol{\sigma}_{dh}$ in Y_h such that

$$(4.12)\qquad \Phi(\mathbf{0},\mathbf{0})=\inf_{\mathbf{w}_h\in X_h}\Phi(\mathbf{w}_h,\mathbf{0})=\Phi^*(\mathbf{0},-\boldsymbol{\sigma}_{dh})=\sup_{\boldsymbol{\tau}_h\in Y_h}\Phi^*(\mathbf{0},\boldsymbol{\tau}_h).$$

From Theorem 1.10 of Chapter 2, it follows that (4.12) can be expressed as

$$(4.13)\qquad (\mathbf{0},-\boldsymbol{\sigma}_{dh})\in\partial\Phi(\mathbf{0},\mathbf{0}),$$

that is, by construction of Φ and $\partial\Phi$,

$$(4.14)\qquad \begin{cases} -\int_\Omega \boldsymbol{\sigma}_{dh}\cdot\mathbf{H}_h\,d\mathbf{x}\le F[\mathbf{D}(\dot{\mathbf{u}}_h+\mathbf{w}_h)-\mathbf{H}_h]-F(\mathbf{D}(\dot{\mathbf{u}}_h))+G(\mathbf{w}_h)\\ \qquad\qquad\qquad\qquad\qquad\qquad \forall\{\mathbf{w}_h,\mathbf{H}_h\}\in X_h\times Y_h.\end{cases}$$

If we first express (4.14) with $\mathbf{w}_h$ arbitrary and $\mathbf{H}_h=\mathbf{D}(\mathbf{w}_h)$, we obtain

$$\int_\Omega \boldsymbol{\sigma}_{dh}\cdot\mathbf{D}(\mathbf{w}_h)\,d\mathbf{x}\ge -G(\mathbf{w}_h)\quad \forall\mathbf{w}_h\in X_h.$$

By linearity, this implies that the linear application L_h defined on V_h by

$$L_h(\mathbf{w}_h)=\int_\Omega \boldsymbol{\sigma}_{dh}\cdot\mathbf{D}(\mathbf{w}_h)\,d\mathbf{x}+G(\mathbf{w}_h)$$

belongs to the orthogonal of X_h in V_h. However, because X_h is the kernel of the divergence operator in the finite-dimensional space V_h, it follows that the orthogonal of X_h is the image of Im (div) = Im (trace $\circ\, B)\subset P_h$ by the transposition of the divergence operator. Therefore, there exists p_h in P_h such that

$$L_h(\mathbf{w}_h)=+\int_\Omega p_h\,\operatorname{div}\mathbf{w}_h\,d\mathbf{x}\quad\forall\mathbf{w}_h\in V_h,$$

that is,

$$(4.15)\qquad \int_\Omega(\boldsymbol{\sigma}_{dh}-p_h\mathbf{Id})\cdot\mathbf{D}(\mathbf{w}_h)\,d\mathbf{x}=-G(\mathbf{w}_h)\quad\forall\mathbf{w}_h\in V_h.$$

On the other hand, we can express (4.14) with $\mathbf{w}_h=\mathbf{0}$ and $\mathbf{H}_h$ arbitrary in Y_h. Since F takes infinite values in Σ_h-Y_h, (4.14) implies

$$\boldsymbol{\sigma}_{dh}\in\partial F(\mathbf{D}(\dot{\mathbf{u}}_h))\quad\text{in }\Sigma_h.$$

Because ∂F is invariant by translation along the tensor **Id**, this yields

$$\boldsymbol{\sigma}_{dh}-p_h\mathbf{Id}\in\partial F(\mathbf{D}(\dot{\mathbf{u}}_h)),$$

that is, by duality,

$$(4.16)\qquad \mathbf{D}(\dot{\mathbf{u}}_h)\in\partial\psi(\boldsymbol{\sigma}_{dh}-p_h\mathbf{Id}).$$

In summary, from (4.15)–(4.16), $\{\dot{\mathbf{u}}_h,\boldsymbol{\sigma}_{dh}-p_h\mathbf{Id}\}$ can be seen to be a stationary solution of (3.5) and, hence, from Theorem 3.4, $\{\dot{\mathbf{u}}_h,\mathbf{D}(\dot{\mathbf{u}}_h),\mathbf{A}^{-1}(\boldsymbol{\sigma}_{dh}-p_h\mathbf{Id})\}$ is a solution of (3.27), that is, of (4.10).

Step 3. The existence of solutions of the minimization problem (4.8), which, from Step 2, can be seen to be equivalent to (4.10), follows directly from the Weierstrass theorem (Chap. 2, Thm. 1.9). Indeed, by construction, (4.8) consists of a minimization of the convex, coercive, lower semicontinuous function J over the nonempty, real, finite-dimensional vector space X_h. □

Remark 4.3. From the above theorem, the direct augmented Lagrangian treatment of the viscoplastic problem (4.8) leads to the augmented Lagrangian formulation (3.27) of stationary elastoviscoplasticity. As a corollary, stationary elastoviscoplastic problems are equivalent to quasi-static viscoplastic problems and admit solutions.

5. Numerical algorithms.

5.1. Time integration schemes. Let us consider the general problem (3.5) of elastoviscoplasticity, once approximated by finite elements. To obtain its numerical solution, let us first express (3.5) as the equivalent initial-value problem (3.14) as follows.

$$\begin{cases} \text{Find } \boldsymbol{\sigma}_h : [0, t_1] \to \Sigma_h \text{ with } \boldsymbol{\sigma}_h(\mathbf{x}, 0) = \boldsymbol{\sigma}_{0h}(\mathbf{x}) \text{ and} \\ S\dot{\boldsymbol{\sigma}}_h + A_1(\boldsymbol{\sigma}_h) + A_2(\boldsymbol{\sigma}_h) \ni \mathbf{0} \quad \text{a.e. in } (0, t_1), \end{cases} \tag{5.1}$$

where $S = \mathbf{A}^{-1}$, $A_1 = \partial\psi_t$, $A_2 = \partial I_t^h$, $\boldsymbol{\sigma}_h$ denotes the approximate stress field, and Σ_h, ψ_t, and I_t^h are as defined in § 3.

Now, let us integrate (5.1) by one of the alternating-directions schemes introduced in Chapter 3, § 5. Denoting by Δt a given time step and by $\boldsymbol{\sigma}_h^n$ an approximation of $\boldsymbol{\sigma}_h(n\Delta t)$, we then obtain the following algorithms.

ALGORITHM (5.2)-(5.3) (Peaceman-Rachford scheme).

Assume that $\boldsymbol{\sigma}_h^0 = \boldsymbol{\sigma}_{0h}$. Then, for $n \geq 0$ and $\boldsymbol{\sigma}_h^n$ known, determine $\boldsymbol{\sigma}_h^{n+1}$ by solving sequentially

$$\frac{S\boldsymbol{\sigma}_h^{n+1/2} - S\boldsymbol{\sigma}_h^n}{\Delta t/2} + A_1(\boldsymbol{\sigma}_h^{n+1/2}) + A_2(\boldsymbol{\sigma}_h^n) \ni \mathbf{0}, \tag{5.2}$$

$$\frac{S\boldsymbol{\sigma}_h^{n+1} - S\boldsymbol{\sigma}_h^{n+1/2}}{\Delta t/2} + A_1(\boldsymbol{\sigma}_h^{n+1/2}) + A_2(\boldsymbol{\sigma}_h^{n+1}) \ni \mathbf{0}. \tag{5.3}$$

ALGORITHM (5.4)-(5.5) (Douglas-Rachford scheme).

Assume that $\boldsymbol{\sigma}_h^0 = \boldsymbol{\sigma}_{0h}$. Then, for $n \geq 0$ and $\boldsymbol{\sigma}_h^n$ known, determine $\boldsymbol{\sigma}_h^{n+1}$ by solving sequentially

$$\frac{S\hat{\boldsymbol{\sigma}}_h^{n+1} - S\boldsymbol{\sigma}_h^n}{\Delta t} + A_1(\hat{\boldsymbol{\sigma}}_h^{n+1}) + A_2(\boldsymbol{\sigma}_h^n) \ni \mathbf{0}, \tag{5.4}$$

$$\frac{S\boldsymbol{\sigma}_h^{n+1} - S\boldsymbol{\sigma}_h^n}{\Delta t} + A_1(\hat{\boldsymbol{\sigma}}_h^{n+1}) + A_2(\boldsymbol{\sigma}_h^{n+1}) \ni \mathbf{0}. \tag{5.5}$$

ALGORITHM (5.6)–(5.8) (θ-scheme).

Assume that $\boldsymbol{\sigma}_h^0 = \boldsymbol{\sigma}_{0h}$. Then, for $n \geq 0$ and $\boldsymbol{\sigma}_h^n$ known, determine $\boldsymbol{\sigma}_h^{n+1}$ by solving sequentially

$$\frac{S\boldsymbol{\sigma}_h^{n+\theta} - S\boldsymbol{\sigma}_h^n}{\theta\,\Delta t} + A_1(\boldsymbol{\sigma}_h^n) + A_2(\boldsymbol{\sigma}_h^{n+\theta}) \ni \mathbf{0}, \tag{5.6}$$

$$\frac{S\boldsymbol{\sigma}_h^{n+1-\theta} - S\boldsymbol{\sigma}_h^{n+\theta}}{(1-2\theta)\,\Delta t} + A_1(\boldsymbol{\sigma}_h^{n+1-\theta}) + A_2(\boldsymbol{\sigma}_h^{n+\theta}) \ni \mathbf{0}, \tag{5.7}$$

$$\frac{S\boldsymbol{\sigma}_h^{n+1} - S\boldsymbol{\sigma}_h^{n+1-\theta}}{\theta\,\Delta t} + A_1(\boldsymbol{\sigma}_h^{n+1-\theta}) + A_2(\boldsymbol{\sigma}_h^{n+1}) \ni \mathbf{0}. \tag{5.8}$$

Note that we exchanged the roles of A_1 and A_2 in the θ-scheme; otherwise, it would have been impossible to express this scheme under a practical form.

ALGORITHM (5.9) (Backward Euler scheme).

Assume that $\boldsymbol{\sigma}_h^0 = \boldsymbol{\sigma}_{0h}$. Then, for $n \geq 0$ and $\boldsymbol{\sigma}_h^n$ known, determine $\boldsymbol{\sigma}_h^{n+1}$ by solving

$$\frac{S\boldsymbol{\sigma}_h^{n+1} - S\boldsymbol{\sigma}_h^n}{\Delta t} + A_1(\boldsymbol{\sigma}_h^{n+1}) + A_2(\boldsymbol{\sigma}_h^{n+1}) \ni \mathbf{0}. \tag{5.9}$$

5.2. Relation to augmented Lagrangian algorithms. A different numerical approach to elastoviscoplasticity would have been to consider the stationary problem (3.27) and to solve it using one of the augmented Lagrangian algorithms introduced in Chapter 3, § 4 for similar saddle-point problems. Actually, this approach turns out to be identical to the time integration of (5.1), as is shown by the following theorem.

THEOREM 5.1. *Let f, g, $\mathbf{A}$, and ψ be independent of time. Then, the numerical solution of the discrete stationary problem* (3.27) *by the augmented Lagrangian algorithm* ALG1 (*resp.*, ALG2, ALG3, ALG4) *consists, in fact, in the time integration of the full stress evolution problem* (5.1) *by a backward Euler scheme with $\rho_n = r = \Delta t$* (*resp., a Douglas–Rachford scheme with $\rho_n = r = \Delta t$; a Peaceman–Rachford scheme with $\rho_n = r = \Delta t/2$, a θ-scheme*).

Proof. In Chapter 3, through Theorem 5.3 and the construction of ALG4, the augmented Lagrangian algorithms ALG1, ALG2, ALG3, and ALG4 were interpreted, under the same conditions, as time integrators of the multivalued initial-value problem

$$\frac{d\boldsymbol{\lambda}}{dt} + \partial F^{-1}(\boldsymbol{\lambda}) - B\,\partial G^{-1}(-B^T\boldsymbol{\lambda}) \ni \mathbf{0}, \tag{5.10}$$

set on the Hilbert space H endowed with the scalar product $(\cdot,\cdot)$. Therefore, our theorem will be proved if, with $H = \Sigma_h$ and F, G, B, and $(\cdot,\cdot)$ defined by (3.7), (3.23), (3.24), and (3.25), the above problem (5.10) turns out to be identical to the stress evolution problem (5.1).

But, by definition, with Σ_h endowed with the scalar product $(\cdot,\cdot)$, we have

$$(5.11)\quad \begin{cases} -B\,\partial G^{-1}(-B^T\boldsymbol{\lambda})=\{-B\mathbf{v},\ -B^T\boldsymbol{\lambda}\in\partial G(\mathbf{v})\} \\ \qquad =\left\{-B\mathbf{v}, -\int_\Omega \mathbf{A}\boldsymbol{\lambda}\cdot B\mathbf{w}\,d\mathbf{x}\le G(\mathbf{v}+\mathbf{w})-G(\mathbf{v})\ \forall\mathbf{w}\in V_h\right\} \\ \qquad =\left\{-\mathbf{D}(\mathbf{v}), \int_\Omega \mathbf{A}\boldsymbol{\lambda}\cdot\mathbf{D}(\mathbf{w})\,d\mathbf{x}\right. \\ \qquad\qquad \left. =\int_\Omega \mathbf{f}\cdot\mathbf{w}\,d\mathbf{x}+\int_{\Gamma_2}\mathbf{g}\cdot\mathbf{w}\,da\ \forall\mathbf{w}\in V_h\right\}, \end{cases}$$

and

$$(5.12)\quad \begin{cases} \partial F^{-1}(\boldsymbol{\lambda})=\{\mathbf{D},\boldsymbol{\lambda}\in\partial F(\mathbf{D})\} \\ \qquad =\left\{\mathbf{D}, \int_\Omega \mathbf{A}\boldsymbol{\lambda}\cdot\mathbf{H}\,d\mathbf{x}\le F(\mathbf{H}+\mathbf{D})-F(\mathbf{D})\ \forall\mathbf{H}\in\Sigma_h\right\}. \end{cases}$$

If we now endow Σ_h with the L^2 scalar product, and if we apply Lemmas 3.1 and 3.2, (5.11) and (5.12) yield

$$-B\,\partial G^{-1}(-B^T\boldsymbol{\lambda})=\partial I^h(\mathbf{A}\boldsymbol{\lambda}),$$
$$\partial F^{-1}(\boldsymbol{\lambda})=\{\mathbf{D},\mathbf{A}\boldsymbol{\lambda}\in\partial F(\mathbf{D})\}=\partial\psi(\mathbf{A}\boldsymbol{\lambda}).$$

Therefore, (5.10) is indeed identical to the stress evolution problem (5.1), with $\boldsymbol{\sigma}=\mathbf{A}\boldsymbol{\lambda}$, and the proof is complete. □

Remark 5.1. Theorem 3.4 proves the equivalence of the stationary problem associated with (5.1) and the saddle-point problem (3.27). Theorem 5.1 proves that the algorithms proposed for solving these two problems not only lead to the same solutions but in fact correspond to the same sequence of numerical computations.

Remark 5.2. The above equivalence result is particularly interesting for the following reasons.

(i) It gives a practical meaning to the formal time-integration schemes of § 5.1 when applied to the multivalued stress initial-value problem (5.1).

(ii) It gives a physical interpretation of all of the values computed during the numerical solution of the stationary elastoviscoplastic problem (3.27) by an augmented Lagrangian algorithm. $\mathbf{A}^{-1}\boldsymbol{\lambda}^n$, $\dot{\mathbf{u}}^n$, and $\mathbf{D}_1^n$, respectively, approximate the values at time nr of the stresses, the velocities, and the plastic strains as they can be observed in the real physical process.

(iii) It gives tools for studying and generalizing the augmented Lagrangian algorithms of Chapter 3 by considering their associated time-integration schemes. More generally, it justifies the use of augmented Lagrangian methods for the solution of (3.27) or (4.8).

5.3. Implementation of the Peaceman–Rachford, Douglas–Rachford, and θ-scheme for elastoviscoplasticity. From the definition of S, A_1, and A_2 in (5.2)-(5.3), it follows that the Peaceman–Rachford scheme (5.2)-(5.3) can be expressed as

$$\text{(5.13)}\qquad \begin{cases} 2\mathbf{A}^{-1}\dfrac{\boldsymbol{\sigma}_h^{n+1/2}-\boldsymbol{\sigma}_h^n}{\Delta t}+\mathbf{D}_1^{n+1/2}+\mathbf{D}_2^n=0, \\ \mathbf{D}_1^{n+1/2}\in\partial\psi_{n+1/2}(\boldsymbol{\sigma}_h^{n+1/2}), \qquad \mathbf{D}_2^n\in\partial I_h^n(\boldsymbol{\sigma}_h^n), \end{cases}$$

$$\text{(5.14)}\qquad \begin{cases} 2\mathbf{A}^{-1}\dfrac{\boldsymbol{\sigma}_h^{n+1}-\boldsymbol{\sigma}_h^{n+1/2}}{\Delta t}+\mathbf{D}_1^{n+1/2}+\mathbf{D}_2^{n+1}=0, \\ \mathbf{D}_2^{n+1}\in\partial I_h^{n+1}(\boldsymbol{\sigma}_h^{n+1}). \end{cases}$$

In (5.13), the condition $\mathbf{D}_2^n\in\partial I_h^n(\boldsymbol{\sigma}_h^n)$ can be eliminated, because it will already be satisfied from the writing of (5.14) at the previous time-step. If, in addition, we compute $\partial\psi$ and ∂I by Lemmas 3.1 and 3.2 and change the order in (5.13)-(5.14), we obtain

$$\text{(5.15)}\qquad \begin{cases} \boldsymbol{\sigma}_h^{n+1/2}\in\partial F_{n+1/2}(\mathbf{D}_1^{n+1/2}), \\ \boldsymbol{\sigma}_h^{n+1/2}=\boldsymbol{\sigma}_h^n-\dfrac{\Delta t}{2}\mathbf{A}(\mathbf{D}_1^{n+1/2}+\mathbf{D}_2^n), \end{cases}$$

$$\text{(5.16)}\qquad \begin{cases} \boldsymbol{\sigma}_h^{n+1}\in S_h((n+1)\,\Delta t), \qquad \mathbf{D}_2^{n+1}=-\mathbf{D}(\mathbf{v}^{n+1}), \\ \boldsymbol{\sigma}_h^{n+1}=\boldsymbol{\sigma}_h^n-\dfrac{\Delta t}{2}\mathbf{A}(\mathbf{D}_1^{n+1/2}+\mathbf{D}_2^{n+1}). \end{cases}$$

After replacement of (5.2)-(5.3) by (5.15)-(5.16) and elimination of $\boldsymbol{\sigma}_h^{n+1/2}$ and $\boldsymbol{\sigma}_h^{n+1}$ in the first line of (5.15) and (5.16), respectively, we obtain the following practical form of the Peaceman–Rachford scheme for elastoviscoplasticity.

ALGORITHM (5.17)-(5.21).

$$\text{(5.17)}\qquad \boldsymbol{\sigma}_h^0=\boldsymbol{\sigma}_{0h},\ \dot{\mathbf{u}}_h^0 \text{ arbitrary in } V_h;$$

then, for $n\geq 0$, $\boldsymbol{\sigma}_h^n$ and $\dot{\mathbf{u}}_h^n$ being known, solve successively

$$\text{(5.18)}\qquad \boldsymbol{\sigma}_h^n+\frac{\Delta t}{2}\mathbf{A}(\mathbf{D}(\dot{\mathbf{u}}_h^n)-\mathbf{D}_1^{n+1/2})\in\partial F_{n+1/2}(\mathbf{D}_1^{n+1/2}),$$

$$\text{(5.19)}\qquad \boldsymbol{\sigma}_h^{n+1/2}=\boldsymbol{\sigma}_h^n+\frac{\Delta t}{2}\mathbf{A}(\mathbf{D}(\dot{\mathbf{u}}_h^n)-\mathbf{D}_1^{n+1/2}),$$

$$\text{(5.20)}\quad \begin{cases} \displaystyle\int_\Omega \left(\boldsymbol{\sigma}_h^{n+1/2}+\frac{\Delta t}{2}\,\mathbf{A}(\mathbf{D}(\dot{\mathbf{u}}_h^{n+1})-\mathbf{D}_1^{n+1/2})\right)\cdot \mathbf{D}(\mathbf{w}_h)\,d\mathbf{x} \\ \displaystyle\quad =\int_\Omega \mathbf{f}^{n+1}\cdot \mathbf{w}_h\,d\mathbf{x}+\int_{\Gamma_2}\mathbf{g}^{n+1}\cdot \mathbf{w}_h\,da \qquad \forall \mathbf{w}_h\in V_h,\quad \dot{\mathbf{u}}_h^{n+1}\in V_h, \end{cases}$$

$$\text{(5.21)}\qquad \boldsymbol{\sigma}_h^{n+1}=\boldsymbol{\sigma}_h^{n+1/2}+\frac{\Delta t}{2}\,\mathbf{A}(\mathbf{D}(\dot{\mathbf{u}}_h^{n+1})-\mathbf{D}_1^{n+1/2}).$$

Let us now transform the Douglas–Rachford scheme (5.4)–(5.5). Using the definitions of S, A_1, and A_2 in (5.4)–(5.5) together with Lemmas 3.1 and 3.2, we obtain

$$\begin{cases} \hat{\boldsymbol{\sigma}}_h^{n+1}\in \partial F_{n+1}(\mathbf{D}_1^{n+1}), \\ \hat{\boldsymbol{\sigma}}_h^{n+1}=\boldsymbol{\sigma}_h^{n}-\Delta t\ \mathbf{A}(\mathbf{D}_1^{n+1}+\mathbf{D}_2^{n}), \\ \boldsymbol{\sigma}_h^{n+1}\in S_h((n+1)\,\Delta t), \qquad \mathbf{D}_2^{n+1}=-\mathbf{D}(\mathbf{v}^{n+1}), \\ \boldsymbol{\sigma}_h^{n+1}=\boldsymbol{\sigma}_h^{n}-\Delta t\ \mathbf{A}(\mathbf{D}_1^{n+1}+\mathbf{D}_2^{n+1}). \end{cases}$$

After elimination of $\hat{\boldsymbol{\sigma}}_h^{n+1}$ and of $\boldsymbol{\sigma}_h^{n+1}$ in the relation $\boldsymbol{\sigma}_h^{n+1}\in S_h((n+1)\Delta t)$, we obtain the following practical form for the Douglas–Rachford scheme for elastoviscoplasticity.

ALGORITHM (5.22)–(5.25).

$$\text{(5.22)}\qquad \boldsymbol{\sigma}_h^{0}=\boldsymbol{\sigma}_{0h},\ \dot{\mathbf{u}}_h^{0}\ \text{arbitrary in}\ V_h;$$

then, for $n\geq 0$, $\boldsymbol{\sigma}_h^{n}$ and $\dot{\mathbf{u}}_h^{n}$ being known, solve successively

$$\text{(5.23)}\qquad \boldsymbol{\sigma}_h^{n}+\Delta t\ \mathbf{A}(\mathbf{D}(\dot{\mathbf{u}}_h^{n})-\mathbf{D}_1^{n+1})\in \partial F_{n+1}(\mathbf{D}_1^{n+1}),$$

$$\text{(5.24)}\quad \begin{cases} \displaystyle\int_\Omega (\boldsymbol{\sigma}_h^{n}+\Delta t\ \mathbf{A}(\mathbf{D}(\dot{\mathbf{u}}_h^{n+1})-\mathbf{D}_1^{n+1}))\cdot \mathbf{D}(\mathbf{w}_h)\,d\mathbf{x} \\ \displaystyle\quad =\int_\Omega \mathbf{f}^{n+1}\cdot \mathbf{w}_h\,d\mathbf{x}+\int_{\Gamma_2}\mathbf{g}^{n+1}\cdot \mathbf{w}_h\,da \qquad \forall \mathbf{w}_h\in V_h,\quad \dot{\mathbf{u}}_h^{n+1}\in V_h, \end{cases}$$

$$\text{(5.25)}\qquad \boldsymbol{\sigma}_h^{n+1}=\boldsymbol{\sigma}_h^{n}+\Delta t\ \mathbf{A}(\mathbf{D}(\dot{\mathbf{u}}_h^{n+1})-\mathbf{D}_1^{n+1}).$$

Let us finally transform the θ-scheme (5.6)–(5.8). Using the definitions of S, A_1, and A_2 in (5.6)–(5.8) and transforming the subgradients by Lemmas 3.1 and 3.2 yields

$$\text{(5.26)}\quad \begin{cases} \boldsymbol{\sigma}_h^{n+\theta}\in S_h((n+\theta)\,\Delta t), \qquad \mathbf{D}_2^{n+\theta}=-\mathbf{D}(\mathbf{v}^{n+\theta}), \\ \boldsymbol{\sigma}_h^{n+\theta}=\boldsymbol{\sigma}_h^{n}-\theta\ \Delta t\ \mathbf{A}(\mathbf{D}_1^{n}+\mathbf{D}_2^{n+\theta}), \end{cases}$$

$$(5.27)\qquad \begin{cases} \boldsymbol{\sigma}_h^{n+1-\theta} \in \partial F_{n+1-\theta}(\mathbf{D}_1^{n+1-\theta}), \\ \boldsymbol{\sigma}_h^{n+1-\theta} = \boldsymbol{\sigma}_h^{n+\theta} - (1-2\theta)\,\Delta t\,\mathbf{A}(\mathbf{D}_1^{n+1-\theta} + \mathbf{D}_2^{n+\theta}), \end{cases}$$

$$(5.28)\qquad \begin{cases} \boldsymbol{\sigma}_h^{n+1} \in S_h((n+1)\,\Delta t), \qquad \mathbf{D}_2^{n+1} = -\mathbf{D}(\mathbf{v}^{n+1}), \\ \boldsymbol{\sigma}_h^{n+1} = \boldsymbol{\sigma}_h^{n+1-\theta} - \theta\,\Delta t\,\mathbf{A}(\mathbf{D}_1^{n+1-\theta} + \mathbf{D}_2^{n+1}). \end{cases}$$

After elimination of $\boldsymbol{\sigma}_h^{n+\theta}$, $\boldsymbol{\sigma}_h^{n+1-\theta}$, and $\boldsymbol{\sigma}_h^{n+1}$ in the first lines of (5.26), (5.27), and (5.28), respectively, we finally obtain the practical form of the θ-scheme for elastoviscoplasticity.

ALGORITHM (5.29)–(5.36).

$$(5.29)\qquad \boldsymbol{\sigma}_h^0 = \boldsymbol{\sigma}_{0h};$$

then, for $n \geq 0$, $\boldsymbol{\sigma}_h^n$ being known, solve successively

$$(5.30)\qquad \boldsymbol{\sigma}_h^n \in \partial F_n(\mathbf{D}_1^n), \qquad \mathbf{D}_1^n \in \Sigma_h,$$

$$(5.31)\qquad \begin{cases} \displaystyle\int_\Omega (\boldsymbol{\sigma}_h^n + \theta\,\Delta t\,\mathbf{A}(\mathbf{D}(\dot{\mathbf{u}}_h^{n+\theta}) - \mathbf{D}_1^n)) \cdot \mathbf{D}(\mathbf{w}_h)\,d\mathbf{x} \\ \displaystyle\quad = \int_\Omega \mathbf{f}^{n+\theta} \cdot \mathbf{w}_h\,d\mathbf{x} + \int_{\Gamma_2} \mathbf{g}^{n+\theta} \cdot \mathbf{w}_h\,d\mathbf{a} \qquad \forall \mathbf{w}_h \in V_h, \quad \dot{\mathbf{u}}_h^{n+\theta} \in V_h, \end{cases}$$

$$(5.32)\qquad \boldsymbol{\sigma}_h^{n+\theta} = \boldsymbol{\sigma}_h^n + \theta\,\Delta t\,\mathbf{A}(\mathbf{D}(\dot{\mathbf{u}}_h^{n+\theta}) - \mathbf{D}_1^n),$$

$$(5.33)\qquad \boldsymbol{\sigma}_h^{n+\theta} + (1-2\theta)\,\Delta t\,\mathbf{A}(\mathbf{D}(\dot{\mathbf{u}}_h^{n+\theta}) - \mathbf{D}_1^{n+1-\theta}) \in \partial F_{n+1-\theta}(\mathbf{D}_1^{n+1-\theta}),$$

$$(5.34)\qquad \boldsymbol{\sigma}_h^{n+1-\theta} = \boldsymbol{\sigma}_h^{n+\theta} + (1-2\theta)\,\Delta t\,\mathbf{A}(\mathbf{D}(\dot{\mathbf{u}}_h^{n+\theta}) - \mathbf{D}_1^{n+1-\theta}),$$

$$(5.35)\qquad \begin{cases} \displaystyle\int_\Omega (\boldsymbol{\sigma}_h^{n+1-\theta} + \theta\,\Delta t\,\mathbf{A}(\mathbf{D}(\dot{\mathbf{u}}_h^{n+1}) - \mathbf{D}_1^{n+1-\theta})) \cdot \mathbf{D}(\mathbf{w}_h)\,d\mathbf{x} \\ \displaystyle\quad = \int_\Omega \mathbf{f}^{n+1} \cdot \mathbf{w}_h\,d\mathbf{x} + \int_{\Gamma_2} \mathbf{g}^{n+1} \cdot \mathbf{w}_h\,d\mathbf{a} \qquad \forall \mathbf{w}_h \in V_h, \quad \dot{\mathbf{u}}_h^{n+1} \in V_h, \end{cases}$$

$$(5.36)\qquad \boldsymbol{\sigma}_h^{n+1} = \boldsymbol{\sigma}_h^{n+1-\theta} + \theta\,\Delta t\,\mathbf{A}(\mathbf{D}(\dot{\mathbf{u}}_h^{n+1}) - \mathbf{D}_1^{n+1-\theta}).$$

The equivalence of the different alternating-direction algorithms and their augmented Lagrangian counterparts is clearly evident in the above. Now, from the numerical point of view, there are three types of steps involved in each of these algorithms.

(i) Explicit updating of the stress field $\boldsymbol{\sigma}$ in (5.19), (5.21), (5.25), (5.32), (5.34), or (5.36).

(ii) Solution of the linear elasticity problem associated with the tensor $\mathbf{A}$ in step (5.20), (5.24), (5.31), or (5.35). This will determine the velocity $\dot{\mathbf{u}}$ when $\boldsymbol{\sigma}$ and $\mathbf{D}_1$ are known, and, if we expand both $\dot{\mathbf{u}}$ and the test functions $\mathbf{w}$ on a given basis $(\mathbf{w}_i)$ of V_h, these elasticity problems reduce to linear systems associated with the same sparse, symmetric positive-definite matrix whose coefficients are given by

$$\mathcal{A}_{ij}=\int_\Omega \mathbf{AD}(\mathbf{w}_i)\cdot\mathbf{D}(\mathbf{w}_j)\,d\mathbf{x}.$$

(iii) Solution of the convex minimization problem (5.18), (5.23), (5.30), or (5.33), for which we will propose a solution procedure in § 6. These problems determine the plastic strains $\mathbf{D}_1$ when $\dot{\mathbf{u}}$ and $\boldsymbol{\sigma}$ are known.

5.4. Implementation of the backward Euler scheme for elastoviscoplasticity. The situation is more complicated for the backward Euler scheme (5.9), which, when applied to (5.1), requires at each step the inversion of the operator $\mathbf{A}^{-1}/\Delta t+\partial\psi+\partial I_t^h$. More precisely, rewriting (5.9) using the definitions of S, A_1, and A_2 and Lemma 3.2, we obtain

$$(5.37)\quad\begin{cases}-\left\{\mathbf{A}^{-1}\dfrac{\boldsymbol{\sigma}_h^{n+1}-\boldsymbol{\sigma}_h^n}{\Delta t}+\mathbf{D}(\dot{\mathbf{u}}_h^{n+1})\right\}\in\partial\psi_{n+1}(\boldsymbol{\sigma}_h^{n+1}),\\ \displaystyle\int_\Omega \boldsymbol{\sigma}_h^{n+1}\cdot\mathbf{D}(\mathbf{w}_h)\,d\mathbf{x}=\int_\Omega \mathbf{f}^{n+1}\cdot\mathbf{w}_h\,d\mathbf{x}+\int_{\Gamma_2}\mathbf{g}^{n+1}\cdot\mathbf{w}_h\,d\mathbf{a}\quad\forall\mathbf{w}_h\in V_h.\end{cases}$$

To solve (5.37), let us express it as a minimization problem by introducing the dual Lagrangian $\mathcal{L}_{n+1}^*$ defined by

$$(5.38)\quad\begin{cases}\mathcal{L}_{n+1}^*(\mathbf{w}_h,\boldsymbol{\tau}_h)=-\displaystyle\int_\Omega \mathbf{f}^{n+1}\cdot\mathbf{w}_h\,d\mathbf{x}-\int_{\Gamma_2}\mathbf{g}^{n+1}\cdot\mathbf{w}_h\,d\mathbf{a}+\int_\Omega\mathbf{D}(\mathbf{w}_h)\cdot\boldsymbol{\tau}_h\,d\mathbf{x}\\ -\dfrac{1}{2\,\Delta t}\displaystyle\int_\Omega \mathbf{A}^{-1}\boldsymbol{\tau}_h\cdot\boldsymbol{\tau}_h\,d\mathbf{x}+\dfrac{1}{\Delta t}\int_\Omega A^{-1}\boldsymbol{\sigma}_h^n\cdot\boldsymbol{\tau}_h\,d\mathbf{x}-\psi_{n+1}(\boldsymbol{\tau}_h).\end{cases}$$

Then, (5.37) can be expressed as

$$(5.39)\quad\begin{cases}\mathcal{L}_{n+1}^*(\dot{\mathbf{u}}_h^{n+1},\boldsymbol{\sigma}_h^{n+1})\ge\mathcal{L}_{n+1}^*(\dot{\mathbf{u}}_h^{n+1},\boldsymbol{\tau}_h)\qquad\forall\boldsymbol{\tau}_h\in\Sigma_h,\quad\boldsymbol{\sigma}_h^{n+1}\in\Sigma_h,\\ \mathcal{L}_{n+1}^*(\dot{\mathbf{u}}_h^{n+1},\boldsymbol{\sigma}_h^{n+1})\le\mathcal{L}_{n+1}^*(\mathbf{w}_h,\boldsymbol{\sigma}_h^{n+1})\qquad\forall\mathbf{w}_h\in V_h,\quad\dot{\mathbf{u}}_h^{n+1}\in V_h.\end{cases}$$

From the convexity of $\mathcal{L}_{n+1}^*(\cdot,\boldsymbol{\tau}_h)$ and $-\mathcal{L}_{n+1}^*(\mathbf{w}_h,\cdot)$ on V_h and Σ_h, respectively, the saddle-point problem (5.39) is classically equivalent to

$$\mathcal{L}_{n+1}^*(\dot{\mathbf{u}}_h^{n+1},\boldsymbol{\sigma}_h^{n+1})=\inf_{\mathbf{w}_h\in V_h}\sup_{\boldsymbol{\tau}_h\in\Sigma_h}\mathcal{L}_{n+1}^*(\mathbf{w}_h,\boldsymbol{\tau}_h),$$

that is,

$$(5.40)\quad\boxed{J_{n+1}(\dot{\mathbf{u}}_h^{n+1})=\inf_{\mathbf{w}_h\in V_h}J_{n+1}(\mathbf{w}_h),}$$

where the function J_{n+1} in (5.40) is defined by

$$(5.41)\quad J_{n+1}(\mathbf{w}_h)=\sup_{\boldsymbol{\tau}_h\in\Sigma_h}\mathcal{L}_{n+1}^*(\mathbf{w}_h,\boldsymbol{\tau}_h).$$

Therefore, the generic step (5.37) of the backward Euler scheme reduces to the minimization problem (5.40), which generalizes in some way the dual formulation encountered in the quadratic programming of Chapter 3, § 2.4. As such, (5.40) can be solved numerically by a nonlinear version of the conjugate-gradient algorithm introduced in that section, where the line search (the determination of ρ_k) is made by one iteration of the secant method applied to the equation $((\partial J(\mathbf{v}-\rho\mathbf{z}), \mathbf{z}))=0$. This version consists of

(i) choosing $\mathbf{v}^0$ in V_h;
(ii) taking $\mathbf{d}^0$ in $\partial J_{n+1}(\mathbf{v}^0)$;
(iii) setting $\mathbf{z}^0=\mathbf{d}^0$ and $\rho_{-1}=1$;
(iv) computing iteratively, for $k\geq 0$, with $\mathbf{v}^k$, $\mathbf{d}^k$, $\mathbf{z}^k$, and ρ_{k-1} known, and until $((\mathbf{d}^k, \mathbf{d}^k))$ is sufficiently small,

$$\rho_k=-\rho_{k-1}((\mathbf{d}^k, \mathbf{z}^k))/((\partial J_{n+1}(\mathbf{v}^k-\rho_{k-1}\mathbf{z}^k)-\partial J(\mathbf{v}^k), \mathbf{z}^k)),$$

$$\mathbf{v}^{k+1}=\mathbf{v}^k-\rho_k\mathbf{z}^k,$$

$$\mathbf{d}^{k+1}\in\partial J_{n+1}(\mathbf{v}^{k+1}),$$

$$\mathbf{z}^{k+1}=\mathbf{d}^{k+1}+((\mathbf{d}^{k+1}, \mathbf{d}^{k+1}-\mathbf{d}^k))/((\mathbf{d}^k, \mathbf{d}^k))\mathbf{z}^k.$$

The practical implementation of the above algorithm still requires the choice of an adequate scalar product $((\cdot,\cdot))$ on V_h and the computation of ∂J_{n+1}. By analogy with § 5.3, in which we saw the linear elasticity tensor $\mathbf{A}$ operating on V_h, it is natural to define

$$((\mathbf{v}, \mathbf{w}))=\int_\Omega \mathbf{A}\mathbf{D}(\mathbf{v})\cdot\mathbf{D}(\mathbf{w})\,d\mathbf{x}. \tag{5.42}$$

On the other hand, from saddle-point theory (Ekeland and Temam [1976]), we have

$$\begin{cases}\partial\{\sup\limits_{\boldsymbol{\tau}_h}\mathscr{L}^*_{n+1}(\mathbf{v}, \boldsymbol{\tau}_h)\}=\partial_w\mathscr{L}^*_{n+1}(\mathbf{v}, \boldsymbol{\sigma}_h(\mathbf{v}))\\ \quad=\{\mathbf{d}, ((\mathbf{d}, \mathbf{w}_h))\leq\mathscr{L}^*_{n+1}(\mathbf{v}+\mathbf{w}_h, \boldsymbol{\sigma}_h(\mathbf{v}))-\mathscr{L}^*_{n+1}(\mathbf{v}, \boldsymbol{\sigma}_h(\mathbf{v}))\ \forall\mathbf{w}_h\in V_h\},\end{cases} \tag{5.43}$$

where $\boldsymbol{\sigma}_h(\mathbf{v})$ is the solution of

$$\mathscr{L}^*_{n+1}(\mathbf{v}, \boldsymbol{\sigma}_h(\mathbf{v}))=\sup_{\boldsymbol{\tau}_h\in\Sigma_h}\mathscr{L}^*_{n+1}(\mathbf{v}, \boldsymbol{\tau}_h). \tag{5.44}$$

Therefore, from (5.38), the calculation of $\partial J_{n+1}(\mathbf{v})$ reduces to the successive solution of

$$-\mathbf{A}^{-1}\frac{\boldsymbol{\sigma}_h-\boldsymbol{\sigma}_h^n}{\Delta t}+\mathbf{D}(\mathbf{v})\in\partial\psi_{n+1}(\boldsymbol{\sigma}_h), \tag{5.45}$$

$$(5.46)\quad \begin{cases} \int_{\Omega} \mathbf{AD}(\mathbf{d}) \cdot \mathbf{D}(\mathbf{w}_h)\, d\mathbf{x} = \int_{\Omega} \boldsymbol{\sigma}_h \cdot \mathbf{D}(\mathbf{w}_h)\, d\mathbf{x} - \int_{\Omega} \mathbf{f}^{n+1} \cdot \mathbf{w}_h\, d\mathbf{x} \\ \qquad - \int_{\Gamma_2} \mathbf{g}^{n+1} \cdot \mathbf{w}_h\, da \qquad \forall \mathbf{w}_h \in V_h, \quad \mathbf{d} \in V_h, \end{cases}$$

and we have

$$(5.47)\qquad \partial J(\mathbf{v}) = \{\mathbf{d}\},$$

$$(5.48)\quad ((\partial J(\mathbf{v}), \mathbf{w}_h)) = \int_{\Omega} \boldsymbol{\sigma}_h \cdot \mathbf{D}(\mathbf{w}_h)\, d\mathbf{x} - \int_{\Omega} \mathbf{f}^{n+1} \cdot \mathbf{w}_h\, d\mathbf{x} - \int_{\Gamma_2} \mathbf{g}^{n+1} \cdot \mathbf{w}_h\, da.$$

Thus, finally, if we substitute (5.40) for (5.37) and solve (5.40) by the above conjugate-gradient algorithm, calculating $\partial J(v)$ by (5.45)–(5.48), we obtain the following practical form of the backward Euler scheme for elastoviscoplasticity.

ALGORITHM (5.49)–(5.61).

$$(5.49)\qquad \boldsymbol{\sigma}_h^0 = \boldsymbol{\sigma}_{0h}, \ \dot{\mathbf{u}}_h^0 \text{ arbitrary in } V_h;$$

then, for $n \geq 0$, with $\{\dot{\mathbf{u}}_h^n, \boldsymbol{\sigma}_h^n\}$ known, determine $\{\dot{\mathbf{u}}_h^{n+1}, \boldsymbol{\sigma}_h^{n+1}\}$ by setting

$$(5.50)\qquad \mathbf{v}^0 = \dot{\mathbf{u}}_h^n, \qquad \rho_{-1} = 1,$$

then by solving successively

$$(5.51)\qquad -\mathbf{A}^{-1}\frac{\boldsymbol{\sigma}^0 - \boldsymbol{\sigma}_h^n}{\Delta t} + \mathbf{D}(\mathbf{v}^0) \in \partial\psi_{n+1}(\boldsymbol{\sigma}^0),$$

$$(5.52)\quad \begin{cases} \int_{\Omega} \mathbf{AD}(\mathbf{d}^0) \cdot \mathbf{D}(\mathbf{w}_h)\, d\mathbf{x} = \int_{\Omega} \{\boldsymbol{\sigma}^0 \cdot \mathbf{D}(\mathbf{w}_h) - \mathbf{f}^{n+1} \cdot \mathbf{w}_h\}\, d\mathbf{x} \\ \qquad - \int_{\Gamma_2} \mathbf{g}^{n+1} \cdot \mathbf{w}_h\, da \qquad \forall \mathbf{w}_h \in V_h, \quad \mathbf{d}^0 \in V_h, \end{cases}$$

$$(5.53)\qquad \mathbf{z}^0 = \mathbf{d}^0,$$

then by computing iteratively, for $k \geq 0$, with $\mathbf{v}^k$, $\boldsymbol{\sigma}^k$, $\mathbf{d}^k$, $\mathbf{z}^k$, ρ_{k-1} known,

$$(5.54)\qquad -\mathbf{A}^{-1}\frac{\boldsymbol{\sigma}_1 - \boldsymbol{\sigma}_h^n}{\Delta t} + \mathbf{D}(\mathbf{v}^k - \rho_{k-1}\mathbf{z}^k) \in \partial\psi_{n+1}(\boldsymbol{\sigma}_1),$$

$$(5.55)\qquad \rho_k = -\rho_{k-1}\frac{\int_{\Omega} \mathbf{AD}(\mathbf{d}^k) \cdot \mathbf{D}(\mathbf{z}^k)\, d\mathbf{x}}{\int_{\Omega} (\boldsymbol{\sigma}_1 - \boldsymbol{\sigma}^k) \cdot \mathbf{D}(\mathbf{z}^k)\, d\mathbf{x}},$$

$$(5.56)\qquad \mathbf{v}^{k+1} = \mathbf{v}^k - \rho_k \mathbf{z}^k,$$

$$(5.57)\qquad -\mathbf{A}^{-1}\frac{\boldsymbol{\sigma}^{k+1} - \boldsymbol{\sigma}_h^n}{\Delta t} + \mathbf{D}(\mathbf{v}^{k+1}) \in \partial\psi_{n+1}(\boldsymbol{\sigma}^{k+1}),$$

$$(5.58)\quad \begin{cases} \displaystyle\int_\Omega \mathbf{AD}(\mathbf{d}^{k+1})\cdot \mathbf{D}(\mathbf{w}_h)\,d\mathbf{x} = \int_\Omega \{\boldsymbol{\sigma}^{k+1}\cdot \mathbf{D}(\mathbf{w}_h) - \mathbf{f}^{n+1}\cdot \mathbf{w}_h\}\,d\mathbf{x} \\ \displaystyle\qquad - \int_{\Gamma_2} \mathbf{g}^{n+1}\cdot \mathbf{w}_h\,da \qquad \forall \mathbf{w}_h \in V_h, \quad \mathbf{d}^{k+1}\in V_h, \end{cases}$$

$$(5.59)\qquad \gamma_k = \frac{\int_\Omega \mathbf{AD}(\mathbf{d}^{k+1})\cdot \mathbf{D}(\mathbf{d}^{k+1}-\mathbf{d}^k)\,d\mathbf{x}}{\int_\Omega \mathbf{AD}(\mathbf{d}^k)\cdot \mathbf{D}(\mathbf{d}^k)\,d\mathbf{x}},$$

$$(5.60)\qquad \mathbf{z}^{k+1} = \mathbf{d}^{k+1} + \gamma_k \mathbf{z}^k,$$

and then by setting, once $((\mathbf{d}^{k+1}, \mathbf{d}^{k+1}))$ is sufficiently small,

$$(5.61)\qquad \{\dot{\mathbf{u}}_h^{n+1}, \boldsymbol{\sigma}_h^{n+1}\} = \{\mathbf{v}^{k+1}, \boldsymbol{\sigma}^{k+1}\}.$$

From the numerical point of view, the steps involved in Algorithm (5.49)–(5.61), in addition to explicit updatings and scalar-product computations, are as follows. First, solve the linear elasticity problems (5.52) and (5.58), which are identical to those encountered in § 5.3 and, therefore, can be solved by the same procedure, and, second, solve the convex minimization problems (5.51), (5.54), and (5.57), which, within the replacement of $\boldsymbol{\sigma}$ by $\mathbf{D}$, are also identical to those encountered in § 5.3 and whose solution will be described in § 6.

Remark 5.3. Inertia terms can easily be substituted into this scheme simply by adding the term

$$\int_\Omega \rho \frac{|\mathbf{w}_h|^2/2 - \mathbf{w}_h\cdot \dot{\mathbf{u}}_h^n}{\Delta t}\,d\mathbf{x}$$

to the dual Lagrangian $\mathscr{L}_{n+1}^*$ and consequently updating the computation of ∂J_{n+1} in (5.46), (5.52), (5.55), and (5.58).

Remark 5.4. If $\partial\psi$ is invertible, the above implicit scheme can be used with $\Delta t = +\infty$, that is, for solving the stationary problem (3.27) of elastoviscoplasticity. It then corresponds to a direct treatment of the minimization problem

$$\inf_{\mathbf{w}_h \in V_h} \{F(B\mathbf{w}_h) + G(\mathbf{w}_h)\}$$

by a nonlinear conjugate-gradient algorithm.

Remark 5.5. In perfect plasticity ($q = +\infty$), Algorithm (5.49)–(5.61) is a variant of the so-called return-mapping algorithm (Zienkiewicz [1977]) that is widely used in engineering. Indeed, (5.57) maps $\boldsymbol{\sigma}_h^n + \Delta t\,\mathbf{A}(\mathbf{D}(\mathbf{v}^{k+1}))$ back into the set $\mathbf{C}$ of locally admissible stresses.

6. The problem in plastic strain rates.

6.1. Localization. We now turn to the study of the most specific step of the previous algorithms, that is, study of the convex minimization problems

(5.18), (5.23), (5.30), (5.33), (5.51), (5.54), and (5.57). All of these problems are of the following form.

$$\text{(6.1)} \qquad \begin{cases} \text{Find } \mathbf{D}_1 \text{ in } \Sigma_h \text{ such that} \\ \boldsymbol{\sigma}_h^n + r\mathbf{A}(\mathbf{D}(\dot{\mathbf{u}}_h^n) - \mathbf{D}_1) \in \partial F(\mathbf{D}_1), \end{cases}$$

with Σ_h being the finite-element space defined in (3.1) and F the function defined in (3.7) by

$$\text{(6.2)} \qquad F(\mathbf{H}) = \int_\Omega \sup_{\boldsymbol{\tau} \in S^9} \left\{ \boldsymbol{\tau} \cdot \mathbf{D} - \frac{\lambda}{q} |j_c(\mathbf{x}, \boldsymbol{\tau})|^q \right\} d\mathbf{x}.$$

To study (6.1), we introduce the space $\mathbb{R}_{\text{sym}}^{M \times M}$ of symmetric $M \times M$ real matrices, recalling that a real function j defined on $\mathbb{R}_{\text{sym}}^{M \times M}$ is said to be isotropic if it is a symmetric function of the eigenvalues of its argument (we will then note $j(\mathbf{H}) = j(H_i)$, where H_i are the eigenvalues of $\mathbf{H}$). We have the following lemma.

LEMMA 6.1. *Let j be isotropic and let $\mathbf{A}$ be given in $\mathbb{R}_{\text{sym}}^{M \times M}$ with eigenvalues $A_1 \geq A_2 \cdots \geq A_M$ and with $\mathbf{Q}$ being the orthogonal matrix whose columns are the eigenvectors of $\mathbf{A}$. Then*

$$\text{(6.3)} \qquad \sup_{\mathbf{H} \in \mathbb{R}_{\text{sym}}^{M \times M}} \{\mathbf{A} \cdot \mathbf{H} - j(\mathbf{H})\} = \sup_{(H_i) \in \mathbb{R}^M} \left\{ \sum_{i=1}^M A_i H_i - j(H_i) \right\},$$

$$\text{(6.4)} \qquad \min_{\mathbf{H} \in \mathbb{R}_{\text{sym}}^{M \times M}} \{j(\mathbf{H}) - \mathbf{A} \cdot \mathbf{H}\} = j(\mathbf{Q}\mathbf{D}^d\mathbf{Q}^T) - \mathbf{A} \cdot \mathbf{Q}\mathbf{D}^d\mathbf{Q}^T,$$

where $\mathbf{D}^d$ is the diagonal matrix with diagonal terms (D_i) such that

$$\text{(6.5)} \qquad j(D_i) - \sum_{i=1}^M A_i D_i = \min_{(H_i) \in \mathbb{R}^M} \left\{ j(H_i) - \sum_{i=1}^M A_i H_i \right\}.$$

Proof. Step 1. Let $\mathbf{H}$ be an arbitrary element of $\mathbb{R}_{\text{sym}}^{M \times M}$ with eigenvalues $H_1 \geq H_2 \cdots \geq H_M$. From a well-known result of von Neumann [1937], the product $\mathbf{A} \cdot \mathbf{H}$ verifies

$$\mathbf{A} \cdot \mathbf{H} = \operatorname{tr}(\mathbf{A}\mathbf{H}) \leq \sum_{i=1}^M A_i H_i.$$

Thus, taking into account the isotropy of j, we obtain

$$\mathbf{A} \cdot \mathbf{H} - j(\mathbf{H}) \leq \sum_{i=1}^M A_i H_i - j(H_i) \quad \forall \mathbf{H} \in \mathbb{R}_{\text{sym}}^{M \times M},$$

which implies

$$\text{(6.6)} \qquad \sup_{\mathbf{H} \in \mathbb{R}_{\text{sym}}^{M \times M}} \{\mathbf{A} \cdot \mathbf{H} - j(\mathbf{H})\} \leq \sup_{(H_i) \in \mathbb{R}^M} \left\{ \sum_{i=1}^M A_i H_i - j(H_i) \right\}.$$

Step 2. Let (H_i) be arbitrary in $\mathbb{R}^M$, to which we associate the diagonal matrix $\mathbf{H}^d$ with diagonal terms (H_i) and the matrix $\mathbf{H}=\mathbf{Q}\mathbf{H}^d\mathbf{Q}^T$. Due to the isotropy of j, we have

$$\sum_{i=1}^{M} A_iH_i - j(H_i) = \operatorname{tr}(\mathbf{Q}^T\mathbf{A}\mathbf{Q}\mathbf{H}^d) - j(\mathbf{H}^d) = \operatorname{tr}(\mathbf{A}\mathbf{Q}\mathbf{H}^d\mathbf{Q}^T) - j(\mathbf{Q}\mathbf{H}^d\mathbf{Q}^T)$$
$$= \mathbf{A}\cdot\mathbf{H} - j(\mathbf{H}) \quad \forall (H_i)\in\mathbb{R}^M,$$

which, combined with (6.6), yields (6.3).

Step 3. Let (D_i) in $\mathbb{R}^M$ satisfy (6.5), with which, as in Step 2, we associate $\mathbf{D}^d$ and $\mathbf{D}=\mathbf{Q}\mathbf{D}^d\mathbf{Q}^T$. From (6.3) and (6.5), we have

$$\inf_{\mathbf{H}\in\mathbb{R}^{M\times M}_{\text{sym}}} \{j(\mathbf{H}) - \mathbf{A}\cdot\mathbf{H}\} = \inf_{(H_i)\in\mathbb{R}^M} \left\{ j(H_i) - \sum_{i=1}^{M} A_iH_i \right\} = j(D_i) - \sum_{i=1}^{M} A_iD_i$$
$$= j(\mathbf{D}) - \operatorname{tr}(\mathbf{Q}^T\mathbf{A}\mathbf{Q}\mathbf{D}^d) = j(\mathbf{D}) - \mathbf{A}\cdot\mathbf{D}.$$

This is precisely (6.4), and our proof is complete. □

With Lemma 6.1, Problem (6.1) reduces to the solution in parallel of N_h minimization problems on $\mathbb{R}^3$ (or $\mathbb{R}^2$).

We have the following theorem.

THEOREM 6.2. *For isotropic materials, employing $M=2$ in plane stresses and $M=3$ otherwise, the solution of* (6.1) *reduces to the following sequence of computations.*

For $\ell = 1, \cdots, N_h$,

(6.7) *compute* $\mathbf{A}_\ell = (\boldsymbol{\sigma}_h^n + r\mathbf{A}\mathbf{D}(\dot{\mathbf{u}}_h^n))|_{\Omega_\ell}$,

(6.8) *diagonalize* $\mathbf{A}_\ell$ *into* $\mathbf{A}_\ell = \mathbf{Q}^T \begin{pmatrix} A_1 & \cdot & 0 \\ \cdot & \ddots & \cdot \\ 0 & \cdot & A_M \end{pmatrix} \mathbf{Q}$,

(6.9) *solve* $J_\ell(D_i) - \sum_{i=1}^{M} A_iD_i \le J_\ell(H_i) - \sum_{i=1}^{M} A_iH_i \quad \forall (H_i)\in\mathbb{R}^M,$

$(D_i)\in\mathbb{R}^M,$

(6.10) *set* $\mathbf{D}_{1|\Omega_\ell} = \mathbf{Q} \begin{pmatrix} D_1 & \cdot & 0 \\ \cdot & \ddots & \cdot \\ 0 & \cdot & D_M \end{pmatrix} \mathbf{Q}^T.$

Above, the function J_q is defined by

$$J_\ell(H_i) = \frac{\nu r E}{2(1-2\nu)(1+\nu)}\left(\sum_{i=1}^{M} H_i\right)^2 + \frac{rE}{2(1+\nu)}\sum_{i=1}^{M} H_i^2 + \sup_{(\tau_i)\in\mathbb{R}^M}\left\{\left(\sum_{i=1}^{M}\tau_i H_i\right) - \frac{\lambda}{q}|j_c(\tau_i)|^q\right\}. \tag{6.11}$$

Proof. By introducing the convex function W that is defined on Σ_h by

$$W(\mathbf{H}) = \int_\Omega \left\{\frac{1}{2} r\mathbf{A}\mathbf{H}\cdot\mathbf{H} - \boldsymbol{\sigma}_h^n\cdot\mathbf{H}\right\} d\mathbf{x} \tag{6.12}$$

and whose gradient is $r\mathbf{A}\mathbf{H} - \boldsymbol{\sigma}_h^n$, (6.1) can be expressed as

$$\mathbf{0} \in \partial F(\mathbf{D}_1) + \partial W(\mathbf{D}_1 - \mathbf{D}(\dot{\mathbf{u}}_h^n)). \tag{6.13}$$

Because W is continuous on Σ_h, we have

$$\partial F(\mathbf{D}_1) + \partial W(\mathbf{D}_1 - \mathbf{D}(\dot{\mathbf{u}}_h^n)) = \partial[F(\mathbf{D}_1) + W(\mathbf{D}_1 - \mathbf{D}(\dot{\mathbf{u}}_h^n))].$$

Thus, by definition of the subgradient, (6.13) can be reduced to

$$F(\mathbf{D}_1) + W(\mathbf{D}_1 - \mathbf{D}(\dot{\mathbf{u}}_h^n)) \le F(\mathbf{H}) + W(\mathbf{H} - \mathbf{D}(\dot{\mathbf{u}}_h^n)) \quad \forall \mathbf{H}\in\Sigma_h,\ \mathbf{D}_1\in\Sigma_h. \tag{6.14}$$

However, for any matrix field $\mathbf{H}$ of Σ_h, the values of $\mathbf{H}$, $F(\mathbf{H})$, and $W(\mathbf{H})$ are constant on each finite element Ω_ℓ and are independent of their values on the other elements. Because the minimum value of the sum of independent terms is equal to the sum of the minimum value of each term, (6.14) can then be expressed as

$$\begin{cases} F|_{\Omega_\ell}(\mathbf{D}_1|_{\Omega_\ell}) + W|_{\Omega_\ell}(\mathbf{D}_1|_{\Omega_\ell} - \mathbf{D}(\dot{\mathbf{u}}_h^n)) \le F|_{\Omega_\ell}(\mathbf{H}) + W|_{\Omega_\ell}(\mathbf{H} - \mathbf{D}(\dot{\mathbf{u}}_h^n)) \\ \qquad \forall \mathbf{H}\in S^9, \quad \mathbf{D}_1|_{\Omega_\ell}\in S^9, \quad \text{and} \quad \ell = 1, N_h. \end{cases} \tag{6.15}$$

By definition of F and W, where S^9 is identified by $\mathbb{R}^{M\times M}_{\text{sym}}$, $\mathbf{A}_\ell$ is defined by (6.7), and $\mathbf{D}_1^\ell$ is the restriction of $\mathbf{D}_1$ to Ω_ℓ, (6.15) becomes

$$\begin{cases} \sup_{\boldsymbol{\tau}\in\mathbb{R}^{M\times M}_{\text{sym}}}\left\{\boldsymbol{\tau}\cdot\mathbf{D}_1^\ell - \frac{\lambda}{q}|j_c(\boldsymbol{\tau})|^q\right\} + \frac{r}{2}\mathbf{A}\mathbf{D}_1^\ell\cdot\mathbf{D}_1^\ell - \mathbf{A}_\ell\cdot\mathbf{D}_1^\ell \\ \le \sup_{\boldsymbol{\tau}\in\mathbb{R}^{M\times M}_{\text{sym}}}\left\{\boldsymbol{\tau}\cdot\mathbf{H} - \frac{\lambda}{q}|j_c(\boldsymbol{\tau})|^q\right\} + \frac{r}{2}\mathbf{A}\mathbf{H}\cdot\mathbf{H} - \mathbf{A}_\ell\cdot\mathbf{H} \\ \qquad \forall \mathbf{H}\in\mathbb{R}^{M\times M}_{\text{sym}}, \quad \mathbf{D}_1^\ell\in\mathbb{R}^{M\times M}_{\text{sym}}, \quad \text{and} \quad \ell = 1, N_h. \end{cases} \tag{6.16}$$

Now, for isotropic materials, j_c is an isotropic function of $\boldsymbol{\tau}$, and

$$\frac{1}{2}\mathbf{A}\mathbf{H}\cdot\mathbf{H} = \frac{E\nu}{2(1-2\nu)(1+\nu)}(\operatorname{tr}(\mathbf{H}))^2 + \frac{E}{2(1+\nu)}\operatorname{tr}\mathbf{H}^2,$$

where E and ν are the Young modulus and the Poisson coefficient, respectively. Using this definition of $\mathbf{A}$, (6.3), and (6.11), (6.1) is finally equivalent to

$$J_\ell(\mathbf{D}_1^\ell) - \mathbf{A}_\ell\cdot\mathbf{D}_1^\ell = \inf_{\mathbf{H}\in\mathbb{R}^{M\times M}_{\text{sym}}}\{J_\ell(\mathbf{H}) - \mathbf{A}_\ell\cdot\mathbf{H}\} \quad \forall \ell = 1, N_h. \tag{6.17}$$

Applying Lemma 6.1 to (6.17) then directly yields the desired result. □

Remark 6.1. For j_c given by (2.6), we have from Lemma 6.1 and Remark 3.1

$$\sup_{(\tau_i)\in\mathbb{R}^M}\left\{\sum_{i=1}^{M}\tau_i H_i-\frac{\lambda}{q}|j_c(\tau_i)|^q\right\}=\sup_{\boldsymbol{\tau}\in S^9}\left\{\boldsymbol{\tau}\cdot\mathbf{H}^d-\frac{\lambda}{q}|j_c(\boldsymbol{\tau})|^q\right\}$$
$$(6.18)\qquad =\frac{\lambda^{1-s}}{s}|\sup_{\boldsymbol{\tau}\in\mathbf{C}(\mathbf{x})\cap S^9}\{\boldsymbol{\tau}\cdot\mathbf{H}^d\}|^s,$$

where $s=q/(q-1)$, $\mathbf{H}^d$ is the diagonal matrix with diagonal terms (H_i), and $\mathbf{C}(\mathbf{x})$ is the closed convex set of locally admissible elastic stresses.

6.2. Maxwell–Norton materials. Obviously, the complexity of (6.9) strongly depends on the choice of j_c, that is, on the material considered. Below, we detail the solution of (6.9) for Maxwell–Norton, Camclay, and Tresca materials.

First, for Maxwell–Norton materials in plane strains or in three dimensions, where $M=3$, we have

$$\mathbf{C}(\mathbf{x})=\{\boldsymbol{\tau}\in\mathbb{R}^{3\times 3}_{\text{sym}}, |\boldsymbol{\tau}_D|=|\boldsymbol{\tau}-\text{tr}\,(\boldsymbol{\tau})\mathbf{Id}/3|\le k\sqrt{2}\}.$$

From Remark 6.1, we then have

$$\begin{cases} J_\ell(H_i)=\dfrac{rE}{2(1+\nu)}|\mathbf{H}^d|^2+\dfrac{\lambda^{1-s}}{s}(k\sqrt{2}|\mathbf{H}^d|)^s & \text{if tr}\,(\mathbf{H}^d)=0,\\ \qquad =+\infty \quad \text{if not.} \end{cases}$$

The extremality conditions associated with the minimization problem (6.9) are therefore

$$\left(\frac{rE}{(1+\nu)}+\lambda^{1-s}(k\sqrt{2})^s|\mathbf{D}^d|^{s-1}\right)D_i-A_i+p=0,\qquad D_1+D_2+D_3=0,$$

and have for their solution

$$\begin{cases} \mathbf{D}^d=|\mathbf{D}^d|\mathbf{A}^D/|\mathbf{A}^D|,\\ (\mathbf{A}^D)_{ij}=(A_i-(A_1+A_2+A_3)/3)\delta_{ij},\\ |\mathbf{D}^d|=\text{argmin}\left\{\dfrac{rE}{2(1+\nu)}x^2+\dfrac{\lambda^{1-s}}{s}(xk\sqrt{2})^s-|\mathbf{A}^D|x\right\}. \end{cases}$$

In this case, (6.8)-(6.10) finally reduce to

$$(6.19)\qquad \begin{cases} \mathbf{A}^D_\ell=\mathbf{A}_\ell-\text{tr}\,(\mathbf{A}_\ell)\mathbf{Id}/3,\\ |\mathbf{D}^\ell_d|=\text{argmin}\left\{\dfrac{rE}{2(1+\nu)}x^2+\dfrac{\lambda^{1-s}}{s}(xk\sqrt{2})^s-|\mathbf{A}^D_\ell|x\right\},\\ \mathbf{D}_1|_{\Omega_\ell}=|\mathbf{D}^\ell_d|\mathbf{A}^D_\ell/|\mathbf{A}^D_\ell|. \end{cases}$$

6.3. Camclay materials. Camclay-type materials are plastically compressible materials that behave differently in compression than they do in traction.

Concrete, of course, is a very good example of such a material, and many other examples have been studied in soil mechanics. Actually, the name Camclay denotes a specific clay extensively studied by the department of soil mechanics of Cambridge University.

For those materials in plane strains or in three dimensions, $M=3$, and the convex $\mathbf{C}$ is the ellipsoid defined by

$$\mathbf{C}(\mathbf{x})=\left\{\boldsymbol{\tau}\in\mathbb{R}^{3\times3}_{\text{sym}}, \left(\frac{1}{3}\operatorname{tr}\boldsymbol{\tau}+\alpha p_0\right)^2 + \frac{1}{2\beta^2}((\tau_1-\tau_2)^2+(\tau_2-\tau_3)^2+(\tau_3-\tau_1)^2)\le p_0^2\right\}, \tag{6.20}$$

with $|\alpha|<1$ and $\beta>0$. From Remark 6.1, we then have

$$\begin{cases} J_\ell(H_i)=\dfrac{\nu rE}{2(1+\nu)(1-2\nu)}(H_1+H_2+H_3)^2+\dfrac{rE}{2(1+\nu)}(H_1^2+H_2^2+H_3^2) \\ \quad +\dfrac{\lambda^{1-s}}{s}(p_0)^s\Big[-\alpha(H_1+H_2+H_3)+\Big((H_1+H_2+H_3)^2 \\ \quad +\dfrac{2\beta^2}{9}((H_1-H_2)^2+(H_2-H_3)^2+(H_3-H_1)^2)\Big)^{1/2}\Big]^s. \end{cases} \tag{6.21}$$

Here, (6.9) can be solved by a standard Newton method operating on $\mathbb{R}^3$.

Remark 6.2. Maxwell–Norton materials in plane stresses lead also to a function J_ℓ given by (6.21) with $\alpha=H_3=0$, $p_0=2k\sqrt{2}/\sqrt{3}$, and $\beta=3/2\sqrt{2}$. Thus they can be considered as a particular case of Camclay materials at least from the numerical point of view.

6.4. Tresca materials in plane stresses. In this case, we have $M=2$, and the set $\mathbf{C}(\mathbf{x})$ is defined by

$$\mathbf{C}(\mathbf{x})=\left\{\boldsymbol{\tau}\in\mathbb{R}^{3\times3}_{\text{sym}}, \max_{i\neq j}|\tau_i-\tau_j|<\sigma_0\right\}.$$

For simplicity, we will assume that $\nu=0$ and, thus, from Remark 6.1, the function J_ℓ is given by

$$J_\ell(H_1,H_2)=\frac{rE}{2}(H_1^2+H_2^2)+\frac{\lambda^{1-s}}{s}\sigma_0^s[\max\{|H_1|,|H_2|,|H_1+H_2|\}]^s. \tag{6.22}$$

This function is strictly convex but not differentiable, a fact that is clearly evident from Fig. 6.1, where the isocontours of J_ℓ are drawn.

To solve (6.9), we first observe that its solution belongs to the half-plane $H_1\ge H_2$, which we partition into seven regions K_i (Fig. 6.2) corresponding to regions where J_ℓ is differentiable and to their boundaries. The solution of (6.9) can then be obtained via the following algorithm.

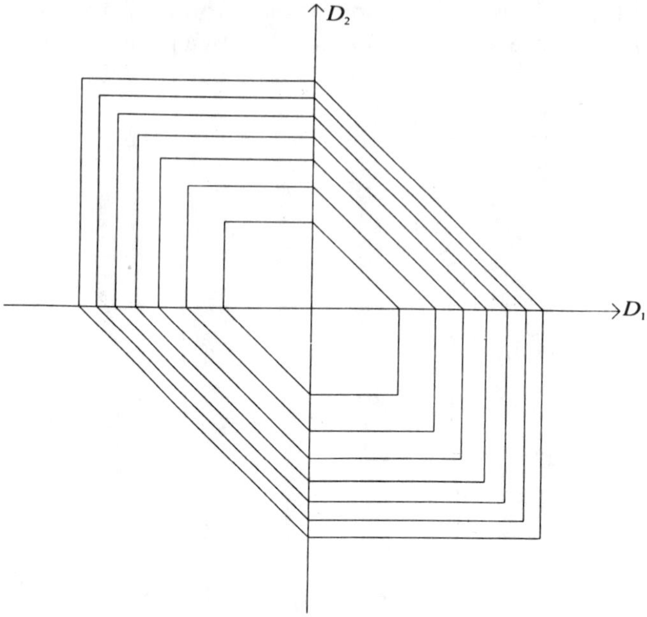

FIG. 6.1. *Isocontours of* J_ℓ $(s=2, E=0)$.

ALGORITHM.

For $i=1$ to 7

test if there exists $\{D_1, D_2\}$ in K_i with $(A_1, A_2)\in\partial J_\ell(D_1, D_2)$;

if yes, solve $(A_1, A_2)\in\partial J_\ell(D_1, D_2)$ in K_i and stop;

if not, continue.

The subgradient of J_ℓ on K_i is very easy to construct. It either contains only the gradient of J_ℓ if J_ℓ is differentiable on K_i, or it contains all of the values between the gradient of J_ℓ on K_{i-1} and the gradient of J_ℓ on K_{i+1} if K_i separates two regions where J_ℓ is differentiable. Having completed all calculations, the solution (D_1, D_2) of (6.9) is finally given by

$$\text{for } A_2>[(A_1-A_2)/rE]^{s-1}\qquad \begin{cases} D_1=\frac{1}{2}[z+(A_1-A_2)/rE],\\ D_2=\frac{1}{2}[z-(A_1-A_2)/rE],\\ z=\operatorname*{argmin}_{t\in\mathbb{R}^+}\left\{\dfrac{2\lambda^{1-s}}{s}\sigma_0^s t^s+\dfrac{rE}{2}t^2-(A_1+A_2)t\right\}; \end{cases}$$

$$\text{for } [(A_1-A_2)/rE]^{s-1}\geq A_2\geq 0\qquad \begin{cases} D_1=\operatorname{argmin}\left\{\dfrac{\lambda^{1-s}}{s}\sigma_0^s t^s+\dfrac{rE}{2}t^2-A_1 t\right\},\\ D_2=0; \end{cases}$$

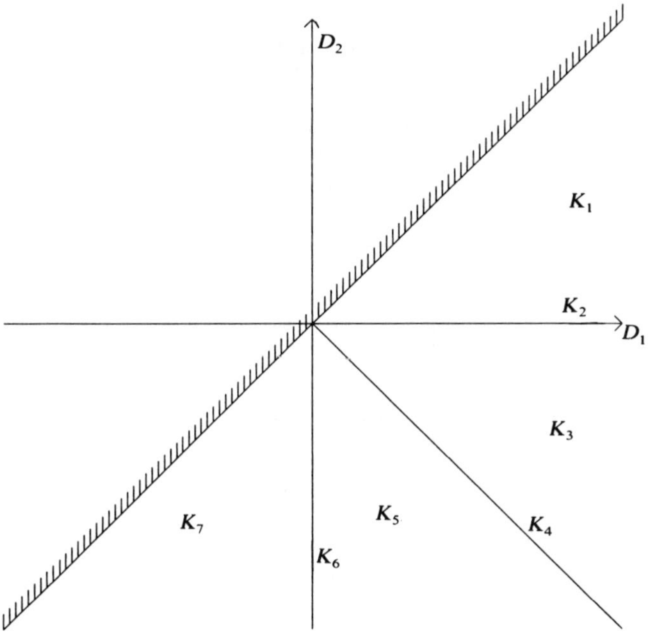

FIG. 6.2. *The half-plane $H_1 > H_2$ of J_ℓ partitioned into regions.*

for $A_1 > [-A_2 + (-A_2/rE)^{s-1}] > 0$
$$\begin{cases} D_1 = \operatorname{argmin}\left\{\dfrac{\lambda^{1-s}}{s}\sigma_0^s t^s + \dfrac{rE}{2}t^2 - A_1 t\right\}, \\ D_2 = A_2/rE; \end{cases}$$

for $-A_2 + (-A_2/rE)^{s-1} \geq A_1 > 0$
and $(A_1/rE)^{s-1} + A_1 \geq -A_2$
$$\begin{cases} D_1 = \operatorname{argmin}\left\{\dfrac{\lambda^{1-s}}{s}\sigma_0^s t^s + rEt^2 - (A_1 - A_2)t\right\}, \\ D_2 = -D_1; \end{cases}$$

for $-A_2 > A_1 + (A_1/rE)^{s-1} > 0$
$$\begin{cases} D_1 = A_1/rE, \\ -D_2 = \operatorname{argmin}\left\{\dfrac{\lambda^{1-s}}{s}\sigma_0^s t^s + \dfrac{rE}{2}t^2 + A_2 t\right\}; \end{cases}$$

for $0 \geq A_1 \geq -[(A_1 - A_2)/rE]^{s-1}$
$$\begin{cases} D_1 = 0, \\ -D_2 = \operatorname{argmin}\left\{\dfrac{\lambda^{1-s}}{s}\sigma_0^s t^s + \dfrac{rE}{2}t^2 + A_2 t\right\}; \end{cases}$$

for $-[(A_1 - A_2)/rE]^{s-1} > A_1$
$$\begin{cases} D_1 = \frac{1}{2}[z + (A_1 - A_2)/rE], \\ D_2 = \frac{1}{2}[z - (A_1 - A_2)/rE], \\ -z = \operatorname{argmin}\left\{\dfrac{2\lambda^{1-s}}{s}\sigma_0^s t^s + \dfrac{rE}{2}t^2 + (A_1 + A_2)t\right\} \end{cases}$$

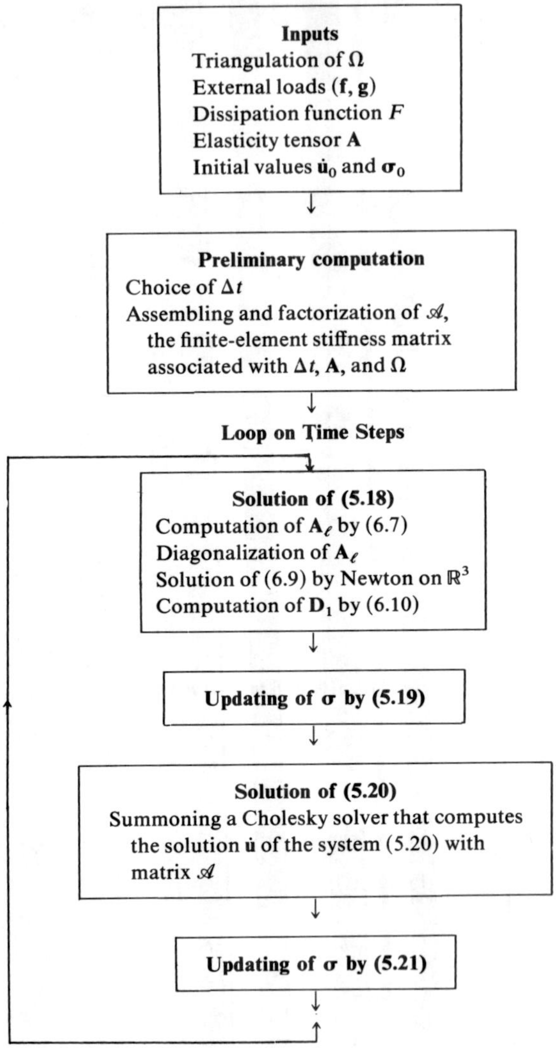

FIG. 7.1. *Computer flow chart for the Peaceman–Rachford algorithm* (5.17)–(5.21).

7. Numerical results. The different algorithms presented in § 5 are easy to implement on computer, as indicated by the computer flow chart for the Peaceman–Rachford algorithm (5.17)–(5.21) for elastoviscoplasticity presented in Fig. 7.1. In what follows, we present three examples of numerical applications.

7.1. Example. The first example corresponds to a problem with a known analytical stationary solution. The domain Ω is described in Fig. 7.2, together

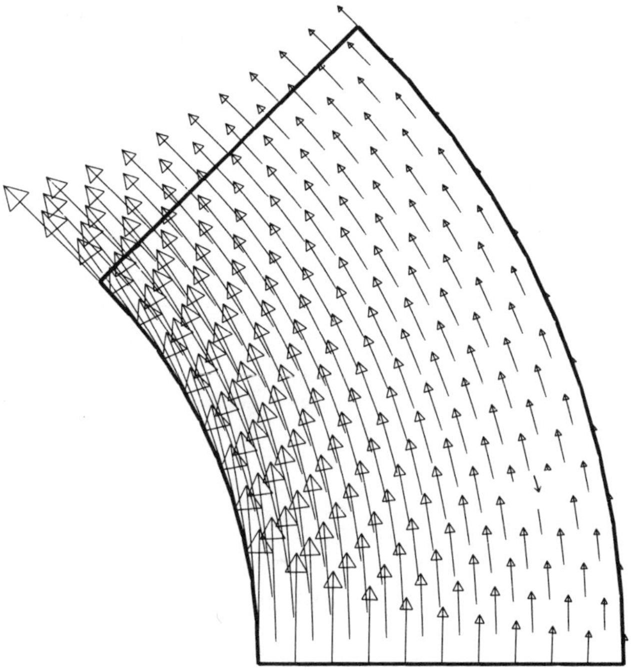

FIG. 7.2. *Analytical solution of Example* 7.1.

with the computed stationary velocity field; it is filled by a Maxwell–Norton material (see § 6.2) with $k\sqrt{2}=1$ MPa, $\lambda=1$ MPa/sec, $q=3$, $E=1.5\ 10^5$ MPa, and $\nu \simeq 0.5$. This material flows in plane strains, and its velocity has an imposed value $\dot{\mathbf{u}}_1=(\rho\sqrt{2})^{-3}\mathbf{e}_y$ on the line $\phi=0$ ($\{\rho, \phi\}$ being the polar coordinates of $\mathbf{x}$). It is subjected to surface tractions $\mathbf{g}=-(2\rho^4)^{-1/2}(\mathbf{n}\cdot\mathbf{e}_\phi\mathbf{e}_r+\mathbf{n}\cdot\mathbf{e}_r\mathbf{e}_\phi)$ on the remaining part Γ_2 of the boundary. Using the Peaceman–Rachford scheme, after 30 times steps, starting from the elastic solution with $\Delta t=0.75\ 10^{-5}$ sec, the relative L^2 error between the computed velocity field and the stationary solution $\dot{\mathbf{u}}=(\rho\sqrt{2})^{-3}\mathbf{e}_\phi$ was equal to 0.008, which is small for a nonlinear problem with this many boundary conditions of the Neumann type. At this time step, the computed solution was almost stationary, since we have

$$\|\mathbf{A}^{-1}(\boldsymbol{\sigma}^{n+1}-\boldsymbol{\sigma}^n)/\Delta t\|/\|\mathbf{D}(\dot{\mathbf{u}}^n)\|=0.8\ 10^{-5}.$$

7.2. Example. The second numerical example considers a perforated thick square plate with a width of 0.20 m that is subjected for positive times to a uniform traction of 0.52 MPa applied on two of its opposite faces. This plate is made of a Maxwell-Norton material with $q=3$, $\lambda=1$ MPa/sec, $k\sqrt{2}=1$ MPa, $E=2\ 10^5$ MPa, and $\nu=0.3$. For this case, we used the Douglas–Rachford scheme (5.22)–(5.25) with $\Delta t=0.5\ 10^{-5}$ sec. For symmetry reasons, we restricted ourselves to one fourth of the plate, as indicated in Fig. 7.3; the initial

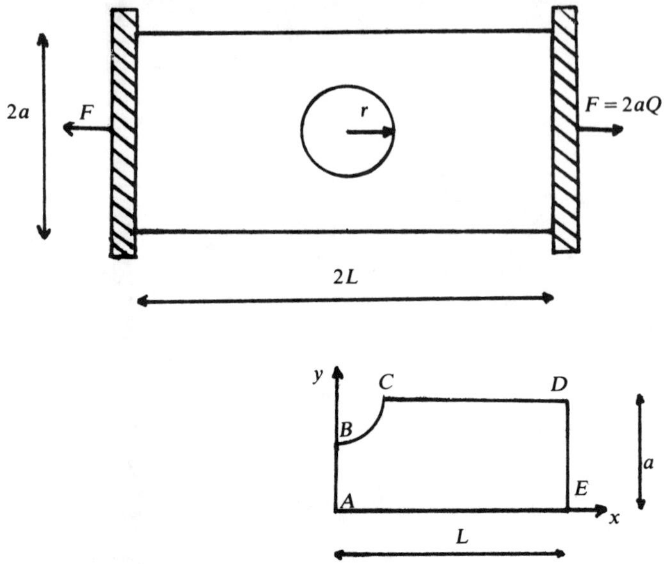

FIG. 7.3. *The perforated square plate problem.*

triangulation and that after 0.2 sec of flow with the computed stationary velocity field are represented on Fig. 7.4. On this problem, we also compared the speed of convergence of the computed velocity field toward the stationary solution for different values of Δt and for the different algorithms presented in § 5. The results are summarized by Fig. 7.5, which plots $\int_\Omega \dot{\sigma}_{11}/E\,d\mathbf{x}$ as a function of time step for the different algorithms, and by Table 7.1, which gives the value

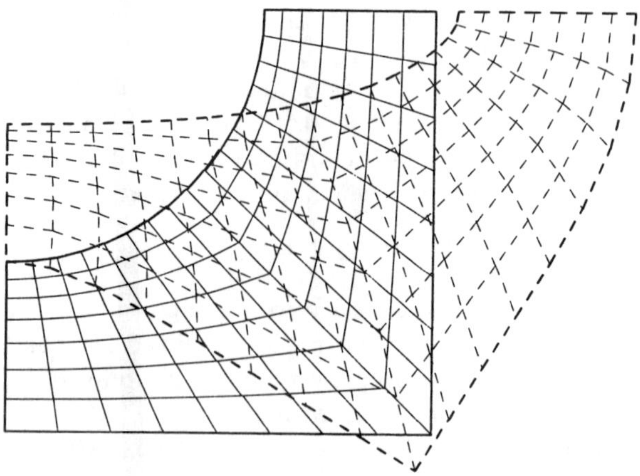

FIG. 7.4. *The solution for the perforated square plate problem* (*Example* 7.2).

TABLE 7.1
The value of $\|A^{-1}\dot{\sigma}\|/\|D(\dot{u})\|$ *for* $n=30$ (*Douglas-Rachford scheme* (5.22)-(5.25) *and Peaceman-Rachford scheme* (5.17)-(5.21)) *and* $n=15$ (θ-*scheme* (5.29)-(5.36)).

$2\times10^5\times\Delta t$	ν	Douglas-Rachford scheme	Peaceman-Rachford scheme $\rho=R/2$	Peaceman-Rachford scheme $\rho=R$	θ-scheme $\theta=0.01$	θ-scheme $\theta=0.1$	θ-scheme $\theta=0.33$
0.1	0.3	0.27×10^{-1}	0.7×10^{-1}	0.27×10^{-1}	0.8×10^{-1}	0.8×10^{-1}	0.7×10^{-1}
0.1	0.4999	0.15×10^{-1}	0.53×10^{-1}	0.15×10^{-1}	0.5×10^{-1}	0.5×10^{-1}	0.45×10^{-1}
1.0	0.3	0.6×10^{-4}	0.55×10^{-3}	0.14×10^{-3}	0.6×10^{-3}	0.6×10^{-3}	0.5×10^{-3}
1.0	0.4999	0.7×10^{-5}	0.2×10^{-3}	0.1×10^{-3}	0.27×10^{-3}	0.12×10^{-3}	0.6×10^{-4}
2.0	0.3	0.29×10^{-4}	0.6×10^{-4}	0.1×10^{-4}	0.7×10^{-4}	0.7×10^{-4}	0.55×10^{-4}
2.0	0.4999	0.6×10^{-4}	0.7×10^{-5}	0.5×10^{-4}	0.2×10^{-4}	0.7×10^{-5}	0.2×10^{-5}
4.0	0.3	0.12×10^{-3}	0.8×10^{-5}	0.5×10^{-2}	explos.	explos.	explos.
4.0	0.4999	0.14×10^{-2}	0.2×10^{-5}	2.0	explos.		
10.0	0.3	0.13×10^{-3}	0.4×10^{-3}	0.4			
10.0	0.4999	0.61×10^{-2}	0.25×10^{-3}	50.0			

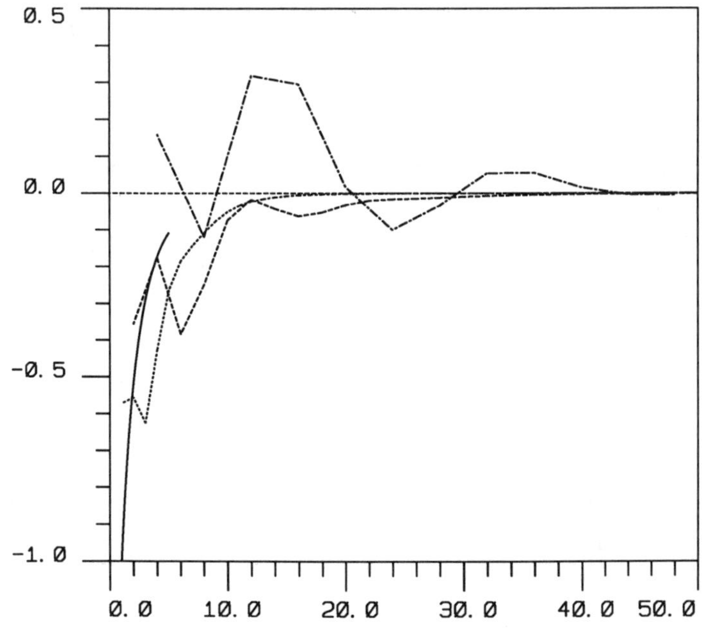

FIG. 7.5a. ALG2.

FIG. 7.5. *Graphs showing* $\int_\Omega \dot{\sigma}_{11}/E\,d\mathbf{x}$ *as a function of the time step* ($\Delta t = 0.1, 1.0, 2.0, 4.0, 10.0$) *for the solution of Example* 7.2 *using the Douglas-Rachford algorithm* (5.22)-(5.25), *Peaceman-Rachford algorithm* (5.17)-(5.21), *and* θ-*algorithm* (5.29)-(5.36). a. *Douglas-Rachford*, $\nu=0.3$. b. *Douglas-Rachford*, $\nu=0.4999$. c. *Peaceman-Rachford*, $\nu=0.3$, $\rho=R$. d. *Peaceman-Rachford*, $\nu=0.4999$, $\rho=R$. e. *Peaceman-Rachford*, $\nu=0.3$, $\rho=R/2$. f. *Peaceman-Rachford*, $\nu=0.4999$, $\rho=R/2$. g. θ, $\nu=0.3$, $\theta=0.01$. h. θ, $\nu=0.4999$, $\theta=0.01$. i. θ, $\nu=0.4999$, $\theta=0.33$. j. θ, $\nu=0.3$, $\theta=0.33$.

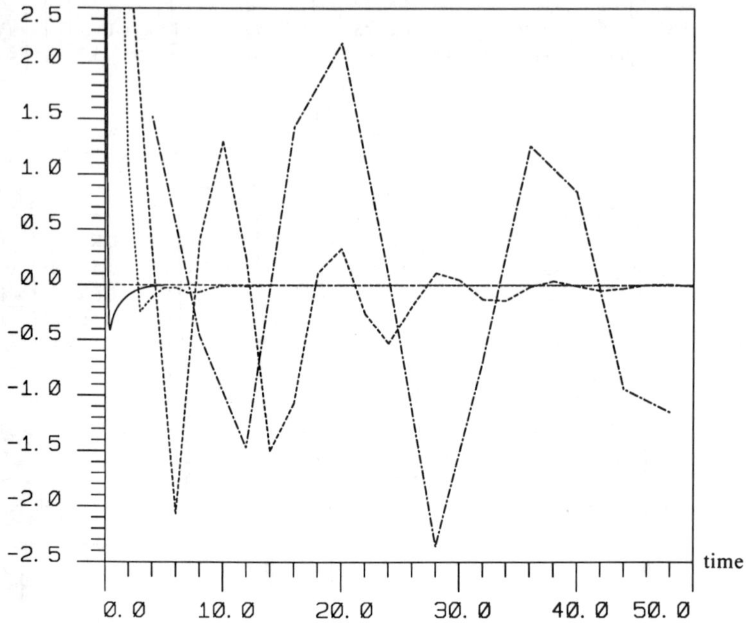

FIG. 7.5b. ALG2.

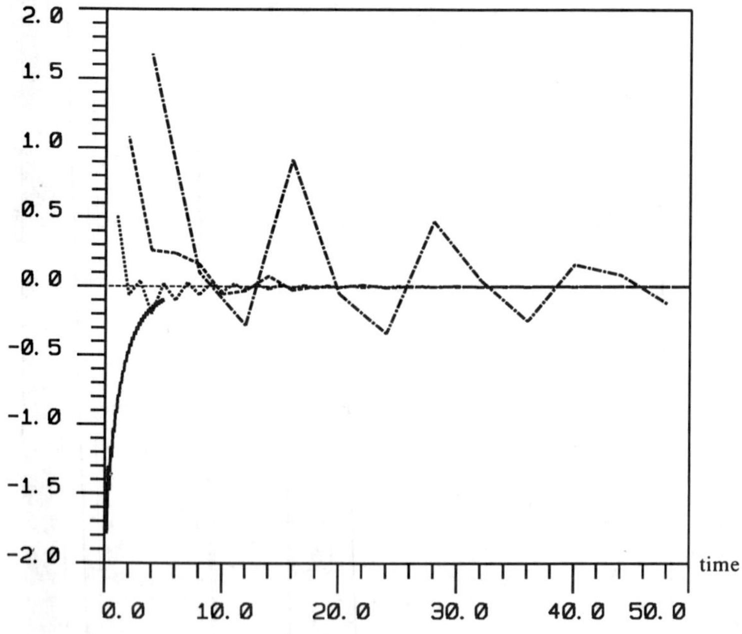

FIG. 7.5c. ALG3, $\rho = R$.

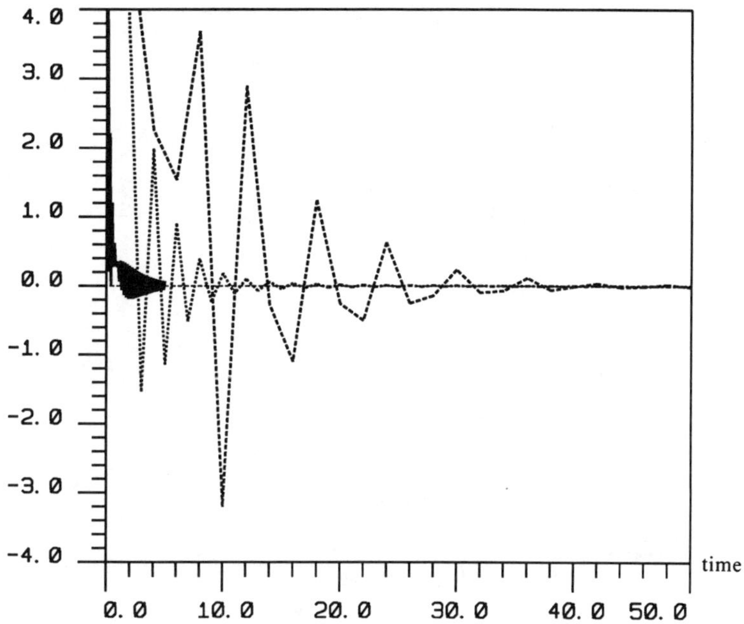

FIG. 7.5d. ALG3, $\rho = R$.

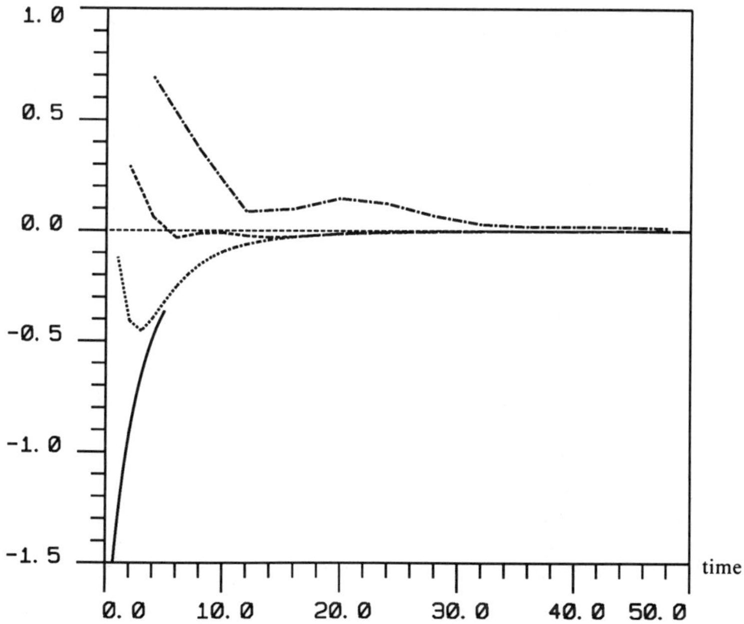

FIG. 7.5e. ALG3, $\rho = R/2$.

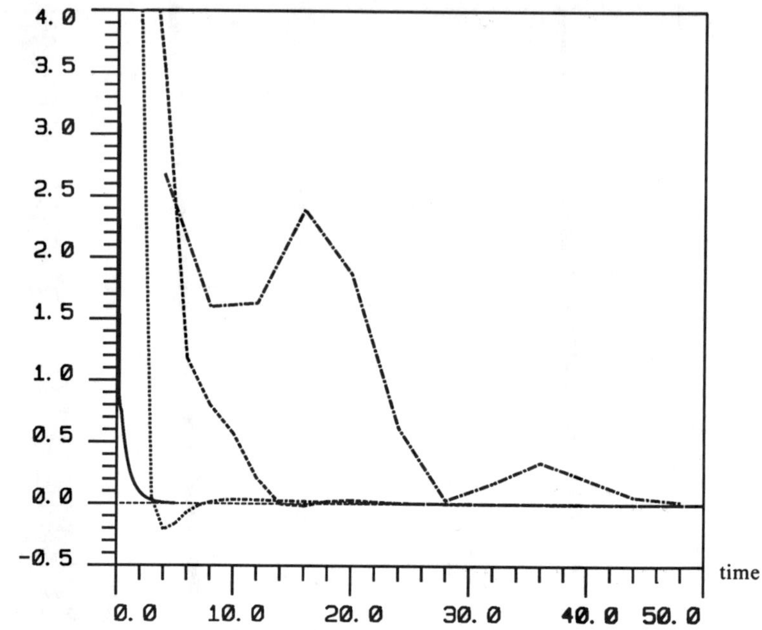

FIG. 7.5f. ALG3, $\rho = R/2$.

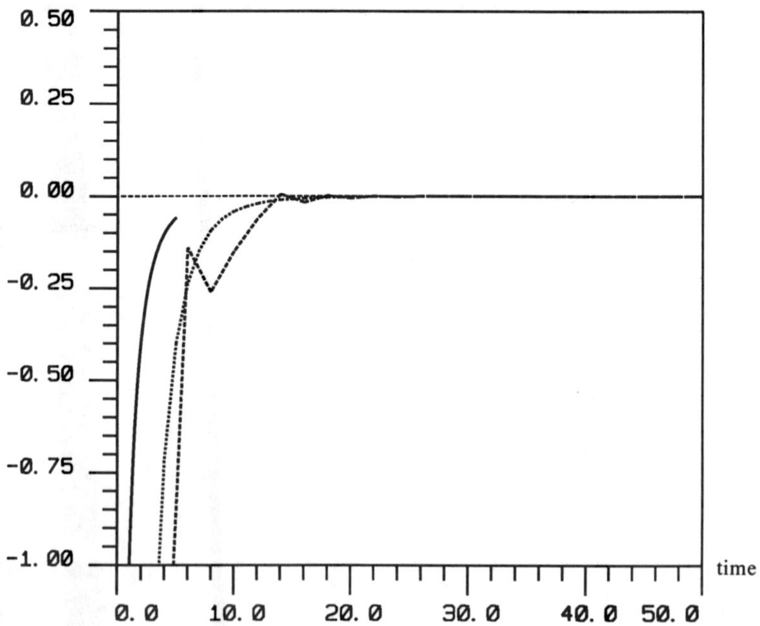

FIG. 7.5g. ALG4, $\theta = 0.01$.

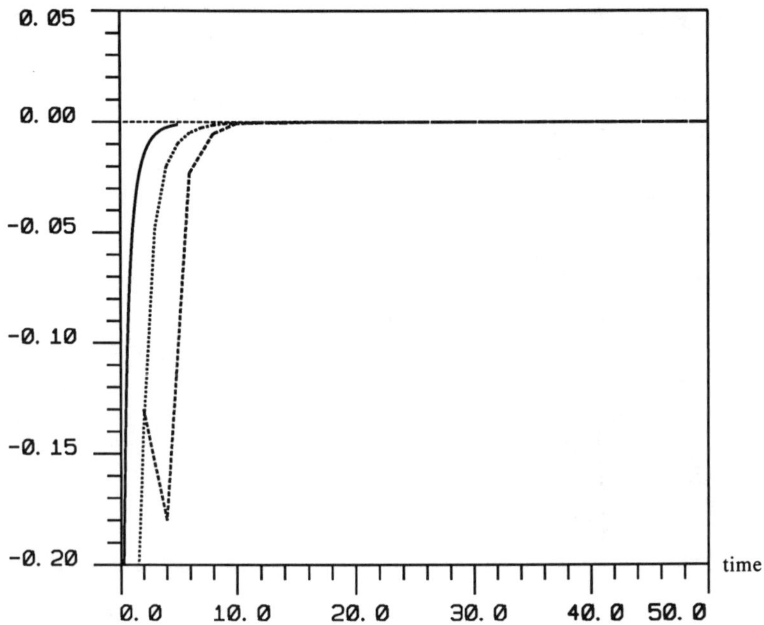

FIG. 7.5h. ALG4, $\theta = 0.01$.

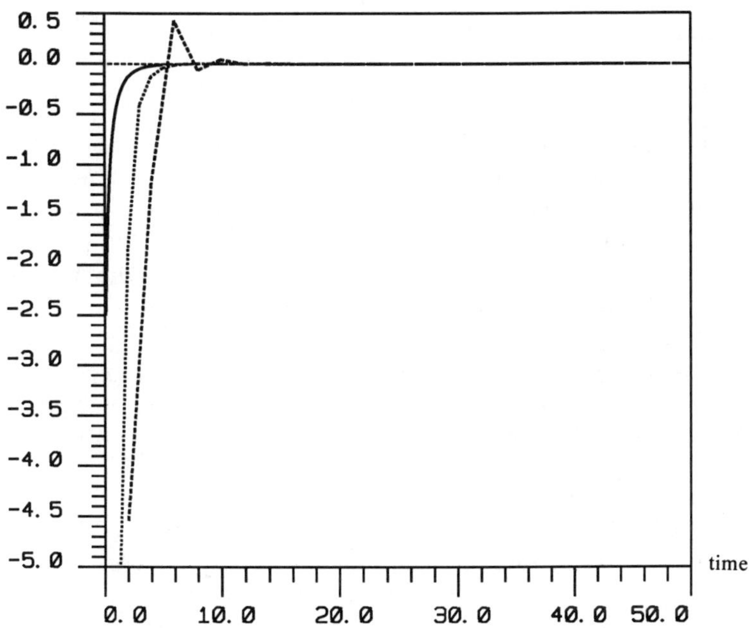

FIG. 7.5i. ALG4, $\theta = 0.33$.

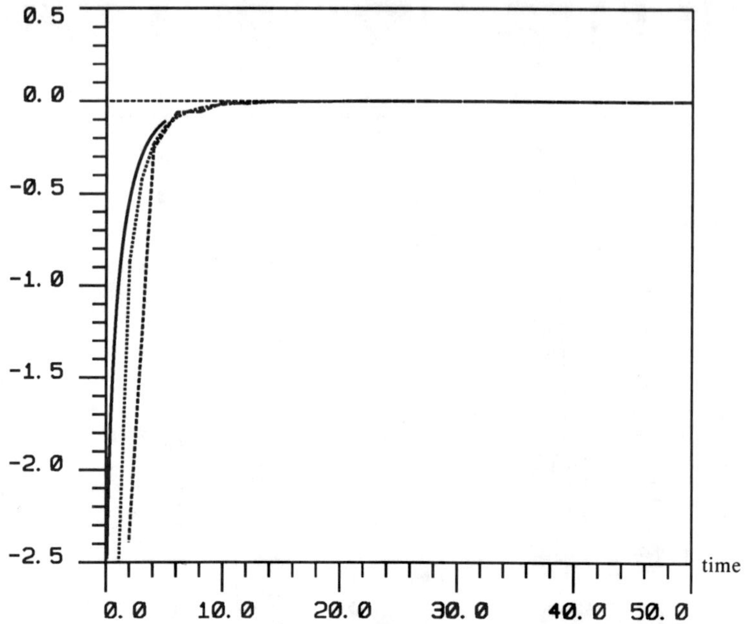

FIG. 7.5j. ALG4, $\theta = 0.33$.

of

$$\|\mathbf{A}^{-1}\dot{\boldsymbol{\sigma}}(n\Delta t)\| / \|\mathbf{D}(\dot{\mathbf{u}}(n\Delta t))\|$$

for $n = 30$ (Peaceman–Rachford or Douglas–Rachford) or for $n = 15$ (θ-scheme (5.29)–(5.36)). We recall that the same amount of computing time is required for 15 iterations of the θ-scheme as for 30 steps of the Peaceman or the Douglas–Rachford algorithm. The same stationary solution can also be obtained after one time step of the backward Euler algorithm (5.49)–(5.61), setting $\Delta t = 0.5 \times 10^{+4}$ sec and using 27 iterations of the conjugate-gradient algorithm (5.54)–(5.60).

In any case, observe the very fast convergence of the θ-scheme when the time step is properly chosen. Unfortunately, this scheme is also the first to diverge when the time step gets too large.

7.3. Example. The final numerical example deals with a nondifferentiable, ill-conditioned problem. It considers a cracked thin plate of Tresca material under plane stresses, with $s = 1.003$, $\sigma_0 = 1$, $\lambda = 1$, $E = 10^5$ MPa, and $\nu = 0$. As above, for symmetry reasons, we considered only one fourth of the plate (Fig. 7.6). The final numerical solution, obtained after 100 time steps of the Douglas–Rachford algorithm, with $\Delta t = 10^{-5}$ s, is represented by Fig. 7.7, where the triangulation after 1.25 sec of flow is indicated.

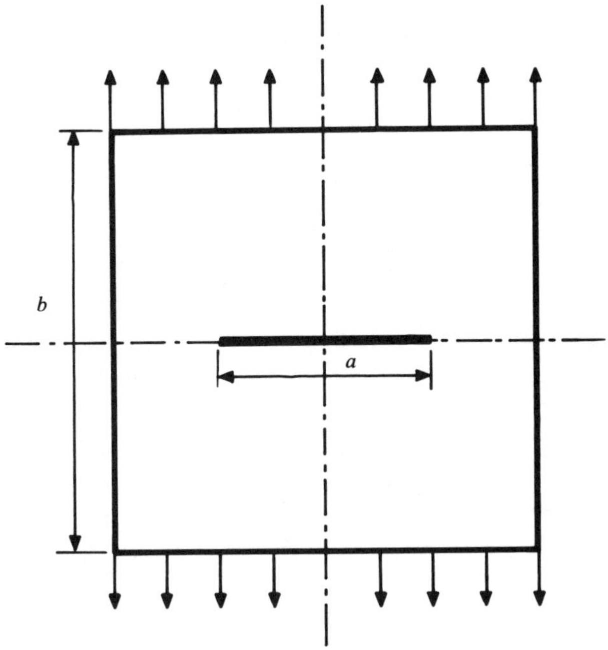

FIG. 7.6. *The cracked plate problem* (*Example* 7.3).

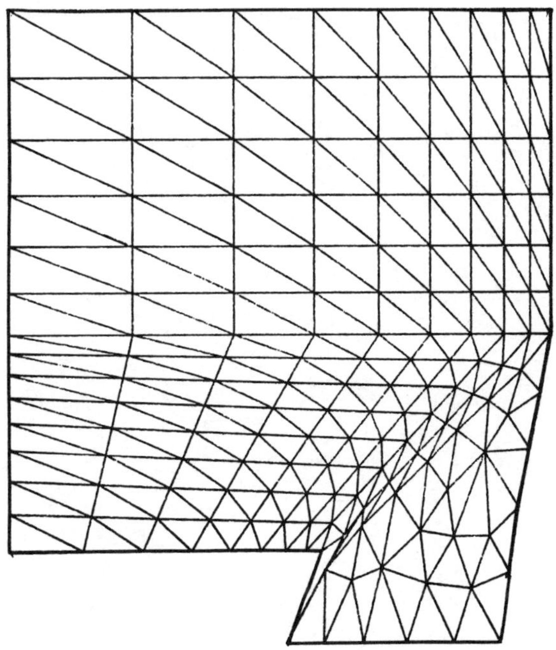

FIG. 7.7. *Solution of the cracked plate problem* (*Example* 7.3).

Chapter 5

Limit Load Analysis

1. Limit loads in plasticity.

1.1. Perfectly elastoplastic materials. Perfect elastoplasticity is a model often used in structural design (Zienkiewicz [1977]) in which the materials composing the structure under study are considered to be subjected to small strains only, and to behave like linearly elastic solids whenever the internal stresses are below a certain limit. If the stresses inside the body reach this limit, called the yield stress, the body begins to flow in an irreversible way. Perfect elastoplasticity also supposes that a characteristic of the considered material is that the internal stresses can never pass this limit.

Typically, the loading and unloading of a straight bar made of a perfectly elastoplastic material corresponds to the stress-strain curve of Fig. 1.1. In the first phase of the loading, stresses and strains increase simultaneously. Then the stresses reach a limit, and only strains continue to increase. If unloading occurs, stresses and strains decrease together. Once the bar is unloaded, the stresses vanish but not the strains. Remanent plastic strains can be observed.

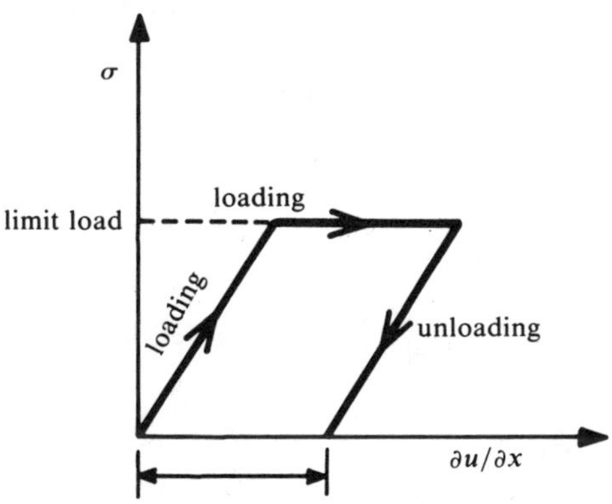

FIG. 1.1. *Loading and unloading of a straight bar.*

In practice, the model of perfect elastoplasticity appears to be a reasonable way to describe steel or concrete structures. From the mathematical point of view, the standard materials introduced in Chapter 1, § 4, provided the necessary tools to describe this model. In that framework, a perfect elastoplastic material is defined by

$$\text{state variables: } \begin{cases} \mathbf{E} & \text{(linearized strain tensor)}, \\ \mathbf{E}^{an} & \text{(anelastic part of } \mathbf{E}); \end{cases}$$

kinematic constraint: $\operatorname{tr} \mathbf{E}^{an} = 0$ (for plastically incompressible bodies only);

$$\text{free-energy potential: } \mathcal{W} = \frac{1}{2\rho}(\mathbf{E}-\mathbf{E}^{an}) \cdot \mathbf{A}(\mathbf{E}-\mathbf{E}^{an});$$

$$\text{internal dissipation potential: } \mathcal{D}_1(\dot{\mathbf{E}}^{an}) = \sup_{\boldsymbol{\sigma}\in\mathbf{C}} \{\boldsymbol{\sigma} \cdot \dot{\mathbf{E}}^{an}\}.$$

In this formulation, $\mathbf{C}$ denotes the set of stresses that can be undergone locally by the material. For the stresses in the interior of $\mathbf{C}$, the material behaves like an elastic solid; for those on the boundary of $\mathbf{C}$, yielding occurs. This set $\mathbf{C}$, which may depend on the material point $\mathbf{x}$, is usually supposed to be closed, to be convex, and to contain the null stress tensor. Basic examples of such a set are

$$\text{(1.1)} \quad \mathbf{C} = \{\boldsymbol{\sigma} \in \mathbb{R}^9_{\text{sym}}, |\boldsymbol{\sigma} - \tfrac{1}{3}(\operatorname{tr} \boldsymbol{\sigma})\mathbf{Id}| \le k\sqrt{2}\} \qquad \text{(Von Mises)},$$

$$\text{(1.2)} \quad \mathbf{C} = \{\boldsymbol{\sigma} \in \mathbb{R}^9_{\text{sym}}, |\sigma_i - \sigma_j| \le \sigma_0, 1 \le i, j \le N\} \qquad \text{(Tresca)},$$

$$\text{(1.3)} \quad \mathbf{C} = \{\boldsymbol{\sigma} \in \mathbb{R}^9_{\text{sym}}, \sigma_1(1+\sin\phi) - \sigma_3(1-\sin\phi) - 2c\cos\phi \le 0\} \quad \text{(Coulomb)},$$

where k, ϕ, σ_0, and c are material constants and where $(\sigma_i)_{i=1,3}$ denotes the eigenvalues of the stress tensor $\boldsymbol{\sigma}(\sigma_1 \le \sigma_2 \le \sigma_3)$. Moreover, for plastically incompressible materials, $\mathbf{C}$ is invariant by translation along the line of the diagonal matrices.

By writing the constitutive laws associated with such a constitutive model, we obtain, for the plastically incompressible case,

$$\begin{cases} \boldsymbol{\sigma} = \rho \dfrac{\partial \mathcal{W}}{\partial \mathbf{E}}(\mathbf{E}-\mathbf{E}^{an}) = \mathbf{A}(\mathbf{E}-\mathbf{E}^{an}) \\ \boldsymbol{\sigma}_D = (\boldsymbol{\sigma} - \tfrac{1}{3}(\operatorname{tr} \boldsymbol{\sigma})\mathbf{Id}) \in \partial\mathcal{D}_1(\dot{\mathbf{E}}^{an}). \end{cases}$$

The anelastic part $\mathbf{E}^{an}$ of the strain tensor can be eliminated by differentiating the first equation with respect to time and by inverting the second equation. Since, as stated earlier, $\mathbf{C}$ is invariant by translation along a diagonal matrix, the elimination of $\mathbf{E}^{\mathrm{an}}$ leads to the following constitutive law for perfectly elastoplastic materials, valid for both the compressible and the incompressible case

$$(1.4)\qquad \boxed{(\dot{\mathbf{E}}-\mathbf{A}^{-1}\dot{\boldsymbol{\sigma}})\in\partial I_c(\boldsymbol{\sigma}),}$$

where I_c is the indicator of $\mathbf{C}$, i.e., the function with value 0 if $\boldsymbol{\sigma}$ belongs to $\mathbf{C}$ and with value $+\infty$ if not. By definition of the subgradient, (1.4) can also be expressed as

$$(1.5)\qquad \boxed{(\dot{\mathbf{E}}-\mathbf{A}^{-1}\dot{\boldsymbol{\sigma}})\cdot(\boldsymbol{\tau}-\boldsymbol{\sigma})\leq 0 \qquad \forall\boldsymbol{\tau}\in\mathbf{C},\quad \boldsymbol{\sigma}\in\mathbf{C},}$$

which is the classical Prandtl–Reuss flow rule for perfectly plastic materials.

1.2. A basic problem in perfect elastoplasticity. An important issue in structural design is to determine whether a given structure can sustain a certain distribution of loads without damage. In that respect, it is less important to compute the final shape of the structure under the specified loading than it is to ensure that the structure can indeed reach an admissible state of equilibrium under that loading. In mathematical words, designers are more interested in the existence of a solution in small strains than its computation.

Three steps are required to solve this existence problem. In the first step, equations that can satisfy the stresses and the displacement field inside the structure during the loading must be derived. In the second step, discussed in the next section, we will derive a general existence theory for the solution of these equations. Finally, in the last step, we will check, in each particular case, the assumptions introduced in the general existence theory. This last operation is precisely the purpose of limit load analysis and is described in detail in the remaining parts of the chapter.

Let us first introduce the equations that define the problem. We consider the quasi-static evolution of a given structure that occupies a domain Ω of $\mathbb{R}^N$ ($N=2$ or 3). This structure is subjected to external body forces $\mathbf{f}$ exerted throughout the volume, and to surface tractions $\mathbf{g}$ applied on a part Γ_2 of the boundary Γ of Ω. In addition, given displacements $\mathbf{u}_1(t)$ are imposed on the complementary part Γ_1 of Γ_2 in Γ. We suppose that this structure is made of a (possibly nonhomogeneous) perfectly elastoplastic material characterized at each point $\mathbf{x}$ of Ω by its elasticity tensor $\mathbf{A}(\mathbf{x})$ and by the set $\mathbf{C}(\mathbf{x})$ of locally admissible stresses. In that context, the equations that satisfy the stresses $\boldsymbol{\sigma}$ and the displacement field $\mathbf{u}$ are the constitutive law (1.5), the boundary conditions on $\mathbf{u}$, and the law of force balance, that is,

$$(1.6)\qquad \begin{cases} (\mathbf{D}(\dot{\mathbf{u}})-\mathbf{A}^{-1}\dot{\boldsymbol{\sigma}})\cdot(\boldsymbol{\tau}-\boldsymbol{\sigma})\le 0 & \forall \boldsymbol{\tau}\in \mathbf{C}(\mathbf{x}) \quad \text{a.e. in } \Omega,\\ \boldsymbol{\sigma}(\mathbf{x})\in \mathbf{C}(\mathbf{x}) \quad \text{a.e. in } \Omega, \\ \mathbf{u}=\mathbf{u}_1 \quad \text{on } \Gamma_1, \\ \operatorname{div}\boldsymbol{\sigma}+\mathbf{f}=0 \quad \text{in } \Omega, \\ \boldsymbol{\sigma}\boldsymbol{\nu}=\mathbf{g} \quad \text{on } \Gamma_2. \end{cases}$$

In (1.6), as in Chapter 4, § 2.4, $N=3$ in the three-dimensional case and $N=2$ in plane strains or plane stresses situations, $\Omega\subset\mathbb{R}^N$ is the interior of the body ($N=3$) or its cross-section ($N=2$), the body being in its reference configuration, and

$$(\mathbf{D}(\mathbf{w}))_{ij}=\frac{1}{2}\left(\frac{\partial \mathbf{w}_j}{\partial x_i}+\frac{\partial \mathbf{w}_i}{\partial x_j}\right) \qquad \text{for } 1\le i, \quad j\le N,$$

$$=0 \qquad \text{for } i>N, \quad j>N,$$

$$\operatorname{div}\boldsymbol{\sigma}=\sum_{i,j=1}^{N}\frac{\partial \sigma_{ij}}{\partial x_j}\mathbf{e}_i.$$

Our basic problem can now be stated as follows.

Problem 1. In perfect elastoplasticity, does there exist a solution $\mathbf{u}(t)$, $\boldsymbol{\sigma}(t)$ to the following quasi-static evolution problem (Suquet [1982, p. 95])?

Find $\mathbf{u}$, $\boldsymbol{\sigma}$, $\mathbf{w}$ such that, almost everywhere in time,

$$(1.7)\qquad \boldsymbol{\sigma}(t)\in \mathbf{C}=\{\boldsymbol{\tau}:\Omega\to S^9,\ \boldsymbol{\tau}(\mathbf{x})\in\mathbf{C}(\mathbf{x}) \text{ a.e. in } \Omega\},$$

$$(1.8)\qquad \int_\Omega \{\mathbf{A}^{-1}\dot{\boldsymbol{\sigma}}\cdot(\boldsymbol{\tau}-\boldsymbol{\sigma})+\dot{\mathbf{u}}\cdot\operatorname{div}(\boldsymbol{\tau}-\boldsymbol{\sigma})\}\,d\mathbf{x} \ge \int_{\Gamma_1}(\boldsymbol{\tau}-\boldsymbol{\sigma})\boldsymbol{\nu}\cdot\dot{\mathbf{u}}_1\,da+\int_{\Gamma_2}(\boldsymbol{\tau}-\boldsymbol{\sigma})\boldsymbol{\nu}\cdot\mathbf{w}\,da \quad \forall\boldsymbol{\tau}\in\mathbf{C},$$

$$(1.9)\qquad \int_\Omega(\operatorname{div}\boldsymbol{\sigma}+\mathbf{f})\cdot\mathbf{v}\,d\mathbf{x}=0 \quad \forall \mathbf{v}\in V,$$

$$(1.10)\qquad \int_{\Gamma_2}(\boldsymbol{\sigma}\boldsymbol{\nu}-\mathbf{g})\cdot\mathbf{z}\,da=0 \quad \forall \mathbf{z}\in W_0(\Gamma_2),$$

$$(1.11)\qquad \boldsymbol{\sigma}(0)=\boldsymbol{\sigma}_0, \qquad \mathbf{u}(0)=\mathbf{u}_0,$$

$$(1.12)\qquad \boldsymbol{\sigma}\in W^{1,2}(0,T;\Sigma_{\text{div}}).$$

The above equations correspond to the weakest possible formulation of (1.6) with respect to the displacements, that is, the formulation that requires the least amount of regularity for the displacement field. The spaces V and W defining the required regularity on $\mathbf{u}$ and $\mathbf{w}$ will be specified later in the existence theory; our only supposition at present is that $\mathbf{H}_0^1(\Omega)$ and $W_0(\Gamma_2)$ are included in V and W, respectively, and that $\mathbf{f}$ belongs to $\mathbf{L}^N(\Omega)$. As for S^9, Σ_{div}, and $W_0(\Gamma_2)$, they are defined by

$$\begin{aligned} S^9 &= \mathbb{R}^9_{\text{sym}} \quad &&\text{(space of symmetric, second-order tensors on } \mathbb{R}^3 \text{ in the three-dimensional or plane strains case)}\\ &= \{\boldsymbol{\tau} \in \mathbb{R}^9_{\text{sym}}, \tau_{i3} = 0\ \forall i = 1, 3\} \quad &&\text{(plane stresses case);}\\ \Sigma_{\text{div}} &= \{\boldsymbol{\tau} : \Omega \to S^9, \boldsymbol{\tau} \in (L^2(\Omega))^9, \operatorname{div} \boldsymbol{\tau} \in \mathbf{L}^N(\Omega)\}; \\ W_0(\Gamma_2) &= \{\mathbf{w} : \Gamma_2 \to \mathbb{R}^N, \exists \mathbf{v} \in (H^1(\Omega))^N \text{ with } \mathbf{v} = \mathbf{0} \text{ on } \Gamma_1 \text{ and } \mathbf{v} = \mathbf{w} \text{ on } \Gamma_2\}. \end{aligned} \tag{1.13}$$

Remark 1.1. The above problem is very similar to the elastoviscoplastic problem studied in Chapter 4; it corresponds to the same constitutive law, now written with $q = +\infty$, and to the same equilibrium equations. However, here we are only interested in existence results and, since $q = +\infty$, the existence result that was valid in Chapter 4 (Chap. 2, Thm. 4.1) is no longer valid. Therefore, we must introduce a different variational formulation and new existence results for the study of Problem 1.

1.3. Existence results. The first existence result is mainly negative. Nevertheless, it does not take into account the constitutive equation (1.8) and, therefore, goes far beyond the framework of perfect elastoplasticity. Moreover, it introduces the basic notion of limit loads. Its statement is particularly simple. If there is no stress field that satisfies both the equilibrium equations (1.9)-(1.10) and the admissibility requirement (1.7), then the evolution problem has no solution. We have the following theorem.

THEOREM 1.1 (First theorem of limit load analysis). *If there is no stress field* $\boldsymbol{\sigma}$ *such that*

$$\begin{cases} \boldsymbol{\sigma}(t) \in \Sigma_{\text{div}}, \\ \operatorname{div} \boldsymbol{\sigma} + \mathbf{f} = \mathbf{0} \quad \textit{in } \mathbf{L}^2(\Omega), \\ \boldsymbol{\sigma}\boldsymbol{\nu} = \mathbf{g} \quad \textit{in } W_0^*(\Gamma_2), \\ \boldsymbol{\sigma}(t) \in \mathbf{C}, \end{cases} \tag{1.14}$$

almost everywhere in time, then the evolution problem (1.7)-(1.12) *has no solution. At time t, a loading* $\{\mathbf{f}, \mathbf{g}\}$ *such that there exists a stress tensor field* $\boldsymbol{\sigma}(t)$ *which satisfies* (1.14) *is said to be potentially admissible. The loading* $\{\lambda\mathbf{f}, \lambda\mathbf{g}\}$, *where* λ *is the supremum of the positive numbers* μ *such that* $\{\mu\mathbf{f}, \mu\mathbf{g}\}$ *is potentially admissible, is called the limit load for the given structure in the direction* $\{\mathbf{f}, \mathbf{g}\}$.

Proof. Suppose there exists a solution $\{\mathbf{u}, \boldsymbol{\sigma}, \mathbf{w}\}$ to the evolution problem (1.7)–(1.12). Then, from (1.12), $\boldsymbol{\sigma}(t)$ belongs to Σ_{div} for almost any t. Moreover, since V contains $\mathbf{H}_0^1(\Omega)$, (1.9) implies

$$\operatorname{div} \boldsymbol{\sigma} + \mathbf{f} = \mathbf{0} \quad \text{in } \mathbf{H}^{-1}(\Omega),$$

which, by density of $\mathbf{H}_0^1(\Omega)$ into $\mathbf{L}^2(\Omega)$, and since $\operatorname{div} \boldsymbol{\sigma}$ and $\mathbf{f}$ both belong to $\mathbf{L}^2(\Omega)$, yields

$$\operatorname{div} \boldsymbol{\sigma} + \mathbf{f} = \mathbf{0} \quad \text{in } \mathbf{L}^2(\Omega).$$

Similarly, since W contains $W_0(\Gamma_2)$, (1.10) implies

$$\boldsymbol{\sigma}\boldsymbol{\nu} = \mathbf{g} \quad \text{in } W_0^*(\Gamma_2).$$

Finally, it follows from (1.7) that $\boldsymbol{\sigma}(t) \in \mathbf{C}$ for almost any t. Therefore, for any solution $\{\mathbf{u}, \boldsymbol{\sigma}, \mathbf{w}\}$ of (1.7)–(1.12), $\boldsymbol{\sigma}$ satisfies (1.14) for almost any time t. Consequently, if there is no stress field that satisfies (1.14) almost everywhere in time, there cannot be any solution of (1.7)–(1.12), and this holds whatever the flow rule (1.8) is. □

Remark 1.2. As pointed out by Salençon [1983], the notation of admissibility of external loads is very old and can be found in the works of Galileo [1638], Coulomb [1773], and Culmann [1866], among others. Theorem 1.1 simply expresses this long-standing notion in a Sobolev function framework. Other frameworks could be chosen and would lead to equivalent results by relaxation (Strang and Temam [1980]). The above framework is used in the existence theory developed by Strang and Temam [1980] and by Suquet [1982] and is convenient for a further characterization of potentially admissible loads.

Remark 1.3. The necessary condition introduced in Theorem 1.1 is independent of the flow rule (1.8). The same necessary condition appears in any situation where a global equilibrium equation and a local admissibility requirement are imposed together. In such a general setting, the variable $\boldsymbol{\sigma}$ and the set $\mathbf{C}(\mathbf{x})$ may have many different physical interpretations; however, the necessary condition (1.14) will remain the same and, therefore, all of the tools we will introduce to check (1.14) can be used in all of these different situations (see Fremond and Friaa [1982] for more details).

We now give sufficient conditions for the existence of solutions of the quasi-static evolution problem (1.7)–(1.12). These conditions are summarized in the following theorem.

THEOREM 1.2. *Suppose that the data of the evolution problem* (1.7)–(1.12) *satisfy the following conditions.*

(i) *The elasticity tensor* $\mathbf{A}(\mathbf{x})$ *is measurable in* $\mathbf{x}$, *it is uniformly continuous and uniformly strongly elliptic, and it satisfies the classical symmetry conditions*

$$A_{ijkl} = A_{klij} = A_{jikl}, \qquad 1 \leq i, j, k, l \leq 3.$$

(ii) *For any* $\mathbf{x}$ *in* Ω, *the set* $\mathbf{C}(\mathbf{x})$ *is closed, convex, and contains a fixed ball of radius* δ_0, *centered at the origin.*

(iii) *The projection operator from* $\mathbb{R}^9_{\text{sym}}$ *into* $\mathbf{C}(\mathbf{x})$ *is a measurable function of* $\mathbf{x}$.

(iv) $\mathbf{f} \in W^{1,\infty}(0, T; \mathbf{L}^N(\Omega))$ *and* $\mathbf{g} \in W^{1,\infty}(0, T; \mathbf{C}^0(\bar{\Gamma}_2))$.

(v) $\boldsymbol{\sigma}(0)$ *satisfies the compatibility condition* (1.14).

(vi) $\dot{\mathbf{u}}_1 \in W^{1,2}(0, T; BD(\bar{\Omega}))$.

(vii) *There exists a stress tensor field* $\boldsymbol{\tau}$ *in* $W^{1,\infty}(0, T; (L^\infty(\Omega))^9)$ *that satisfies the compatibility condition* (1.14) *such that, for any t,*

$$(\boldsymbol{\tau}(t) + d\boldsymbol{\tau}) \in \mathbf{C} \qquad \forall d\boldsymbol{\tau} \in L^\infty(\Omega, S^9) \quad \textit{with } \|d\boldsymbol{\tau}\|_\infty < \delta_1.$$

Then, the evolution problem (1.7)–(1.12) *has a solution* $\{\mathbf{u}, \boldsymbol{\sigma}, \mathbf{w}\}$ *where*

$$\dot{\mathbf{u}} \in L^2_\omega(0, T; \mathbf{BD}(\Omega)), \qquad \boldsymbol{\sigma} \in W^{1,2}(0, T; \Sigma_{\text{div}}), \qquad \mathbf{w} \in L^2_\omega(0, T; \mathbf{M}(\Gamma_2)),$$

$$V = \mathbf{BD}(\Omega), \qquad W = \mathbf{M}(\Gamma_2),$$

and $\mathbf{C}$ *is replaced in* (1.8) *by*

$$\mathbf{C}_c = \{\boldsymbol{\tau} \in \mathbf{C} \cap \Sigma_{\text{div}}, \boldsymbol{\tau\nu} \in \mathbf{C}^0(\bar{\Gamma}_2)\}.$$

Proof. See Suquet [1981] for the proof of this theorem. □

In the foregoing solution, the subscript ω denotes weak measurability with respect to time, $\mathbf{M}(\Gamma_2)$ is the topological dual space of $C^0(\bar{\Gamma}_2)$, and $\mathbf{BD}(\Omega)$ represents the space of bounded deformations, that is, the space of vector fields $\mathbf{v}$ of $\mathbf{L}^1(\Omega)$ whose associated linearized deformation tensor $\mathbf{D}(\mathbf{v})$ belongs to the topological dual of the set of continuous functions with compact support in Ω.

Remark 1.4. The conditions imposed by Theorem 1.2 on the external load $\{\mathbf{f}, \mathbf{g}\}$ are the regularity condition (iv) and the safety condition (vii). In most cases, these conditions are satisfied whenever $\{(1+\varepsilon)\mathbf{f}, (1+\varepsilon)\mathbf{g}\}$ is a regular, potentially admissible load, ε being an arbitrarily small, strictly positive number. Indeed, if $\boldsymbol{\tau}_\varepsilon$ denotes a stress tensor field that satisfies the compatibility condition (1.14) with external loads $\{(1+\varepsilon)\mathbf{f}, (1+\varepsilon)\mathbf{g}\}$, the stress tensor $\boldsymbol{\tau} = \boldsymbol{\tau}_\varepsilon/(1+\varepsilon)$ will usually verify (vii).

Remark 1.5. Theorem 1.2 does not guarantee the stability of the obtained solutions, which may quite well be unstable. For example, in the case of a cylindrical pipe subjected to a uniform external pressure, buckling will occur well before the pressure ceases to be potentially admissible, which indicates that the solution obtained for small strains is unstable.

Remark 1.6. It is proved in Temam [1986], under additional regularity assumptions on $\mathbf{C}$, that the solution of (1.7)–(1.12) does in fact satisfy the constitutive law (1.5) in a stronger sense than (1.8). More precisely, for any sufficiently regular element $\boldsymbol{\tau}$ of $\mathbf{C}$, the constitutive law

$$(\mathbf{D}(\dot{\mathbf{u}}) - \mathbf{A}^{-1}\dot{\boldsymbol{\sigma}}, \boldsymbol{\tau} - \boldsymbol{\sigma}) \le 0$$

is satisfied in the sense of measure on $\Omega \times (0, T)$.

1.4. Numerical analysis of the basic evolution problem. Let us come back to the initial Problem 1, looking for the existence of solutions to the quasi-static evolution problem (1.7)–(1.12) in perfect elastoplasticity for small strains cases. This problem can be approached numerically in two ways.

The first way is to ignore the existence results of § 1.3 and compute the solution of the evolution problem (1.7)–(1.12) directly, stopping the computation whenever numerical results can no longer be obtained, and assuming, then, that the limit load has been reached and that the structure cannot sustain the imposed loading. With this approach, the numerical solution of equations (1.7)–(1.12) can be obtained by a finite-element discretization in space and by an implicit integration in time, combined at each time step with the projection of the extrapolated stress $\boldsymbol{\sigma}((n+1)\,\Delta T)=\boldsymbol{\sigma}(n\,\Delta T)+\dot{\boldsymbol{\sigma}}(n\,\Delta T)\,\Delta T$ onto the admissible set $\mathbf{C}$ (Zienkiewicz [1977, p. 467]). Alternatively, in view of Remark 1.1, one can use the numerical time-integration technique described in Chapter 4.

This first approach has two drawbacks. One, the additional information it provides on the internal stresses and on the displacement field is reliable only if the constitutive law (1.8) is accurate. Second, and even more important, since no solution can exist above the limit load, this computation is unstable and fails when the external loads approach the limit load. Therefore, this limit load, which in a way estimates how safe a structure is when subjected to a specified loading $\{\mathbf{f}, \mathbf{g}\}$, usually cannot be estimated accurately by this type of computation.

The second numerical approach to Problem 1 computes the set $\mathcal{K}$ of all potentially admissible external loads that belong to a given family $\mathcal{L}$ of external loads. From the first theorem of limit load analysis, this set $\mathcal{K}$ can be seen to contain all the elements of the family of external loads that can be sustained by the structure. Moreover, in the case of perfect elastoplasticity, Remark 1.4 indicates that the regular elements of $\mathcal{K}$ correspond, within any practical requirements for accuracy, to those external loads for which a solution of the equations (1.7)–(1.12) exists in the sense of Theorem 1.2. Now, two main directions can be followed for computing $\mathcal{K}$.

First is the static approach, which solves, for any direction $\{\mathbf{f}, \mathbf{g}\}$ of $\mathcal{L}$, the following nonlinear programming problem.

Find $\{\lambda, \boldsymbol{\sigma}\}\in\mathbb{R}_+\times\Sigma_{\text{div}}$ that maximizes $F(\lambda, \boldsymbol{\sigma})=\lambda$ under the constraints

$$\begin{cases}\operatorname{div}\boldsymbol{\sigma}+\lambda\mathbf{f}=\mathbf{0} & \text{in } \mathbf{H}^{-1}(\Omega),\\ \boldsymbol{\sigma\nu}=\lambda\mathbf{g} & \text{in } W_0^*(\Gamma_2),\\ \boldsymbol{\sigma}\in\mathbf{C}.\end{cases}$$

When $\mathbf{H}^1(\Omega)$ is replaced by an adequate finite-element discretization and $\mathbf{C}(\mathbf{x})$ is approximated by a convex polytope of $\mathbb{R}^9_{\text{sym}}$, this problem is reduced to a linear programming problem that can be solved, for example, by the primal simplex method. Although this technique, described in detail in Pastor [1978], is very attractive, it still remains incredibly expensive both in core memory requirements and in computer running time—a good approximation of $\boldsymbol{\sigma}$ requires many degrees of freedom (several thousand for $N=2$) and, moreover, the simplex method behaves poorly when dealing with piecewise linear approximations of continuous convex functions.

A second method is the kinematic approach, which introduces a dual characterization of potentially admissible loads and then carries out a computation of $\mathcal{K}$ based on this dual definition. This method is described in detail in the following sections.

1.5. Synopsis of the chapter. The remainder of this chapter describes a numerical method for computing the set $\mathcal{K}$ of potentially admissible external loads using the kinematic approach, a viscoplastic regularization, and augmented Lagrangian techniques for solving the regularized viscoplastic problem. More precisely, as observed by Koiter [1960] and as proved in Fremond and Friaa [1982] or in Strang and Temam [1980], under regularity conditions on $\mathbf{C}(\mathbf{x})$, *it is equivalent* to verify (1.14) and to verify that, for any kinematically admissible velocity field, the dissipated plastic power is greater or equal to the power developed by the external loads (§ 2). Following Friaa [1979] or Casciaro and Cascini [1982], the dissipated plastic power can then be replaced by the limit, as the viscosity goes to zero, of the power dissipated by a viscoplastic material flowing under the action of the external loads. These loads will be potentially admissible if and only if a given function G_s of this viscoplastic flow remains bounded by 1 whatever the viscosity is (§ 3).

The numerical method that follows, then, simply computes this function G_s, determining the viscoplastic flows as was done in Chapter 4, that is, by a finite-element discretization of the continuous problems and by solving the resulting discrete problems by augmented Lagrangian techniques. The convergence of this numerical algorithm for computing G_s is studied in § 4. We then include the algorithm in a global program that determines the limit load in a given direction $\{\mathbf{f}, \mathbf{g}\}$ and give several practical examples of its application to the study of plastically compressible and incompressible materials (§ 5).

2. Limit load analysis.

2.1. Review of the static problem. Under the notation of §§ 1.2 and 1.3, the problem consists of determining whether a given load $\{\mathbf{f}, \mathbf{g}\}$ is potentially admissible for a given structure Ω.

Static problem 2. Does there exist a stress tensor field $\boldsymbol{\sigma}$ that satisfies

$$\left\{\begin{array}{l}\boldsymbol{\sigma}\in\mathbf{C}=\{\boldsymbol{\tau}:\Omega\to S^9,\ \boldsymbol{\tau}(\mathbf{x})\in\mathbf{C}(\mathbf{x})\text{ a.e. in }\Omega\},\\ \boldsymbol{\sigma}\in\Sigma_{\text{div}}=\{\boldsymbol{\tau}:\Omega\to S^9,\ \boldsymbol{\tau}\in(L^2(\Omega))^9,\ \operatorname{div}\boldsymbol{\tau}\in\mathbf{L}^N(\Omega)\},\\ \operatorname{div}\boldsymbol{\sigma}+\mathbf{f}=\mathbf{0}\quad\text{in }\mathbf{H}^{-1}(\Omega),\\ \boldsymbol{\sigma}\boldsymbol{\nu}=\mathbf{g}\quad\text{in }W_0^*(\Gamma_2)?\end{array}\right.\tag{2.1}$$

2.2. The kinematic characterization. In most cases, the admissibility condition (2.1) can be transformed by duality into an equivalent identity that may be easier to check. Introducing

$$V^s=\{\mathbf{w}:\Omega\to\mathbb{R}^N,\ \mathbf{w}=\mathbf{0}\text{ on }\Gamma_1,\ \mathbf{D}(\mathbf{w})\in(L^s(\Omega))^9\},\tag{2.2}$$

$$\mathbf{C}_D(\mathbf{x})=\{\boldsymbol{\tau}\in\mathbf{C}(\mathbf{x}),\ \mathrm{tr}\,(\boldsymbol{\tau})=0\},\tag{2.3}$$

we have the following theorem.

THEOREM 2.1. *Let* $\mathbf{C}(\mathbf{x})\subset\mathbb{R}^9_{\text{sym}}$ *(resp.,* $\mathbf{C}_D(\mathbf{x})\subset\mathbb{R}^9_{\text{sym}}\cap\ker(\mathrm{tr})$ *for the plastically incompressible case) be closed, convex, and contain a fixed ball of radius* δ_0 *and center* $\mathbf{0}$*, and let it be bounded uniformly in* $\mathbf{x}$*. Then the admissibility condition* (2.1) *is equivalent to*

$$\inf_{\mathbf{w}\in V^2}\left\{\int_\Omega\sup_{\boldsymbol{\tau}\in\mathbf{C}}\{\boldsymbol{\tau}\cdot\mathbf{D}(\mathbf{w})\}\,d\mathbf{x}-\int_\Omega\mathbf{f}\cdot\mathbf{w}\,d\mathbf{x}-\int_{\Gamma_2}\mathbf{g}\cdot\mathbf{w}\,da\right\}=0.\tag{2.4}$$

Proof. The proof of this theorem can be found in Fremond and Friaa [1982] for an abstract functional framework or in Strang and Temam [1980, § 3.2] for the above functional framework. The latter uses the same techniques of convex duality as does the proof of Theorem 2.2 in Chapter 2. We will outline this proof in the plastically incompressible case, which is the most difficult case. As in Theorem 2.2 of Chapter 2, we first introduce

$$X=\{\mathbf{w}\in V^2,\ \mathrm{div}\,\mathbf{w}=0\},$$

$$Y=\{\mathbf{D}:\Omega\to S^9,\ \mathbf{D}\in(L^2(\Omega))^9,\ \mathrm{tr}\,(\mathbf{D})=0\},$$

$$\Phi(\mathbf{w},\mathbf{D})=\int_\Omega\sup_{\boldsymbol{\tau}\in\mathbf{C}}\{\boldsymbol{\tau}\cdot(\mathbf{D}(\mathbf{w})-\mathbf{D})\}\,d\mathbf{x}-\int_\Omega\mathbf{f}\cdot\mathbf{w}\,d\mathbf{x}-\int_{\Gamma_2}\mathbf{g}\cdot\mathbf{w}\,da.$$

Moreover, we identify Y to its topological dual through the scalar product

$$\langle\mathbf{D},\mathbf{H}\rangle=\int_\Omega\mathbf{D}\cdot\mathbf{H}\,d\mathbf{x}.$$

By assumption, $\mathbf{C}$ is invariant by translation along the set of diagonal matrices, and $\mathbf{C}_D$ contains the origin and is uniformly bounded. Thus we have

$$0\le\sup_{\boldsymbol{\tau}\in\mathbf{C}}(\boldsymbol{\tau}\cdot\mathbf{D})=\sup_{\boldsymbol{\tau}\in\mathbf{C}_D}(\boldsymbol{\tau}\cdot\mathbf{D})\le C_1|\mathbf{D}|\quad\text{if tr}\,(\mathbf{D})=0,$$

$$\sup_{\boldsymbol{\tau}\in\mathbf{C}}(\boldsymbol{\tau}\cdot\mathbf{D})=+\infty\quad\text{if tr}\,(\mathbf{D})\neq0,$$

where C_1 denotes a positive constant independent of $\mathbf{x}$. Therefore, Φ is bounded and hence continuous on $X\times Y$ and takes on infinite values on V^2-X. Thus we can rewrite (2.4) as

$$\inf\mathcal{P}=0,$$

the primal problem $\mathcal{P}$ being defined by

$$\Phi(\mathbf{v},\mathbf{0})\le\Phi(\mathbf{w},\mathbf{0})\qquad\forall\mathbf{w}\in X,\quad\mathbf{v}\in X.$$

To prove the equivalence of (2.1) and (2.4), let us first assume that (2.4) holds. Then, $\Phi(\mathbf{w}, \mathbf{0})$ is bounded below on X. This implies, applying Theorem 1.10 of Chapter 2 (one of the fundamental theorems of convex analysis), that

$$(2.5) \qquad 0 = \inf \mathcal{P} = \max \mathcal{P}^* = -\Phi^*(\mathbf{0}, -\boldsymbol{\sigma}_D),$$

where $(-\boldsymbol{\sigma}_D)$ is a solution of the dual problem $\mathcal{P}^*$. Moreover, a direct computation characterizes Φ^* as

$$\begin{cases} \Phi^*(\mathbf{0}, \boldsymbol{\tau}) = 0 \quad \text{if } (-\boldsymbol{\tau}) \in \mathbf{C} \cap Y \quad \text{and satisfies} \\ \qquad \displaystyle\int_\Omega (-\boldsymbol{\tau}) \cdot \mathbf{D}(\mathbf{w})\, d\mathbf{x} - \int_\Omega \mathbf{f} \cdot \mathbf{w}\, d\mathbf{x} - \int_{\Gamma_2} \mathbf{g} \cdot \mathbf{w}\, da = 0 \quad \forall \mathbf{w} \in X, \\ \Phi^*(\mathbf{0}, \boldsymbol{\tau}) = +\infty \quad \text{if not.} \end{cases}$$

Then, since $\Phi^*(\mathbf{0}, -\boldsymbol{\sigma}_D) = 0$, we have automatically

$$(2.6) \qquad \begin{cases} \boldsymbol{\sigma}_D \in \mathbf{C} \cap Y, \\ \displaystyle\int_\Omega \boldsymbol{\sigma}_D \cdot \mathbf{D}(\mathbf{w})\, d\mathbf{x} - \int_\Omega \mathbf{f} \cdot \mathbf{w}\, d\mathbf{x} - \int_{\Gamma_2} \mathbf{g} \cdot \mathbf{w}\, da = 0 \quad \forall \mathbf{w} \in X. \end{cases}$$

Using the characterization of X^* obtained in the proof of Theorem 2.2, Chapter 2 and based on the closed range theorem, (2.6) can be expressed

$$\begin{cases} \exists p \in L^2(\Omega) \text{ such that} \\ \displaystyle\int_\Omega (\boldsymbol{\sigma}_D - p\mathbf{Id}) \cdot \mathbf{D}(\mathbf{w})\, d\mathbf{x} = \int_\Omega \mathbf{f} \cdot \mathbf{w}\, d\mathbf{x} + \int_{\Gamma_2} \mathbf{g} \cdot \mathbf{w}\, da \qquad \forall \mathbf{w} \in V^2, \quad \boldsymbol{\sigma}_D \in \mathbf{C} \cap Y. \end{cases}$$

In other words, the tensor $\boldsymbol{\sigma}_D - p\mathbf{Id}$ satisfies (2.1).

Conversely, suppose there exists a stress tensor $\boldsymbol{\sigma}$ that satisfies (2.1). By construction of Φ^*, and by definition of the dual and primal problems, we then have

$$0 = \Phi^*(\mathbf{0}, -\boldsymbol{\sigma} + \tfrac{1}{3} \operatorname{tr}(\boldsymbol{\sigma})\mathbf{Id}) = \sup \mathcal{P}^* \leq \inf \mathcal{P}.$$

Therefore, $\Phi(\mathbf{w}, \mathbf{0})$ is bounded below by zero for any $\mathbf{w}$ in X. We can again apply Theorem 1.10 of Chapter 2, which states that

$$\inf \mathcal{P} = \max \mathcal{P}^* = 0,$$

that is, (2.4).

Remark 2.1. Equation (2.4) expresses in mechanical terms that the plastic power $\sup_{\boldsymbol{\tau}} (\boldsymbol{\tau} \cdot \mathbf{D}(\mathbf{w}))$ which is dissipated inside the body by any kinematically admissible velocity field $\mathbf{w}$ is always greater than or equal to the power developed by the external forces for this velocity field.

Remark 2.2. The assumptions made in Theorem 2.1 on the convex $\mathbf{C}(\mathbf{x})$ of locally admissible stresses are satisfied in the plastically compressible case by materials like concrete, Camclay materials (see Chap. 4, Eq. (6.20)) or by

Coulomb materials with a maximum limit in compression (see Chap. 5, § 5.5). In the plastically incompressible case they are satisfied by Von Mises or by Tresca materials (see (1.1) and (1.2)).

They are not satisfied by standard Coulomb materials. In fact, when $\mathbf{C}(\mathbf{x})$ is not convex or is not bounded, it is very difficult to introduce any kinematic characterization of potentially admissible loads.

3. Viscoplastic regularization and numerical algorithm.

3.1. Associated Norton–Hoff viscoplastic material. The local capacity of resistance of the materials studied in § 2 is characterized by the set $\mathbf{C}(\mathbf{x})$ of locally admissible stresses. To $\mathbf{C}$, one can always associate a rigid, perfectly plastic material which, when subjected to external loads, obeys the constitutive law

$$\boldsymbol{\sigma} \in \partial\Big[\sup_{\boldsymbol{\tau}\in \mathbf{C}(\mathbf{x})} \{\boldsymbol{\tau} \cdot \mathbf{D}(\mathbf{v})\}\Big]. \tag{3.1}$$

Formally, as seen in Chapter 2, § 3, the velocity field of such a material flowing under the action of the load $\{\mathbf{f}, \mathbf{g}\}$ will realize the minimum in (2.4). Thus, Theorem 2.1 expresses that the load $\{\mathbf{f}, \mathbf{g}\}$ is potentially admissible if, in the resulting flow of the above rigid plastic material, the rate of energy dissipation is positive.

However, the constitutive law (3.1) can be considered as the limit, when the viscosity goes to zero (that is, when s goes to 1), of the constitutive law associated with a Norton–Hoff viscoplastic material, given by

$$\boldsymbol{\sigma} \in \partial\mathcal{D}_{1s}(\mathbf{D}(\mathbf{v})), \qquad \mathcal{D}_{1s}(\mathbf{D}) = \frac{1}{s}\Big|\sup_{\boldsymbol{\tau}\in\mathbf{C}} (\boldsymbol{\tau} \cdot \mathbf{D})\Big|^s. \tag{3.2}$$

The idea of Friaa [1979] and of Casciaro and Cascini [1982] consists of replacing the rigid, perfectly plastic material (3.1) by the associated viscoplastic material (3.2) in the computation of the rate of energy dissipation in Theorem 2.1. This approach is perfectly justified, as is proved in the theorem below.

THEOREM 3.1. *Under the assumptions of Theorem* 2.1, *a load* $\{\mathbf{f}, \mathbf{g}\}$ *is potentially admissible if and only if*

$$\lim_{s\to 1}\left\{\inf_{\mathbf{w}\in V^s}\left[\int_\Omega \mathcal{D}_{1s}(\mathbf{D}(\mathbf{w}))\,d\mathbf{x} - \int_\Omega \mathbf{f}\cdot\mathbf{w}\,d\mathbf{x} - \int_{\Gamma_2} \mathbf{g}\cdot\mathbf{w}\,da\right]\right\} = 0. \tag{3.3}$$

Proof. See Friaa [1979] for the proof of Theorem 3.1. In general terms, to prove this theorem one considers the stress field $\boldsymbol{\sigma}_s$ solution of the viscoplastic problem associated with (3.3) and shows that, if (3.3) holds, these fields $\boldsymbol{\sigma}_s$ converge to a limit satisfying (2.1) and conversely. □

There may not be any real materials that obey the constitutive law (3.2). This law is introduced here as a computational device only.

In mathematical terms, since $\mathscr{D}_{1s}$ is a positively homogeneous function of degree s, the replacement of (2.4) by (3.3) amounts formally to the approximation of $|\xi|$ by $1/s|\xi|^s$ in $\mathbb{R}_+$. This regularization is not very good globally but transforms the initial flow problem associated with (3.1), whose solution must be looked for in the awkward space $\mathbf{BD}(\Omega)$, into a strongly elliptic problem set on $W^{1,s}(\Omega)$. Moreover, from Theorem 3.1, this does not significantly affect the rate of energy dissipation.

Numerically, the characterization of potentially admissible loads can now be achieved first by computing the quasi-static flows of viscoplastic materials, as was done in Chapter 4, then by computing the associated rate of energy dissipation, and finally by going to the limit as s goes to 1. These computations must be organized in a specific way to be efficient, and this is described in the next sections.

3.2. Final characterization of admissible loads. From a numerical point of view, the limit in (3.3) cannot be obtained accurately. Therefore, it is better to replace this limit by the characterization below introduced by Friaa [1979].

THEOREM 3.2. *A load* $\{\mathbf{f}, \mathbf{g}\}$ *is potentially admissible if and only if*

$$\lim_{s\to 1} G_s(\mathbf{f}, \mathbf{g}) \le 1 \tag{3.4}$$

where $G_s(\cdot, \cdot)$ *is the convex, positively homogeneous function of degree* 1, *defined by*

$$\begin{cases} G_s(\mathbf{f}, \mathbf{g}) = \left\{ -\dfrac{s}{(s-1)\operatorname{meas}(\Omega)} \inf_{\mathbf{w}\in V^s} [F_s(\mathbf{D}(\mathbf{w})) + G(\mathbf{w})] \right\}^{(s-1)/s}, \\ F_s(\mathbf{D}) = \displaystyle\int_\Omega \mathscr{D}_{1s}(\mathbf{D})\, d\mathbf{x} = \int_\Omega \frac{1}{s} |\sup_{\boldsymbol{\tau}\in\mathbf{C}} (\boldsymbol{\tau}\cdot\mathbf{D})|^s \, d\mathbf{x}, \\ G(\mathbf{w}) = -\displaystyle\int_\Omega \mathbf{f}\cdot\mathbf{w}\, d\mathbf{x} - \int_{\Gamma_2} \mathbf{g}\cdot\mathbf{w}\, da. \end{cases} \tag{3.5}$$

The proof of this result is a variant of the proof of Theorem 3.1 and will not be given here. In essence, $G_s(\mathbf{f}, \mathbf{g})$ is a scaled equivalent of the limit (3.3), which turns out to be easier to compute.

Remark 3.1. It can also be shown that, for a fixed external load $\{\mathbf{f}, \mathbf{g}\}$, $G_s(\mathbf{f}, \mathbf{g})$ is a monotone decreasing function of s. Therefore, if $\mathbf{v}_{ns}$ denotes minimizing sequences of

$$F_s(\mathbf{D}(\mathbf{w})) + G(\mathbf{w})$$

on V^s, (3.4) can be rewritten as

$$\sup_{s>1} \sup_{\mathbf{v}_{ns}} \left\{ -\frac{s}{(s-1)\operatorname{meas}(\Omega)} (F_s(\mathbf{D}(\mathbf{v}_{ns})) + G(\mathbf{v}_{ns})) \right\}^{(s-1)/s} \le 1. \tag{3.6}$$

3.3. Numerical method for characterizing admissible loads. Based on Theorem 3.2, the numerical methods for characterizing potentially admissible loads finally correspond to the following algorithm.

ALGORITHM (3.7)–(3.10).

Inputs

(3.7) The load $\{\mathbf{f}, \mathbf{g}\}$ in $H^{-1}(\Omega) \times W_0^*(\Gamma_2)$;

(3.8) the triangulation $\mathcal{T}_h$ of Ω (see Chap. 4, § 3.1).

Loop on s

For s decreasing toward 1, compute $G_s(\mathbf{f}, \mathbf{g})$ as follows.

(3.9) Obtain an approximate value $\mathbf{v}_{ns}$ of the quasi-static flow of the viscoplastic material (3.2) subjected to the load $\{\mathbf{f}, \mathbf{g}\}$ using a finite-element discretization of V^s and the augmented Lagrangian techniques of Chapter 4 for solving the discrete problems.

$$\text{(3.10)} \qquad \text{Set } G_s(\mathbf{f}, \mathbf{g}) = \left\{ -\frac{s}{(s-1) \text{ meas } (\Omega)} (F_s(\mathbf{D}(\mathbf{v}_{ns})) + G(\mathbf{v}_{ns})) \right\}^{(s-1)/s}$$

If $G_s(\mathbf{f}, \mathbf{g}) > 1$, go to the Output procedure; if not, decrease the value of s and go to the first instruction of the loop.

Outputs

Consider the last computed value of $G_s(\mathbf{f}, \mathbf{g})$. The load is potentially admissible if and only if this value is less than or equal to 1.

Remark 3.2. The quasi-static flows of the viscoplastic fluid (3.2) under the action of the load $\{\mathbf{f}, \mathbf{g}\}$ minimize $F_s(\mathbf{D}(\mathbf{w})) + G(\mathbf{w})$ on V^s. Thus, (3.10) does indeed correspond to the definition of G_s given in (3.5).

Remark 3.3. Observe that the numerical method (3.7)–(3.10) is quite general—it works for any loading, in any geometry, in any dimension. Moreover, it does not require any differentiability of the function $\mathcal{D}_{1s}$, that is, any smoothness of the set $\mathbf{C}(\mathbf{x})$. Both the definition of G_s and the numerical methods of (3.9) require $\mathcal{D}_{1s}$ only to be continuous and convex.

Remark 3.4. Many viscoplastic regularizations have been proposed for the kinematic characterization (2.4) of potentially admissible loads. In particular, Mercier [1977] introduced such a technique in a study of viscoplastic Bingham fluids using augmented Lagrangian techniques. In this chapter, we chose the regularization (3.3) not only because it is theoretically justified but because the introduction of G_s leads to practical estimates of the limit loads that appear to be reasonably accurate.

Remark 3.5. For any $s > 1$ and for any $\mathbf{v}$ in V^s, the quantity

$$\text{(3.11)} \qquad \lambda(s, \mathbf{v}) = \left\{ -\frac{s}{(s-1) \text{ meas } (\Omega)} [F_s(\mathbf{D}(\mathbf{v})) + G(\mathbf{v})] \right\}^{(1-s)/s}$$

gives an upper bound of the limit load in the direction $\{\mathbf{f}, \mathbf{g}\}$. Indeed its inverse is a lower bound of $G_s(\mathbf{f}, \mathbf{g})$. Since G_s is a positively homogeneous function of degree 1, we then have $G_s(\mu\mathbf{f}, \mu\mathbf{g})>1$ for any $\mu>\lambda(s, \mathbf{v})$, which, from Remark 3.1, implies that the load $\{\mu\mathbf{f}, \mu\mathbf{g}\}$ is not potentially admissible. (In (3.11), $\lambda(s, \mathbf{v})$ takes the value $+\infty$ if the term between brackets is negative.)

4. Computation of $G_s(\mathbf{f}, \mathbf{g})$ and convergence results.

4.1. Computing strategy. In Algorithm (3.7)-(3.10) the computation of $G_s(\mathbf{f}, \mathbf{g})$ requires the solution of a viscoplastic flow problem associated with the material (3.2). If we solve this last problem by the numerical techniques of Chapter 4, and if we add a subscript s to indicate dependence on the regularizing exponent s, the computation of $G_s(\mathbf{f}, \mathbf{g})$ finally consists of

(i) the introduction of the finite-element spaces

(4.1) $$\Sigma_h=\{\boldsymbol{\tau}_h:\Omega\rightarrow S^9, \boldsymbol{\tau}_h|_{\Omega_\ell}=\text{const}\ \forall \ell=1, \cdots, N_h\},$$

(4.2) $$V_h=\{\mathbf{w}_h\in \mathbf{C}^0(\Omega),\ \mathbf{w}_h=0 \text{ on } \Gamma_1, \mathbf{w}_h|_{\Omega_\ell}\in \mathbf{P}_1(\Omega_\ell)\ \forall \ell=1, \cdots, N_h\};$$

(ii) the approximation of the viscoplastic flow problem by

(4.3) $$F_s(B\mathbf{v}_{hs})+G(\mathbf{v}_{hs})\leq F_s(B\mathbf{w}_h)+G(\mathbf{w}_h) \qquad \forall \mathbf{w}_h\in V_h, \quad \mathbf{v}_h\in V_h,$$

with $B\mathbf{w}_h=\mathbf{D}(\mathbf{w}_h)$, $\mathbf{F}_s$ and G defined in (3.5);

(iii) the transformation of this approximate problem into the equivalent saddle-point problem, using Theorem 4.1 of Chapter 4,

(4.4) $$\begin{cases}\mathscr{L}_r^s(\mathbf{v}_{hs}, \mathbf{D}_{hs}; \boldsymbol{\mu}_h)\leq \mathscr{L}_r^s(\mathbf{v}_{hs}, \mathbf{D}_{hs}; \boldsymbol{\lambda}_{hs})\leq \mathscr{L}_r^s(\mathbf{w}_h, \mathbf{H}_h; \boldsymbol{\lambda}_{hs})\\ \forall\{\mathbf{w}_h, \mathbf{H}_h; \boldsymbol{\mu}_h\}\in V_h\times\Sigma_h\times\Sigma_h, \quad \{\mathbf{v}_{hs}, \mathbf{D}_{hs}; \boldsymbol{\lambda}_{hs}\}\in V_h\times\Sigma_h\times\Sigma_h,\end{cases}$$

with $\mathscr{L}_r^s$ defined by

(4.5) $$\begin{aligned}\mathscr{L}_r^s(\mathbf{w}_h, \mathbf{H}_h; \boldsymbol{\mu}_h)=F_s(\mathbf{H}_h)+G(\mathbf{w}_h)&+\frac{r}{2}\int_\Omega \mathbf{A}(B\mathbf{w}_h-\mathbf{H}_h)\cdot(B\mathbf{w}_h-\mathbf{H}_h)\,d\mathbf{x}\\ &+\int_\Omega \mathbf{A}\boldsymbol{\mu}_h\cdot(B\mathbf{w}_h-\mathbf{H}_h)\,d\mathbf{x};\end{aligned}$$

(iv) the solution of the saddle-point problem (4.4) using one of the algorithms proposed in Chapter 3, § 4, for example, ALG1;

(v) the approximation of $G_s(\mathbf{f}, \mathbf{g})$ by $g_s(\mathbf{v}_s^n, \mathbf{D}_s^n)$ where $\{\mathbf{v}_s^n, \mathbf{D}_s^n\}$ is the result of iteration n of ALG1 and where $g_s(\mathbf{w}, \mathbf{H})$ is defined by

(4.6) $$g_s(\mathbf{w}, \mathbf{H})=\left\{-\frac{s}{(s-1)\text{ meas }(\Omega)}(F_s(\mathbf{H})+G(\mathbf{w}))\right\}^{(s-1)/s}$$

In the above strategy, the positive number r and the symmetric positive tensor $\mathbf{A}$ are arbitrary. The functions $\boldsymbol{\mu}_h$ and $\boldsymbol{\lambda}_{hs}$ are multipliers of the constraint $\mathbf{D}_{hs}=\mathbf{B}\mathbf{v}_{hs}$.

4.2. Convergence results. To evaluate the accuracy of the approximation of $G_s(\mathbf{f}, \mathbf{g})$ by $g_s(\mathbf{v}_s^n, \mathbf{D}_s^n)$, we must first study the truncation error, that is, the difference $g_s(\mathbf{v}_s^n, \mathbf{D}_s^n) - g_s(\mathbf{v}_{hs}, \mathbf{D}_{hs})$, and then the discretization error, that is the difference $g_s(\mathbf{v}_{hs}, \mathbf{D}_{hs}) - G_s(\mathbf{f}, \mathbf{g})$. For this purpose, we denote by C_i positive constants independent of s and h. We then have the following lemma.

LEMMA 4.1. *The Lagrange multiplier $\boldsymbol{\lambda}_{hs}$ satisfies*

$$\|\mathbf{A}\boldsymbol{\lambda}_{hs}\|_{0,2} \leq C_1 g_s(\mathbf{v}_{hs}, \mathbf{D}_{hs}). \tag{4.7}$$

Proof. Our proof is based on the subdifferential calculus of Lemma 3.1 of Chapter 4. First, from Theorem 3.4 of Chapter 4, $\{\mathbf{v}_{hs}, \mathbf{A}\boldsymbol{\lambda}_{hs}\}$ is a stationary solution of the variational system of elastoviscoplasticity, that is,

$$\int_\Omega \mathbf{A}\boldsymbol{\lambda}_{hs} \cdot \mathbf{D}(\mathbf{w}_h)\, d\mathbf{x} = -G(\mathbf{w}_h) \quad \forall \mathbf{w}_h \in V_h; \tag{4.8}$$

$$\mathbf{D}_{hs} = \mathbf{D}(\mathbf{v}_{hs}) \in \partial\psi_q(\mathbf{A}\boldsymbol{\lambda}_{hs}) \quad \text{in } \Sigma_h, \tag{4.9}$$

with ψ_q defined by

$$\psi_q(\boldsymbol{\tau}) = \int_\Omega \frac{1}{q} |\inf\{\mu > 0, \boldsymbol{\tau} \in \mu\mathbf{C}(\mathbf{x})\}|^q\, d\mathbf{x}, \qquad \frac{1}{q} + \frac{1}{s} = 1. \tag{4.10}$$

From Lemma 3.1 of Chapter 4, we see that (4.9) can be expressed as

$$F_s(\mathbf{D}_{hs}) + \psi_q(\mathbf{A}\boldsymbol{\lambda}_{hs}) = \langle \mathbf{A}\boldsymbol{\lambda}_{hs}, \mathbf{D}_{hs}\rangle = \int_\Omega \mathbf{A}\boldsymbol{\lambda}_{hs} \cdot \mathbf{D}(\mathbf{v}_{hs})\, d\mathbf{x},$$

which, from (4.8), yields

$$\psi_q(\mathbf{A}\boldsymbol{\lambda}_{hs}) = -F_s(\mathbf{D}_{hs}) - G(\mathbf{v}_{hs}). \tag{4.11}$$

On the other hand, since $\mathbf{C}(\mathbf{x})$ contains a fixed ball centered in $\mathbf{0}$ with radius δ_0, we have that

$$\psi_q(\boldsymbol{\tau}) \geq \int_\Omega \frac{1}{q} (|\boldsymbol{\tau}|/\delta_0)^q\, d\mathbf{x}. \tag{4.12}$$

Adding (4.12) to (4.11), we obtain

$$\|\mathbf{A}\boldsymbol{\lambda}_{hs}\|_{0,q}^q \leq (g_s(\mathbf{v}_{hs}, \mathbf{D}_{hs}))^q \delta_0^q \text{ meas}(\Omega),$$

which completes the proof. □

The proof of this lemma is in fact quite similar to the proofs of the characterization theorems of § 3. Using Lemma 4.1 and denoting by $\{\mathbf{v}_s^n, \mathbf{D}_s^n\}$ the result of iteration n of ALG1 operating on the saddle-point problem (4.4), with $\rho_n = \rho = r$, we can estimate the associated truncation error by the following theorem.

THEOREM 4.2. *For any $s>1$, when n goes to infinity, $g_s(\mathbf{v}_s^n, \mathbf{D}_s^n)$ converges toward $g_s(\mathbf{v}_{hs}, \mathbf{D}_{hs})$ at least as fast as $(n \log n)^{-1/2}$ goes to zero. The asymptotic constant is independent of s, provided that we have*

$$g_s(\mathbf{v}_{hs}, \mathbf{D}_{hs}) \geq C_2^{(s-1)/(3-2s)} \tag{4.13}$$

for some positive constant C_2 independent of s and h.

Proof. From (4.9), Lemma 3.1 of Chapter 4, and the definition of the subgradient, we have that

$$\int_\Omega \mathbf{A}\boldsymbol{\lambda}_{hs} \cdot (\mathbf{D}_s^n - \mathbf{D}_{hs})\, d\mathbf{x} \leq F_s(\mathbf{D}_s^n) - F_s(\mathbf{D}_{hs}). \tag{4.14}$$

Adding (4.8) written with $\mathbf{w}_h = \mathbf{v}_s^n - \mathbf{v}_{hs}$ to (4.14) yields

$$F_s(\mathbf{D}_s^n) + G(\mathbf{v}_s^n) - (F_s(\mathbf{D}_{hs}) + G(\mathbf{v}_{hs})) + \int_\Omega \mathbf{A}\boldsymbol{\lambda}_{hs} \cdot (\mathbf{D}(\mathbf{v}_s^n) - \mathbf{D}_s^n)\, d\mathbf{x} \geq 0. \tag{4.15}$$

On the other hand, since $\{\mathbf{v}_s^n, \mathbf{D}_s^n\}$ realizes by construction the minimum of $\mathcal{L}_r^s(\cdot, \cdot; \boldsymbol{\lambda}_s^n)$ over $V_h \times \Sigma_h$, we have

$$\mathcal{L}_r^s(\mathbf{v}_{hs}, \mathbf{D}_{hs}; \boldsymbol{\lambda}_s^n) \geq \mathcal{L}_r^s(\mathbf{v}_s^n, \mathbf{D}_s^n; \boldsymbol{\lambda}_s^n). \tag{4.16}$$

Adding (4.15) to (4.16) yields

$$(\boldsymbol{\lambda}_{hs} - \boldsymbol{\lambda}_s^n, \mathbf{D}(\mathbf{v}_s^n) - \mathbf{D}_s^n)_A \geq r|\mathbf{D}(\mathbf{v}_s^n) - \mathbf{D}_s^n|_A^2 \tag{4.17}$$

under the notation

$$(\mathbf{H}, \mathbf{D})_A = \int_\Omega \mathbf{A}\mathbf{H} \cdot \mathbf{D}\, dx, \qquad |\mathbf{H}|_A^2 = (\mathbf{H}, \mathbf{H}).$$

Moreover, the computation of $\boldsymbol{\lambda}_s^{n+1}$ in ALG1 implies that

$$|\boldsymbol{\lambda}_{hs} - \boldsymbol{\lambda}_s^{n+1}|_A^2 = |\boldsymbol{\lambda}_{hs} - \boldsymbol{\lambda}_s^n|_A^2 + r^2|\mathbf{D}(\mathbf{v}_s^n) - \mathbf{D}_s^n|_A^2 - 2r(\boldsymbol{\lambda}_{hs} - \boldsymbol{\lambda}_s^n, \mathbf{D}(\mathbf{v}_s^n) - \mathbf{D}_s^n)_A. \tag{4.18}$$

Substituting (4.17) into (4.18), we obtain

$$-|\boldsymbol{\lambda}_{hs} - \boldsymbol{\lambda}_s^{n+1}|_A^2 + |\boldsymbol{\lambda}_{hs} - \boldsymbol{\lambda}_s^n|_A^2 \geq r^2|\mathbf{D}(\mathbf{v}_s^n) - \mathbf{D}_s^n|_A^2. \tag{4.19}$$

The sequence $|\boldsymbol{\lambda}_{hs} - \boldsymbol{\lambda}_s^n|_A^2$ is therefore decreasing and converges toward a positive limit. By summing from $n=0$ to $n=+\infty$, inequality (4.19) then yields

$$r^2 \sum_{n=0}^{+\infty} |\mathbf{D}(\mathbf{v}_s^n) - \mathbf{D}_s^n|_A^2 \leq |\boldsymbol{\lambda}_{hs} - \boldsymbol{\lambda}_s^0|_A^2. \tag{4.20}$$

This means that the positive series on the left-hand side of (4.20) is convergent and, therefore, for n sufficiently large, each term is bounded by the generic term of any divergent positive series, and, in particular, by $|\boldsymbol{\lambda}_{hs}|_A^2(n \log n)^{-1}$. We now come back to the inequalities (4.15) and (4.16). Denoting

$$\Delta_n = F_s(\mathbf{D}_s^n) + G(\mathbf{v}_s^n) - (F_s(\mathbf{D}_{hs}) + G(\mathbf{v}_{hs})),$$

these inequalities can be expressed as

$$(\boldsymbol{\lambda}_{hs}, \mathbf{D}_s^n - \mathbf{D}(\mathbf{v}_s^n))_A \leq \Delta_n \leq (\boldsymbol{\lambda}_s^n, \mathbf{D}_s^n - \mathbf{D}(\mathbf{v}_s^n))_A - r|\mathbf{D}_s^n - \mathbf{D}(\mathbf{v}_s^n)|_A^2.$$

From (4.19), and assuming, for example, that ALG1 starts with $\boldsymbol{\lambda}_s^0 = 0$, this implies that

$$|\Delta_n| \leq 2|\boldsymbol{\lambda}_{hs}|_A|\mathbf{D}_s^n - \mathbf{D}(\mathbf{v}_s^n)|_A \tag{4.21}$$

which, from the above-boundedness of $|\mathbf{D}_s^n - \mathbf{D}(\mathbf{v}_s^n)|_A$ and Lemma 4.1, yields

$$|\Delta_n| \leq 2C_1^2 g_s^2(\mathbf{v}_{hs}, \mathbf{D}_{hs})(n \log n)^{-1/2}/r. \tag{4.22}$$

We are now ready to estimate the truncation error

$$\delta_n = g_s(\mathbf{v}_s^n, \mathbf{D}_s^n) - g_s(\mathbf{v}_{hs}, \mathbf{D}_{hs}).$$

Indeed, we can write

$$\delta_n = g_s(\mathbf{v}_{hs}, \mathbf{D}_{hs})\{(1 + \Delta_n/(F_s(\mathbf{D}_{hs}) + G(\mathbf{v}_{hs})))^{(s-1)/s} - 1\}.$$

But, for n sufficiently large, we have, from (4.22), that

$$|\Delta_n| \leq \tfrac{1}{2}|F_s(\mathbf{D}_{hs}) + G(\mathbf{v}_{hs})|;$$

therefore, δ_n can be estimated by a Taylor expansion of $(1+x)^q$ around 1, which gives

$$|\delta_n| \leq C_3 g_s(\mathbf{v}_{hs}, \mathbf{D}_{hs})|\Delta_n|(s-1)/[s|F_s(\mathbf{D}_{hs}) + G(\mathbf{v}_{hs})|].$$

Using (4.22) and the definition of g_s, we finally obtain

$$\boxed{|\delta_n| \leq C_5 g_s(\mathbf{v}_{hs}, \mathbf{D}_{hs})^{(2s-3)/(s-1)}(n \log n)^{-1/2}/(r \operatorname{meas}(\Omega)).} \qquad \square$$

Remark 4.1. Condition (4.13) is satisfied by $C_2 = 1$ when the load $\{\mathbf{f}, \mathbf{g}\}$ is not potentially admissible at the finite-element level, that is, when $g_s(\mathbf{v}_{hs}, \mathbf{D}_{hs})$ is strictly greater than 1. In that case, Theorem 4.2 guarantees that the speed of convergence of the sequence $g_s(\mathbf{v}_s^n, \mathbf{D}_s^n)$ does not deteriorate when s approaches 1.

We now turn to the study of the discretization error. We have the following theorem.

THEOREM 4.3. *For s given, $g_s(\mathbf{v}_{hs}, \mathbf{D}_{hs})$ converges from above toward $G_s(\mathbf{f}, \mathbf{g})$ when the diameter h of the triangulation $\mathcal{T}_h$ goes to zero. More precisely, if h is sufficiently small, we have*

$$1 \leq g_s(\mathbf{v}_{hs}, \mathbf{D}_{hs})/G_s(\mathbf{f}, \mathbf{g}) \leq 1 + C_s \inf_{\mathbf{w}_h \in W_h} \|\mathbf{v}_s - \mathbf{w}_h\|_{1,s}, \tag{4.23}$$

where C_s is a positive constant that is dependent on s and independent of h, and $\mathbf{v}_s \in V^s$ is the solution of the continuous viscoplastic problem associated with the constitutive law (3.2) and load $\{\mathbf{f}, \mathbf{g}\}$ and W_h standing for V_h (resp., for the space X_h of divergence-free elements of V_h) in the plastically compressible (resp., incompressible) case.

Proof. Let us consider the incompressible case and denote $\mathbf{z}_h$ the element of X_h such that

$$\|\mathbf{v}_s-\mathbf{z}_h\|_{1,s}=\inf_{\mathbf{w}_h\in X_h}\|\mathbf{v}_s-\mathbf{w}_h\|_{1,s}.$$

Since, by construction, $\mathbf{v}_s$ and $\mathbf{v}_{hs}$ realize the minimum of $F_s(\mathbf{D}(\cdot))+G(\cdot)$ on V^s and X_h, respectively, and since $F_s(\mathbf{D}(\cdot))+G(\cdot)$ is locally Lipschitz on $V^s\cap\ker(\mathrm{div})$, we have

$$\begin{aligned}F_s(\mathbf{D}(\mathbf{v}_s))+G(\mathbf{v}_s)&\le F_s(\mathbf{D}(\mathbf{v}_{hs}))+G(\mathbf{v}_{hs})\le F_s(\mathbf{D}(\mathbf{z}_h))+G(\mathbf{z}_h)\\&\le F_s(\mathbf{D}(\mathbf{v}_s))+G(\mathbf{v}_s)+C_s\|\mathbf{v}_s-\mathbf{z}_h\|_{1,s}.\end{aligned}\tag{4.24}$$

Now, to obtain (4.23) from (4.24), one proceeds as in the last part of the proof of Theorem 4.2, replacing (4.22) by (4.24). □

4.3. Comments on the numerical computation of $G_s(\mathbf{f},\mathbf{g})$. The convergence results of the preceding section indicate three possible weaknesses of Algorithm (3.7)–(3.10). The convergence of the sequence $(\mathbf{v}_s^n)$ is not proved and in fact can be very poor when s is close to 1; the discretization error worsens when s approaches 1; and the proposed method uses ALG1, which requires, at each step, the minimization of the nonlinear and possibly poorly conditioned problem

$$\begin{aligned}&\mathcal{L}_r^s(\mathbf{v}_s^n,\mathbf{D}_s^n;\boldsymbol{\lambda}_s^n)\le\mathcal{L}_r^s(\mathbf{w}_h,\mathbf{D}_h;\boldsymbol{\lambda}_s^n)\\&\forall\{\mathbf{w}_h,\mathbf{D}_h\}\in V_h\times\Sigma_h,\quad\{\mathbf{v}_s^n,\mathbf{D}_s^n\}\in V_h\times\Sigma_h.\end{aligned}\tag{4.25}$$

In practice, these difficulties can be overcome by the following: working with elasticity tensors $\mathbf{A}(\mathbf{x})$, which improve the conditioning of (4.25) (for example, in plastically incompressible situations, we use almost incompressible, nonhomogeneous elasticity tensors $\mathbf{A}$); solving (4.25) very crudely with few iterations of a block-relaxation algorithm such as Algorithm (4.17)–(4.19) of Chapter 3; and monitoring the numerical behavior not of $(\mathbf{v}_s^n)$ but of the sequence $g_s(\mathbf{v}_s^n,\mathbf{D}_s^n)$.

One can expect reasonable upper bounds of the limit loads using this strategy, since $F_s(\mathbf{D}_{s1}^n)+G(\mathbf{v}_{s1}^n)$ has been proven to converge reasonably well when ALG1 is used with one iteration of block relaxation (Chap. 3, Remark 4.4), and since $g_s(\mathbf{v}_s^n,\mathbf{D}_s^n)$ converges uniformly in s when levels above the limit load are reached. In fact, our numerical experiments did produce errors of less than 5% on the limit loads when $\mathbf{A}$ was properly chosen. This was achieved using values of s going as low as 1.01, with no more than four iterations of block relaxation per step, and with a number of steps in ALG1 being bounded by 30.

5. Examples of computations of limit loads.

5.1. Limit load on a line. Let us consider a given structure subjected to a fixed external load $\{\mathbf{f}_0,\mathbf{g}_0\}$ and to a load $\{\mathbf{f}_1,\mathbf{g}_1\}$ of variable intensity γ. Our problem consists of finding the maximum value of γ in which the load

$\{\mathbf{f}_0+\gamma\mathbf{f}_1, \mathbf{g}_0+\gamma\mathbf{g}_1\}$ is admissible for the structure. Obviously such a problem is of great practical importance in structural dimensioning.

Solving this problem with Algorithm (3.7)–(3.10) and using the computing strategy outlined in § 4.1 with V_h, Σ_h, and $\mathscr{L}_r^s$ defined by (4.1)–(4.5), we obtain the following algorithm.

ALGORITHM (5.1)–(5.11).

Inputs

The loads $\mathbf{f}_0, \mathbf{f}_1, \mathbf{g}_0, \mathbf{g}_1$;
the triangulation $\mathscr{T}_h$ of Ω.

Initialization of γ

Choose γ such that $\{\mathbf{f}_0+\gamma\mathbf{f}_1, \mathbf{g}_0+\gamma\mathbf{g}_1\}$ is not admissible.

Loop on γ (decrease external load)

(5.1) Set $\gamma=\gamma(1-\varepsilon)$, $\mathbf{f}=\mathbf{f}_0+\gamma\mathbf{f}_1$, $\mathbf{g}=\mathbf{g}_0+\gamma\mathbf{g}_1$, $s=s_0$.

Loop on s (decrease regularizing exponent)

(5.2) Initialize $\boldsymbol{\lambda}$ and set $n=1$.

Loop on n (ALG1)

(5.3) Initialize $\mathbf{D}_0$ and set $k=1$.

Loop on k (block relaxation)

(5.4) Solve $\mathscr{L}_r^s(\mathbf{v}_k, \mathbf{D}_{k-1}; \boldsymbol{\lambda}) \le \mathscr{L}_r^s(\mathbf{w}, \mathbf{D}_{k-1}; \boldsymbol{\lambda}) \quad \forall \mathbf{w} \in V_h$;

(5.5) Solve $\mathscr{L}_r^s(\mathbf{v}_k, \mathbf{D}_k; \boldsymbol{\lambda}) \le \mathscr{L}_r^s(\mathbf{v}_k, \mathbf{H}; \boldsymbol{\lambda}) \quad \forall \mathbf{H} \in \Sigma_h$;

(5.6) If $k<m$, set $k=k+1$ and return to (5.4).

End loop on k

(5.7) $\boldsymbol{\lambda}=\boldsymbol{\lambda}+r(\mathbf{D}(\mathbf{v}_m)-\mathbf{D}_m)$;

(5.8) $G_s^n=g_s(\mathbf{v}_m, \mathbf{D}_m)$;

(5.9) if $G_s^n>1$, return to (5.1);

(5.10) if $n<n_{\max}$, and $|G_s^n-G_s^{n-1}|>\eta$, set $n=n+1$ and return to (5.3).

End loop on n

(5.11) If $(s-1)$ is not too small, decrease s and return to (5.2).

End loop on s

Answer

Maximal intensity $=\gamma$ (with an accuracy of $\varepsilon/2$).

Many other procedures can be used. The advantage of Algorithm (5.1)–(5.11) is that Algorithm (3.7)–(3.10), on which it is based, is used mainly for loads that are not admissible, and therefore G_s^n converges well toward $G_s(\mathbf{f}, \mathbf{g})$ even if $(s-1)$ is small (see Remark 4.1). The practical application of (5.1)–(5.11) is illustrated below in several examples.

5.2. Example. The perforated square plate problem. Let us consider a thin square plate with a circular hole in its center. This plate is supposed to be made of a Von Mises homogeneous material, that is, of a material whose set of locally admissible stresses is given by

$$\mathbf{C}=\{\boldsymbol{\sigma}\in\mathbb{R}^9_{\text{sym}},\ |\boldsymbol{\sigma}-\tfrac{1}{3}(\operatorname{tr}\boldsymbol{\sigma})\mathbf{Id}|\le k\sqrt{2}\}.$$

This plate is subjected to two pairs of opposite surface tractions characterized by their surface densities $\mathbf{g}_1$ and $\mathbf{g}_2$. We are interested in determining the maximum tractions that can be supported by the plate.

For symmetry reasons, only one fourth of the plate has to be considered. In addition, we suppose that the stresses remain planar in nature. We are then faced with a well-known problem for which there are many experimental, analytical, and numerical results (Gaydon and MacCrum [1954], Hodge [1959], Belytschko and Hodge [1970]), and we can compare our numerical results with these known results in three basic cases.

$$\text{(i)}\ \mathbf{g}_2\cdot\mathbf{n}_2=\mathbf{g}_1\cdot\mathbf{n}_1,\qquad \text{(ii)}\ \mathbf{g}_2\cdot\mathbf{n}_2=0,\qquad \text{(iii)}\ \mathbf{g}_2\cdot\mathbf{n}_2=-\mathbf{g}_1\cdot\mathbf{n}_1.$$

The geometry of the loading is shown in Fig. 5.1, the finite-element meshes used in Fig. 5.2, and our numerical results in Table 5.1. Note the close agreement of our results with those of Gaydon and MacCrum [1954].

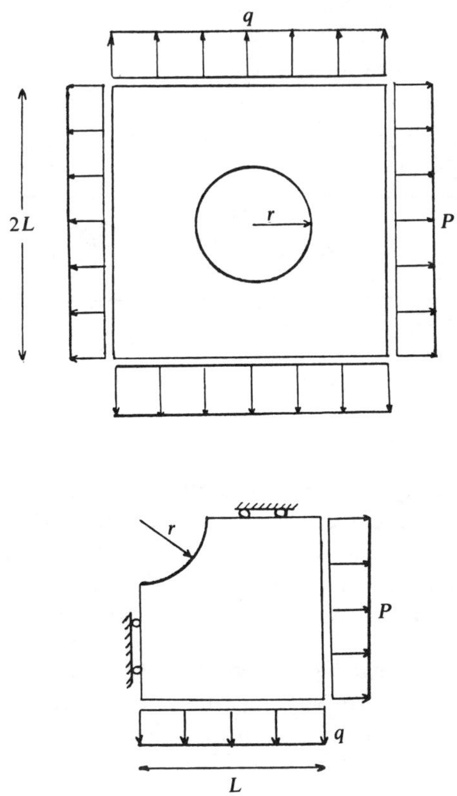

FIG. 5.1. *The perforated square plate problem* (*Example* 5.2).

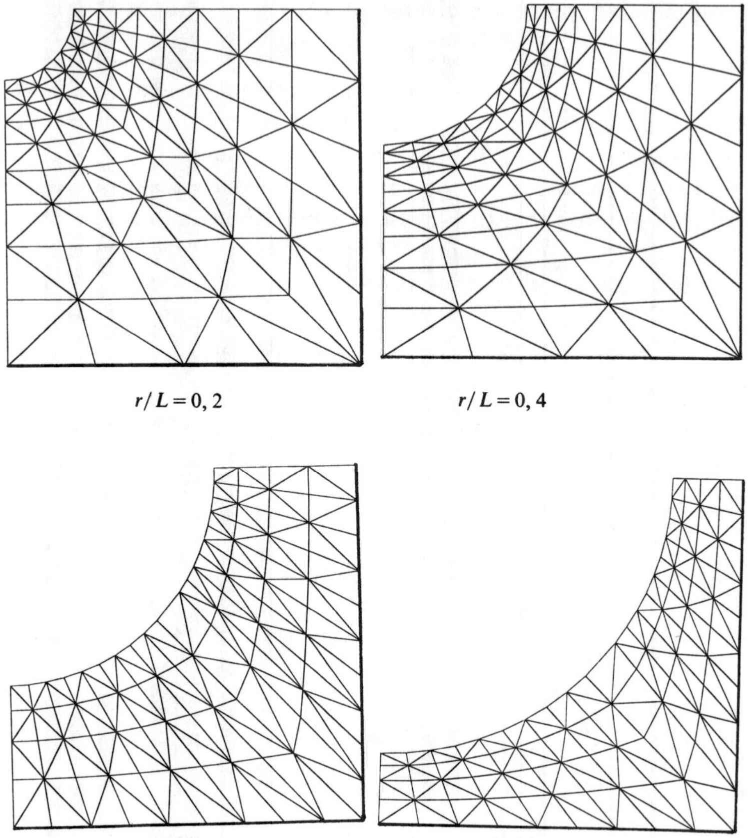

$r/L = 0,2$ $r/L = 0,4$

$r/L = 0,6$ $r/L = 0,8$

FIG. 5.2. *The different finite-element meshes used to solve the perforated square plate problem* (*Example* 5.2).

The computation time for a given geometry and a given ratio of surface tractions was approximately 20 sec on an IBM 360. These results are taken from Guennouni [1982], who used Algorithm (5.1)–(5.11) with $m = 2$ or 4, $n_{\max} = 30$, and values of s ranging from 1.5 to 1.02. Moreover, the augmented Lagrangian was constructed with $r = 1$, $\mathbf{A} = \mathbf{Id}$, and $B\mathbf{w} = \nabla \mathbf{w}$.

5.3. Example. The cracked plate problem. Our second example also considers a thin square plate. This time, the plate has a crack in its center parallel to one of its sides. The plate is made of a Tresca material characterized by the set

$$\mathbf{C}(\mathbf{x}) = \{\boldsymbol{\sigma} \in \mathbb{R}^9_{\text{sym}} \sup_{i \neq j} |\sigma_i - \sigma_j| \leq \sigma_0\},$$

TABLE 5.1
Limit loads of a perforated square plate: Summary of numerical results and comparison with those of Gaydon and MacCrum [1954].

	Case (i)		Case (ii)		Case (iii)	
	G.-LT.	G.-MC.	G.-LT.	G.-MC.	G.-LT.	G.-MC.
r/L	$\lvert g\rvert/k\sqrt{2}$	—	$\lvert g\rvert/k\sqrt{2}$	—	$\lvert g\rvert/k\sqrt{2}$	—
0	—	1.000	1.000	1.000	—	1.000
0.2	0.908	0.910	0.843	0.800	0.860	0.870
0.4	0.704	0.693	0.621	0.621	0.640	0.660
0.5	—	—	0.423	0.423	—	—
0.6	0.461	0.462	0.270	0.259	0.240	0.240
0.8	0.223	0.231	0.059	0.056	0.057	0.050

and it is subjected to a uniform traction perpendicularly to the crack. The problem consists of determining the maximal traction that can be supported by such a plate (Fig. 5.3). In plane stresses, the solution to this problem is known and corresponds to a maximal traction of

$$|\mathbf{g}| = \sigma_0 \frac{b-a}{b},$$

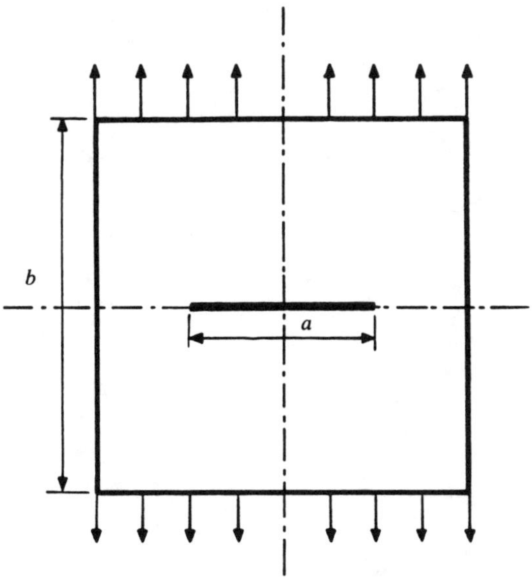

FIG. 5.3. *The cracked square plate problem (Example 5.3).*

with a and b the length of the crack and the width of the plate, respectively (Hodge [1959]).

Our numerical computation was done in plane stresses for a ratio $(b-a)/b = 0.4$. Recall that the potential $\mathscr{D}_{1s}$ associated with a Tresca material in plane stresses, which was computed in Chapter 4, § 6, is not differentiable and is given by

$$\mathscr{D}_{1s}(\mathbf{D}) = \frac{1}{s}(\sigma_0 \sup \{|D_1|, |D_2|, |D_1 + D_2|\})^s.$$

Restricting ourselves to one fourth of the plate for symmetry reasons, we used Algorithm (5.1)–(5.11) on the finite-element mesh shown in Fig. 5.4 (169 nodes) with $r/\sigma_0 = 10$, $\mathbf{A} = \mathbf{Id}$, $m = 1$, $n_{\max} = 100$, and s ranging from 1.05 to 1.003. The maximal traction obtained numerically after 23 min of CPU time on a VAX 780 was equal to $|\mathbf{g}| = 0.4189\sigma_0$ as compared with the theoretical value $|\mathbf{g}| = 0.4\sigma_0$. The velocity field $\mathbf{v}_{1.003}^{100}$ corresponding to a traction $|\mathbf{g}| = 0.4165\sigma_0$ is represented in Fig. 7.7 of Chapter 4, where the aspect of the mesh is shown after 1.25 sec of flow. Observe that this velocity field looks qualitatively like the piecewise continuous mechanisms used to analytically determine limit loads by the kinematic method of Theorem 2.1.

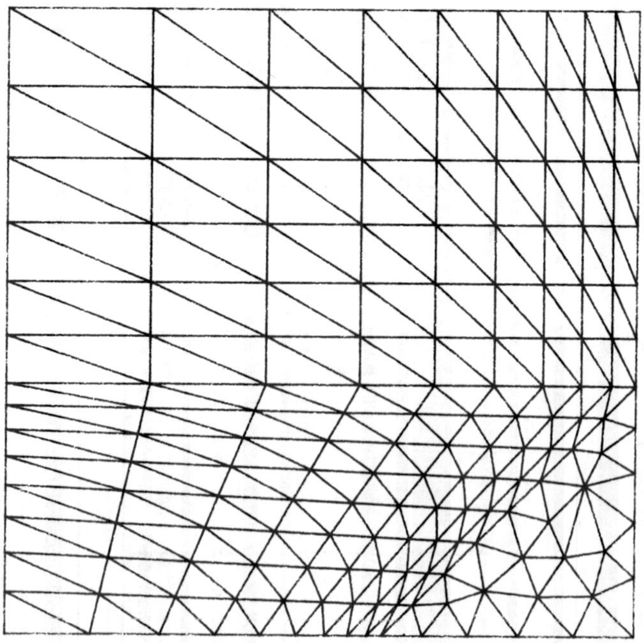

FIG. 5.4. *The finite-element mesh used to discretize the cracked square plate problem* (*Example* 5.3).

5.4. Example. The vertical bank problem. We now turn to a problem involving incompressible materials in plane strains. More specifically, we consider the problem of determining the maximum height of a vertical bank made of a Tresca material (Salençon [1983]). From dimensional analysis, this problem is equivalent to the determination of the maximum density allowable for the material, so that a vertical bank of unit height made of this material can sustain its own weight. Here, if sufficiently large, the width of the bank does not affect the solution. In plane strains, Tresca materials and Von Mises materials correspond to the same dissipation potential, given by

$$\mathcal{D}_{1s}(\mathbf{D}) = \begin{cases} \frac{1}{s}(k\sqrt{2})^s|\mathbf{D}|^s & \text{if tr}(\mathbf{D})=0, \\ +\infty & \text{if not.} \end{cases}$$

For this material, in this problem, the best estimate of the maximum density obtained through use of a kinematic method is (De Josselin de Jong [1977])

$$\gamma = 3.83k.$$

For this problem, the discretization strategy proposed in Chapter 4, § 3.3(i), to handle plastically incompressible materials turns out to be too stiff. In Algorithm (5.1)–(5.11) this results in a sequence G_s^n that converges poorly toward the correct value $G_s(\mathbf{f}, \mathbf{g})$. To get better results, we used the discretization strategy of Chapter 4, § 3.3(ii), with

(5.12) $\mathcal{T}_{h/2}$ = triangulation obtained by joining the midsides of every edge of any triangle Ω_ℓ of $\mathcal{T}_h$ (Fig. 5.6),

(5.13) $$V_h = \{\mathbf{w}_h \in \mathbf{C}^0(\Omega), \mathbf{w}_h = 0 \text{ on } \Gamma_1, \mathbf{w}_h|_{\Omega_p} \in \mathbf{P}_1(\Omega_p)\ \forall \Omega_p \in \mathcal{T}_{h/2}\},$$

(5.14) $$\Sigma_h = \{\boldsymbol{\tau}_h : \Omega \to S^9, \boldsymbol{\tau}_h|_{\Omega_p} = \text{const}\ \forall \Omega_p \in \mathcal{T}_{h/2}\},$$

(5.15) $$P_h = \{q_h \in C^0(\Omega), q_h|_{\Omega_\ell} \in P_1(\Omega_\ell)\ \forall \Omega_\ell \in \mathcal{T}_h\}.$$

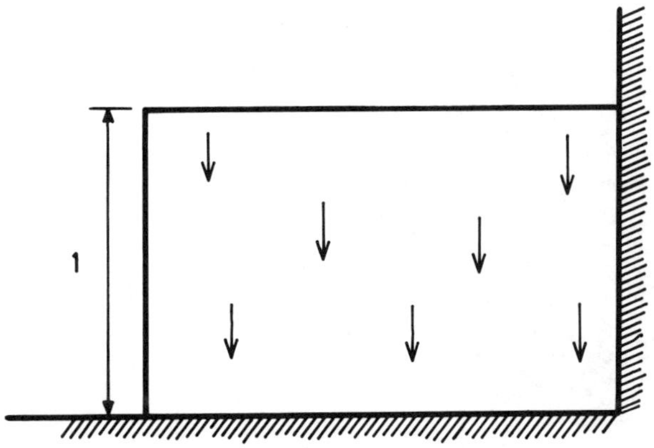

FIG. 5.5. *The vertical bank problem (Example 5.4).*

The following results were obtained with a triangulation $\mathcal{T}_h$ of 162 triangles corresponding to 100 pressure nodes and 361 velocity nodes. With an initial value of $\gamma = 3.96k$, and s having values of {1.2, 1.1, 1.05, 1.02, 1.01, 1.008, 1.005, 1.003}, the computed limit load was $\gamma = 3.94k$. As expected, this value was not affected by a change in the initial density γ_0 or by an increase of the foundation width.

In view of the difficulties already observed by Mercier [1977] in the numerical solution of this problem, this numerical estimate can be considered to be rather good, since it lies within 3% of the best known estimate. Nevertheless, although only one block relaxation was done per Uzawa iteration, and although the initial density γ_0 was close to the final estimate, the computation lasted two hours on a VAX 780. Unfortunately, such lengthy computations seem to be characteristic of incompressible materials.

For completeness, we show in Fig. 5.6 and Fig. 5.7 the appearance of the computed velocity field and of the triangulation $\mathcal{T}_{h/2}$ after 0.4 sec of flow. These results correspond to the case

$$\begin{cases} s = 1.005, \quad \gamma = 3.96k, \quad k = 1/\sqrt{2}, \\ r = 10, \quad \mathbf{A} = \mathbf{Id}. \end{cases}$$

5.5. Example. Foundation on a modified Coulomb material. Our last example, studied in Guennouni and Le Tallec [1982], considers the computation of the bearing capacity of a rigid foundation that lies on the top of an embankment made of a Coulomb material (Fig. 5.8). The embankment is also subjected to its own weight, and we suppose that the compression stresses inside the material cannot exceed a given threshold Q.

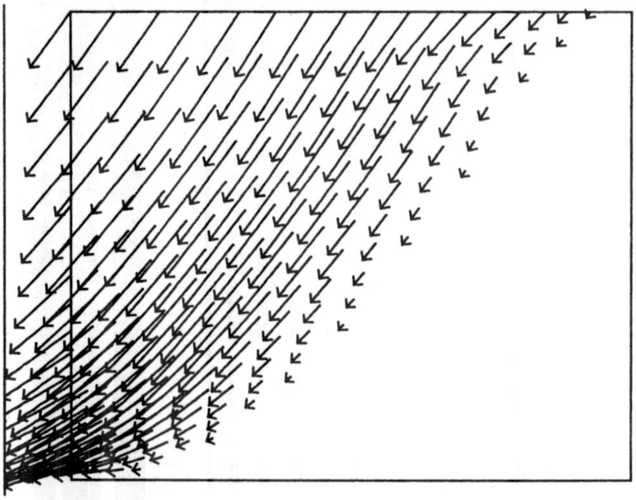

FIG. 5.6. *Vertical bank problem. The computed velocity field after* 0.4 *sec of flow.*

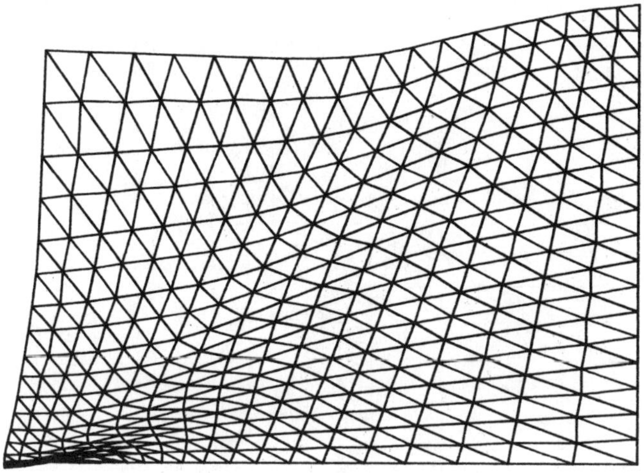

FIG. 5.7. *Vertical bank problem. The triangulation* $\mathcal{T}_{h/2}$ *after* 0.4 *sec of flow.*

Without this last assumption, the set of locally admissible stresses for a Coulomb material is given by

$$\mathbf{C}=\{\boldsymbol{\sigma}\in\mathbb{R}^9_{\text{sym}},\ \sigma_1(1+\sin\phi)-\sigma_3(1-\sin\phi)-2c\cos\phi\le 0\},$$

where σ_1 and σ_3 are, respectively, the largest and the smallest eigenvalues of $\boldsymbol{\sigma}$ and where c and ϕ are material constants denoting the internal cohesion of the material and the angle of internal friction, respectively. This set is not bounded in R^9_{sym}; therefore, the theory of §§ 2 and 3 cannot be applied. Now, if we assume the existence of a maximum compression threshold Q, the set of locally admissible stresses becomes

$$\mathbf{C}=\{\boldsymbol{\sigma}\in\mathbb{R}^9_{\text{sym}},\ \sigma_1(1+\sin\phi)-\sigma_3(1-\sin\phi)-2c\cos\phi\le 0,\ \sigma_1+\sigma_3\ge -Q\}.$$

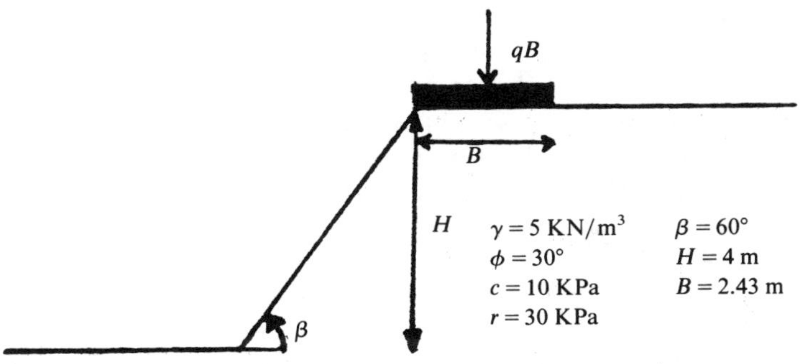

FIG. 5.8. *The bearing capacity of a rigid foundation lying on top of an embankment made of Coulomb material* (*Example* 5.5).

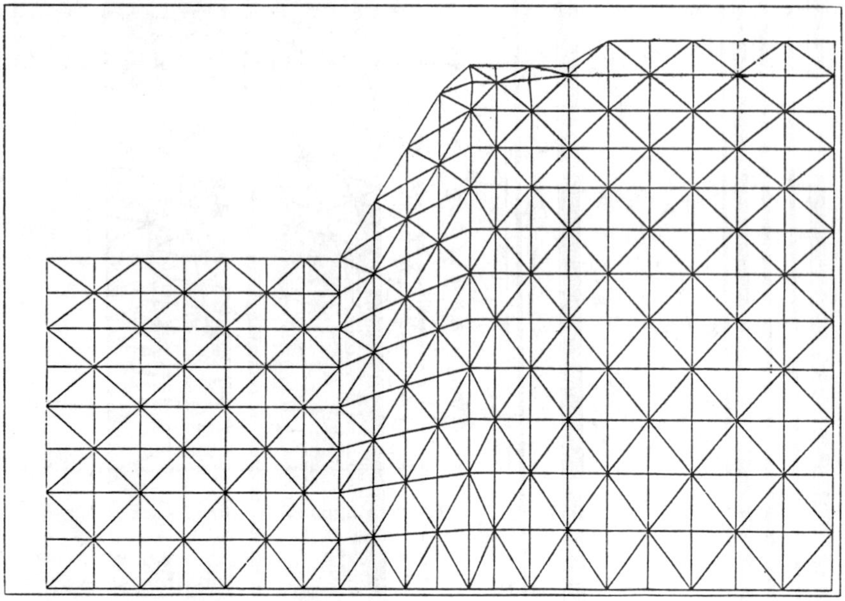

FIG. 5.9. *The finite-element mesh after one second of flow, at* $s = 1.02$ (*Example* 5.5).

This new set is now convex and bounded in $\mathbb{R}^9_{\text{sym}}$. Therefore, the whole theory developed in this chapter applies and leads to the introduction of an associated nondifferentiable material dissipation potential given by

$$\mathscr{D}_{1s}(\mathbf{D}) = \frac{1}{s}\left| c \cot\phi \sup\left\{\operatorname{tr}(\mathbf{D}), \sum_i |D_1| \sin\phi + Q\left(\sum_i |D_i| \sin\phi + \operatorname{tr}(\mathbf{D})\right)\right\}\right|^s.$$

Assuming plane strains and using the above dissipation potential, the application of Algorithm (5.1)–(5.11) to the geometry of Fig. 5.8 gives a limit load corresponding to the ratio $|\mathbf{g}|/\gamma B = 6.45$, where $|\mathbf{g}|$ is the density of the load exerted on the foundation, γ is the volumic weight of the material, and B is the width of the foundation. Engineering curves, given in Kusakabe, Kimura, and Yamaguchi [1981], and based on a completely different method, lead to a ratio of $|\mathbf{g}|/\gamma B = 6.20$ for classical Coulomb materials. From an engineering point of view, the agreement between these two results is quite good. The aspect of the computed flow, corresponding at the limit load to $s = 1.02$, is indicated on Fig. 5.9, where the final shape of the mesh after one second of flow is represented.

Chapter 6

Two-Dimensional Flow of Incompressible Viscoplastic Fluids

1. Classical formulation of the flow problem.

1.1. The physical problem. The present chapter is based largely on the work of Begis [1979], Glowinski, Lions, and Tremolières [1981, App. 6] and Fortin and Glowinski [1983, Chap. 7] on the problem of the unsteady flow of a Bingham fluid in a bounded two-dimensional cavity. To simplify our notation, we will denote by Ω the geometrical domain associated with the cavity, and, by Γ, its boundary. Moreover, we will omit the overbars in all Eulerian quantities. The problem then consists of finding, for all times t in $[0, t_1]$, the in-plane components $\mathbf{v} = \{v_1, v_2\}$ of the fluid velocity where its initial value $\mathbf{v}_0$, its trace $\mathbf{v}_1$ on Γ, and the applied body forces $\mathbf{f}$ are known.

1.2. Variational velocity formulation of the flow problem. In Chapter 2, § 3, we introduced a well-posed mathematical formulation of the time-dependent flow problem for a Bingham fluid using the *virtual work theorem* (Chap. 1, Eq. (5.3)) and the constitutive law of a *Bingham fluid* (Chap. 1, Eqs. (5.4), (5.6)) and neglecting the convection terms. When applied to a plane flow situation, this formulation is as follows.

$$
(1.1)\quad \begin{cases} \text{Find } (\mathbf{v}-\mathbf{v}_1) \in L^2(0, t_1; X), \quad \dfrac{\partial \mathbf{v}}{\partial t} \in L^2(0, t_1; X^*) \text{ such that} \\[2ex] \displaystyle\int_\Omega \rho \frac{\partial \mathbf{v}}{\partial t}(t) \cdot (\mathbf{w}-\mathbf{v}(t))\, d\mathbf{x} + \int_\Omega [\mathcal{D}_1(\mathbf{D}(\mathbf{w})) - \mathcal{D}_1(\mathbf{D}(\mathbf{v}(t)))]\, d\mathbf{x} \\[2ex] \qquad \displaystyle\geq \int_\Omega \mathbf{f} \cdot (\mathbf{w}-\mathbf{v}(t))\, d\mathbf{x} \qquad \forall \mathbf{w} \in X \quad \text{a.e. in } (0, t_1), \\[2ex] \mathbf{v}(0) = \mathbf{v}_0, \end{cases}
$$

under the notation

$$X=\{\mathbf{w}\,|\,\mathbf{w}\in(H^1(\Omega))^2,\ \mathbf{w}=\mathbf{0}\text{ on }\Gamma,\ \nabla\cdot\mathbf{w}=0\},\tag{1.2}$$

$$\mathcal{D}_1(\mathbf{D})=\mu|\mathbf{D}|^2+g\sqrt{2}|\mathbf{D}|,\tag{1.3}$$

$$\mathbf{D}(\mathbf{w})=\sum_{1\le i,j\le 2}\frac{1}{2}\left(\frac{\partial w_i}{\partial x_j}+\frac{\partial w_j}{\partial x_i}\right)\mathbf{e}_i\otimes\mathbf{e}_j.\tag{1.4}$$

Above, ρ is the fluid *density*, μ its *viscosity*, and g its *rigidity*.

1.3. Synopsis of the chapter. One possible numerical treatment of the evolution problem (1.1) is to introduce a backward Euler time discretization of (1.1) and to treat the resulting problem at each step as a quasi-static viscoplastic problem to be solved using the techniques of Chapter 4. Although this approach is perfectly legitimate and efficient, in this chapter we will propose a different treatment of (1.1) that is based on the same techniques but operates on a stream function formulation of (1.1). Our main motivation for this approach is that it will permit us to take advantage of the two-dimensionality of the problem and thus to eliminate the difficulties associated with the numerical treatment of the incompressibility condition. More precisely, we shall see that the introduction of a stream function enables (1.1) to be reduced to a parabolic variational inequality of order four with respect to the space variables. We shall then examine the approximation of the above problem using mixed finite elements for the space approximation and finite differences for the time discretization. Next, we will show that, at each time step, these approximate problems can be solved by the augmented Lagrangian methods of Chapter 3, § 4. Finally, we will illustrate the above ideas by several numerical examples.

The reader should note that these ideas in fact introduce two time scales: a real one associated with the evolution problem (1.1), and an artificial one associated at each time step with the augmented Lagrangian treatment of the discrete problem. Indeed, we have seen that augmented Lagrangian methods correspond to time-integration techniques of an associated dual evolution problem. Here, contrary to what we observed in our study of elastoviscoplasticity, the associated dual evolution problem does not correspond to the original problem (1.1) and, thus, rather than being identical, the two time scales are completely independent.

2. Stream function formulation. In this section, we shall make the following two simplifying assumptions

$$\Omega\text{ is simply connected,}\tag{2.1}$$

$$\mathbf{v}_1=\mathbf{0}.\tag{2.2}$$

There are, in fact, no real numerical difficulties in extending the following techniques to situations where (2.1) or (2.2) are not satisfied. If we confine our attention to two-dimensional flows, we can eliminate the condition $\nabla \cdot \mathbf{v}=0$ by introducing a stream function defined to within an additive constant by

$$v_1=\frac{\partial \psi}{\partial x_2}, \qquad v_2=-\frac{\partial \psi}{\partial x_1}. \tag{2.3}$$

The condition $\mathbf{v}=\mathbf{0}$ on Γ implies

$$\psi=\text{const} \quad \text{on } \Gamma, \tag{2.4}$$

$$\frac{\partial \psi}{\partial n}=0 \quad \text{on } \Gamma. \tag{2.5}$$

We shall take $\psi=0$ on Γ, which fixes the above constant.

Let us consider now $\mathbf{w}\in X$; we associate to $\mathbf{w}$ the function $\phi\in H_0^2(\Omega)$, uniquely defined by

$$w_1=\frac{\partial \phi}{\partial x_2}, \qquad w_2=-\frac{\partial \phi}{\partial x_1}. \tag{2.6}$$

Recall that

$$H_0^2(\Omega)=\left\{\phi \mid \phi\in H^2(\Omega),\ \phi=\frac{\partial \phi}{\partial n}=0 \text{ on } \Gamma\right\}. \tag{2.7}$$

In view of (2.3) and (2.6), we can reduce (1.1) to the following parabolic variational inequality.

$$\left\{\begin{aligned}
&\text{Find } \psi\in L^2(0, t_1; H_0^2(\Omega)), \frac{\partial \psi}{\partial t}\in L^2(0, t_1: L^2(\Omega)) \text{ such that}\\
&-\rho\int_\Omega \frac{\partial \psi}{\partial t}(t)\Delta(\phi-\psi(t))\,d\mathbf{x}+\mu a(\psi(t), \phi-\psi(t))\\
&\quad +gj(\phi)-gj(\psi(t))\\
&\quad \geq \int_\Omega\left(\frac{\partial f_2}{\partial x_1}-\frac{\partial f_1}{\partial x_2}\right)(\phi-\psi(t))\,d\mathbf{x} \qquad \forall \phi\in H_0^2(\Omega) \quad \text{a.e. in } (0, t_1),\\
&\psi(\mathbf{x}, 0)=\psi_0(\mathbf{x}),
\end{aligned}\right. \tag{2.8}$$

where

$$\left\{\begin{aligned}
a(\phi_1, \phi_2)=\int_\Omega\Bigg[&\left(2\frac{\partial^2 \phi_1}{\partial x_1\,\partial x_2}\right)\left(2\frac{\partial^2 \phi_2}{\partial x_1\,\partial x_2}\right)\\
&+\left(\frac{\partial^2 \phi_1}{\partial x_1^2}-\frac{\partial^2 \phi_1}{\partial x_2^2}\right)\left(\frac{\partial^2 \phi_2}{\partial x_1^2}-\frac{\partial^2 \phi_2}{\partial x_2^2}\right)\Bigg]\,d\mathbf{x} \qquad \forall \phi_1, \phi_2\in H_0^2(\Omega),
\end{aligned}\right. \tag{2.9}$$

and

$$j(\phi)=\int_\Omega\left[\left(2\frac{\partial^2 \phi}{\partial x_1\,\partial x_2}\right)^2+\left(\frac{\partial^2 \phi}{\partial x_1^2}-\frac{\partial^2 \phi}{\partial x_2^2}\right)^2\right]^{1/2} d\mathbf{x} \qquad \forall \phi\in H_0^2(\Omega). \tag{2.10}$$

Remark 2.1. In fact, we have

$$a(\phi_1, \phi_2)=\int_\Omega \Delta\phi_1\,\Delta\phi_2\,d\mathbf{x} \qquad \forall \phi_1, \phi_2 \in H_0^2(\Omega). \tag{2.11}$$

In the following, we shall be using (2.9) and (2.11) simultaneously. We observe that the function $j(\cdot)$ is nondifferentiable.

3. Approximation of the steady-state problem.

3.1. Synopsis and formulation of the steady-state problem. Before approximating (2.8) by means of a mixed finite-element method, we shall first study the approximation of the corresponding steady-state problem, i.e., the following elliptic variational inequality of order 4.

$$\begin{cases} \text{Find } \psi \in H_0^2(\Omega) \text{ such that, } \forall \phi \in H_0^2(\Omega), \text{ we have} \\ \mu a(\psi, \phi-\psi)+gj(\phi)-gj(\psi) \geq \int_\Omega \left(\dfrac{\partial f_2}{\partial x_1}-\dfrac{\partial f_1}{\partial x_2}\right)(\phi-\psi)\,d\mathbf{x}, \end{cases} \tag{3.1}$$

where $a(\cdot,\cdot)$ and $j(\cdot)$ are defined by (2.9) and (2.10). We note that (3.1) is equivalent to the following minimization problem.

$$\begin{cases} \text{Find } \psi \in H_0^2(\Omega) \text{ such that} \\ J(\psi) \leq J(\phi) \qquad \forall \phi \in H_0^2(\Omega), \end{cases} \tag{3.2}$$

where, in (3.2), we have

$$J(\phi)=\frac{\mu}{2}a(\phi,\phi)+gj(\phi)-(f,\phi). \tag{3.3}$$

From now on, we shall use the notation $f=\partial f_2/\partial x_1-\partial f_1/\partial x_2$, and we shall assume that $f\in H^{-1}(\Omega)$; in fact, there would be no difficulty in treating the case in which the linear function $\phi \to (f, \phi)$ would be defined by

$$(f,\phi)=\int_\Omega f\,\Delta\phi\,d\mathbf{x} \qquad \forall \phi\in H_0^2(\Omega), \quad f\in L^2(\Omega). \tag{3.4}$$

Since the bilinear form $a(\cdot,\cdot)$ is $H_0^2(\Omega)$ elliptic and the function $j(\cdot)$ is convex and continuous on $H_0^2(\Omega)$, with $\phi\to(f,\phi)$ being linear and continuous, then it is a classical result (see, for example, Chap. 2 and Glowinski [1984, Chap. 1]) that (3.1)–(3.2) admits a unique solution.

3.2. Approximation of (3.1)–(3.2) by a mixed finite-element method. We shall approximate (3.1)-(3.2) here by a mixed finite-element method (suggested by Miyoshi [1973] for other fourth-order problems). The objective is to reduce the approximation to that of a problem in which we only have to perform the discretization of $H^1(\Omega)$ and $L^2(\Omega)$ instead of discretizing $H^2(\Omega)$, which is a much more complicated task. To do this, we first introduce a weakened variational formulation of our problem. The new variational problem thus obtained possesses a unique solution which coincides with that of (3.1)-(3.2) under fairly unrestrictive conditions. For a general presentation of this approach, the reader may refer to Brezzi [1979].

Thus, suppose we have that $\phi \in H_0^2(\Omega)$, where $1 \le i, j \le 2$, and

$$z_{ij} = z_{ji} = \frac{\partial^2 \phi}{\partial x_i \, \partial x_j} \in L^2(\Omega). \tag{3.5}$$

We then have, for all test functions $v \in H^1(\Omega)$,

$$\int_\Omega z_{ij} v \, d\mathbf{x} + \frac{1}{2} \int_\Omega \left(\frac{\partial \phi}{\partial x_i} \frac{\partial v}{\partial x_j} + \frac{\partial \phi}{\partial x_j} \frac{\partial v}{\partial x_i} \right) d\mathbf{x} = 0. \tag{3.6}$$

Conversely, if $\phi \in H_0^1(\Omega)$ and $\mathbf{z} = \{z_{ij}\}_{1 \le i, j \le 2}$ satisfies (3.6), then $\phi \in H_0^2(\Omega)$ and $\mathbf{z}$ and ϕ are related by (3.5). If we have

$$\begin{cases} J(\phi, \mathbf{z}) = \alpha \dfrac{\mu}{2} \displaystyle\int_\Omega (z_{11} + z_{22})^2 \, d\mathbf{x} + \beta \dfrac{\mu}{2} \displaystyle\int_\Omega [(2z_{12})^2 + (z_{22} - z_{11})^2] \, d\mathbf{x} \\ \qquad + g j(\mathbf{z}) - (f, \phi), \end{cases} \tag{3.7}$$

where $\alpha, \beta \in (0, 1)$ with $\alpha + \beta = 1$, and

$$j(\mathbf{z}) = \int_\Omega [(2z_{12})^2 + (z_{22} - z_{11})^2]^{1/2} \, d\mathbf{x}, \tag{3.8}$$

and if we define W as

$$W = \{\{\phi, \mathbf{z}\} \mid \phi \in H_0^1(\Omega), \mathbf{z} \in (L^2(\Omega))^4, \phi \text{ and } \mathbf{z} \text{ satisfy (3.6)}\}, \tag{3.9}$$

we can replace problem (3.1)-(3.2) with the following problem.

$$\begin{cases} \text{Find } \{\psi, \mathbf{s}\} \in W \text{ such that} \\ J(\psi, \mathbf{s}) \le J(\phi, \mathbf{z}) \quad \forall \{\phi, \mathbf{z}\} \in W. \end{cases} \tag{3.10}$$

This problem, which is equivalent to the original problem (3.1)-(3.2), offers a considerable advantage as far as the discretization is concerned, since it requires the approximation of the spaces $H^1(\Omega)$ and $L^2(\Omega)$ only. The discrete variables are then related by (3.6), a weak form of (3.5).

We shall assume in the following that Ω is a convex polygonal in $\mathbb{R}^2$; let $\{\mathcal{T}_h\}_h$ be a standard family of triangulations of Ω. We then define the following.

$$\text{(3.11)} \quad \begin{cases} V_h = \{v_h \mid v_h \in C^0(\bar{\Omega}),\ v_{h|T} \in P_k\ \forall T \in \mathcal{T}_h\}, \\ V_{0h} = \{v_h \mid v_h \in V_h,\ v_h = 0 \text{ on } \Gamma\} = V_h \cap H_0^1(\Omega), \\ W_h = \Big\{\{\phi_h, \mathbf{z}_h\} \mid \phi_h \in V_{0h},\ \mathbf{z}_h \in (V_h)^4, \\ 2\displaystyle\int_\Omega z_{ijh} v_h \, d\mathbf{x} + \int_\Omega \left(\frac{\partial \phi_h}{\partial x_i}\frac{\partial v_h}{\partial x_j} + \frac{\partial \phi_h}{\partial x_j}\frac{\partial v_h}{\partial x_i}\right) d\mathbf{x} = 0\ \forall v_h \in V_h,\ 1 \le i, j \le 2\Big\}. \end{cases}$$

In (3.11), P_k is the space of the polynomials in x_1, x_2 of degree $\le k$. We should observe that the approximations of $H^1(\Omega)$ and $L^2(\Omega)$ are performed here using the same space (V_h) of finite elements. While this procedure is well suited to the present situation, other approaches are available. Finally, the approximate problem will obviously be as follows.

$$\text{(3.12)} \quad \begin{cases} \text{Find } \{\psi_s, \mathbf{s}_h\} \in W_h \text{ such that} \\ J(\psi_h, \mathbf{s}_h) \le J(\phi_h, \mathbf{z}_h) \quad \forall \{\phi_h, \mathbf{z}_h\} \in W_h. \end{cases}$$

In conclusion, we note that the use of $z_{ij} = \partial^2\phi/\partial x_i\,\partial x_j$ as an auxiliary variable indicates the process is particularly well adapted for the treatment of the nondifferentiable term appearing in the function to be minimized; it is for this reason also that the above mixed method was chosen.

3.3. Solvability of the approximate problem (3.12). We have the following theorem.

THEOREM 3.1. *The approximate problem* (3.12) *has a unique solution.*

Proof. See Glowinski, Lions, and Tremolières [1981, App. 6] for the proof of this theorem. □

The solution of (3.12) by augmented Lagrangian algorithms is discussed in § 5 of the present chapter.

3.4. Convergence of the approximate solutions. We shall restrict our attention to the cases $k = 1, 2$ (see Remark 3.1 for $k \ge 3$). We have the following theorem concerning the convergence of the approximate solutions when $h \to 0$.

THEOREM 3.2. *Suppose that when $h \to 0$ the angles of $\mathcal{T}_h$ remain bounded below uniformly in h by $\theta_0 > 0$; suppose also that the following condition is satisfied.*

$$\frac{\max_{T\in\mathcal{T}_h} h(T)}{\min_{T\in\mathcal{T}_h} h(T)} \leq \tau \qquad \forall \mathcal{T}_h, \quad \textit{with } \tau \textit{ independent of } h, \tag{3.13}$$

and where $h(T)$ equals the length of the largest side of T. We then have

$$\lim_{h\to 0} \{\psi_h, \mathbf{s}_h\} = \{\psi, \mathbf{s}\} \quad \textit{strongly in } H_0^1(\Omega)\times(L^2(\Omega))^4 \tag{3.14}$$

where $\{\psi_h, \mathbf{s}_h\}$ is the solution of the approximate problem (3.12), *ψ is that of the continuous problem* (3.1)–(3.2), *and*

$$\mathbf{s} = \{s_{ij}\}_{1\leq i,j\leq 2} \quad \textit{with} \quad s_{ij} = \frac{\partial^2 \psi}{\partial x_i\, \partial x_j}.$$

Proof. See Glowinski, Lions, and Tremolières [1981, App. 6] for the proof of this theorem. □

Remark 3.1. We have assumed that $k = 1, 2$; in fact, similar convergence results could be obtained for approximations based on finite elements of order $k \geq 3$, but, given the limited regularity of the solutions ($\psi \notin H^4(\Omega) \cap H_0^2(\Omega)$ in general), the use of elements of such a high order is not justified.

3.5. Approximation using numerical integration. From a practical point of view, it is necessary to use a numerical integration procedure to approximate the function $J(\cdot,\cdot)$ in (3.10) and (3.12); we shall restrict our attention to the case of $k = 1$. Let Σ_h denote the set of the vertices of $\mathcal{T}_h$; we approximate on V_h the inner product induced by $L^2(\Omega)$, i.e., we approximate

$$\{\lambda_h, \mu_h\} \to \int_\Omega \lambda_h \mu_h \, d\mathbf{x}$$

by

$$(\lambda_h, \mu_h)_h = \frac{1}{3} \sum_{P\in\Sigma_h} m(P)\lambda_h(P)\mu_h(P), \tag{3.15}$$

where, in (3.15), $m(P)$ is the sum of the areas of the triangles that have P as a common vertex. In view of (3.15), we shall use in (3.12) the function $J_h(\cdot,\cdot)$ defined (if $k = 1$) by

$$J_h(\phi_h, \mathbf{z}_h) = \frac{\alpha\mu}{2}(z_{11h} + z_{22h}, z_{11h} + z_{22h})_h$$
$$+ \frac{\beta\mu}{2}[(2z_{12h}, 2z_{12h})_h + (z_{22h} - z_{11h}, z_{22h} - z_{11h})_h]$$

$$
(3.16) \qquad +\frac{g}{3}\sum_{P\in\Sigma_h} m(P)[(2z_{12h}(P))^2+(z_{22h}(P)-z_{11h}(P))^2]^{1/2}
$$

$$
-(f_h, \phi_h),
$$

where f_h is an approximation of f. Similarly, instead of using W_h defined by (3.9), we shall, if $k=1$, use W_h defined by

$$
(3.17) \quad \begin{cases} W_h = \Big\{ \{\phi_h, \mathbf{z}_h\} \in V_{0h} \times (V_h)^4, \forall q_h \in V_h \\ 2(z_{ijh}, q_h)_h = -\displaystyle\int_\Omega \left(\frac{\partial \phi_h}{\partial x_i}\frac{\partial q_h}{\partial x_j} + \frac{\partial \phi_h}{\partial x_j}\frac{\partial q_h}{\partial x_i} \right) d\mathbf{x}, 1 \le i, j \le 2 \Big\}. \end{cases}
$$

Using equation (3.17) it is easy to express $z_{ijh}(P)$ for all $P\in\Sigma_h$ explicitly as a function of the values assumed by ϕ_h on Σ_h; in fact, the matrix associated with the discrete inner product $(\cdot,\cdot)_h$ in V_h is diagonal. In the numerical solution, it is therefore possible to eliminate the variable z_h; we refer the reader to Begis [1979] for further details.

4. Approximation of the time-dependent problem (2.8).

4.1. Semi-discretization with respect to time. Let $k=\Delta t\ (>0)$ denote a time-discretization step; we then approximate (2.8) by the following (backward Euler) implicit scheme (where $\psi^n \sim \psi(nk)$)

$$
(4.1) \quad \begin{cases} \text{for } \psi^n \text{ known, compute } \psi^{n+1} \text{ by solving} \\ \displaystyle\int_\Omega \rho\nabla\left(\frac{\psi^{n+1}-\psi^n}{k}\right)\cdot\nabla(\phi-\psi^{n+1})\,d\mathbf{x} + \mu a(\psi^{n+1}, \phi-\psi^{n+1}) \\ \qquad + gj(\phi) - gj(\psi^{n+1}) \\ \qquad \ge (f((n+1)k), \phi-\psi^{n+1}) \qquad \forall\phi\in H_0^2(\Omega), \quad \psi^{n+1}\in H_0^2(\Omega), \\ \qquad\qquad n=0,1,\cdots, \qquad \psi^0=\psi(0)=\psi_0. \end{cases}
$$

The use of the above semi-discrete scheme has thus enabled us to reduce the solution of the evolution problem (2.8) to that of a sequence of elliptic variational inequalities equivalent to the following sequence of minimization problems (with $n \geq 0$).

$$\text{(4.2)} \qquad \begin{cases} \text{Find } \psi^{n+1} \in H_0^2(\Omega) \text{ such that} \\ J_k^{n+1}(\psi^{n+1}) \leq J_k^{n+1}(\phi) \quad \forall \phi \in H_0^2(\Omega), \end{cases}$$

where

$$\text{(4.3)} \qquad \begin{cases} J_k^{n+1}(\phi) = \dfrac{\rho}{2k} \displaystyle\int_\Omega |\nabla \phi|^2 \, d\mathbf{x} + \dfrac{\mu}{2} a(\phi, \phi) + gj(\phi) \\ \qquad\qquad - (f((n+1)k), \phi) - \dfrac{\rho}{k} \displaystyle\int_\Omega \nabla \psi^n \cdot \nabla \phi \, d\mathbf{x}. \end{cases}$$

The discretization of (4.2)–(4.3) by the mixed finite-element method of § 3 is treated in the following section.

4.2. Full discretization of problem (2.8). The notation is the same as that of § 3.2; we approximate $\psi^0 = \psi_0$ by $\psi_h^0 \in V_{0h}$, and the semi-discrete scheme (4.1) by the following. With the function $\psi_h^n \in V_{0h}$ known, obtain $\{\psi_h^{n+1}, s_h^{n+1}\}$ by solving, for $n = 0, 1, \cdots$, the following minimization problem.

$$\text{(4.4)} \qquad \begin{cases} \text{Find } \{\psi_h^{n+1}, \mathbf{s}_h^{n+1}\} \in W_h \text{ such that} \\ J_{kh}(\psi_h^{n+1}, \mathbf{s}_h^{n+1}) \leq J_{kh}(\phi_h, \mathbf{z}_h) \qquad \forall \{\phi_h, \mathbf{z}_h\} \in W_h, \end{cases}$$

where $j(\cdot)$ is still defined by (3.8), and where

$$\text{(4.5)} \qquad \begin{cases} J_{kh}(\phi_h, \mathbf{z}_h) = \dfrac{\rho}{2k} \displaystyle\int_\Omega |\nabla \phi_h|^2 \, d\mathbf{x} + \dfrac{\alpha\mu}{2} \displaystyle\int_\Omega (z_{11h} + z_{22h})^2 \, d\mathbf{x} \\ \qquad + \dfrac{\beta\mu}{2} \displaystyle\int_\Omega [(2z_{12h})^2 + (z_{22h} - z_{11h})^2] \, d\mathbf{x} + gj(\mathbf{z}_h) \\ \qquad - (f((n+1)k), \phi_h) - \dfrac{\rho}{k} \displaystyle\int_\Omega \nabla \psi_h^n \cdot \nabla \phi_h \, d\mathbf{x}. \end{cases}$$

It can easily be shown that Problem (4.4)–(4.5) has a unique solution; furthermore, the comments in § 3.5 concerning the use of numerical integration are still valid for Problem (4.4)–(4.5).

With regard to the convergence, as h and k approach 0, of the above approximate solutions of Problem (2.8), we refer the reader to Glowinski, Lions, and Tremolieres [1981].

5. Solution of Problems (3.1) and (4.2) by augmented Lagrangian methods.

5.1. Synopsis. In this section, we shall show that it is possible to solve the steady-state problem (3.1), or the sequence of problems (4.2) obtained by the semi-discretization in time of Problem (2.8), by means of augmented Lagrangian methods that fall within the general framework defined in Chapter 3, § 4. We shall confine our attention to the case that is continuous with respect to the space variables, but the generalization to fully discrete problems does not present any particular difficulty (apart from the fact that the formalism that has to be constructed is extremely cumbersome).

5.2. The model problem. Introduction of an augmented Lagrangian function. Problems (3.1) and (4.2) lead us to consider the following minimization problem.

$$\text{(5.1)}\qquad \begin{cases} \text{Find } \psi \in H_0^2(\Omega) \text{ such that} \\ J(\psi) \le J(\phi) \quad \forall \phi \in H_0^2(\Omega), \end{cases}$$

where

$$\text{(5.2)}\qquad \begin{cases} J(\phi) = \dfrac{\gamma}{2}\displaystyle\int_\Omega |\nabla \phi|^2\, d\mathbf{x} + \dfrac{\mu}{2}\int_\Omega |\Delta \phi|^2\, d\mathbf{x} \\ \qquad + g\displaystyle\int_\Omega \left[\left(2\frac{\partial^2 \phi}{\partial x_1\, \partial x_2}\right)^2 + \left(\frac{\partial^2 \phi}{\partial x_2^2} - \frac{\partial^2 \phi}{\partial x_1^2}\right)^2\right]^{1/2} d\mathbf{x} - (f, \phi), \end{cases}$$

and $\gamma \ge 0$ ($\gamma = 0$ for the steady-state problem, $\gamma = \rho/k$ if (6.1) arises from problem (4.2)). The principal difficulty in the solution of (5.1)-(5.2) arises from the nondifferentiable function

$$\phi \to \int_\Omega \left[\left(2\frac{\partial^2 \phi}{\partial x_1\, \partial x_2}\right)^2 + \left(\frac{\partial^2 \phi}{\partial x_2^2} - \frac{\partial^2 \phi}{\partial x_1^2}\right)^2\right]^{1/2} d\mathbf{x}.$$

To overcome this difficulty (as well as to simplify the discretization of the problem), we shall adopt the framework of § 3.2 and consider a mixed variational formulation of Problem (5.1)-(5.2).

With $j(\cdot)$ still defined by (3.8), we consider again

$$\begin{cases} W = \Bigg\{\{\phi, \mathbf{z}\} \in H_0^1(\Omega) \times (L^2(\Omega))^4,\ 2\displaystyle\int_\Omega z_{ij} v\, d\mathbf{x} \\ \qquad = -\displaystyle\int_\Omega \left(\frac{\partial \phi}{\partial x_i}\frac{\partial v}{\partial x_j} + \frac{\partial \phi}{\partial x_j}\frac{\partial v}{\partial x_i}\right) d\mathbf{x}\ \forall v \in H^1(\Omega),\ 1 \le i, j \le 2\Bigg\}, \end{cases}$$

and

$$\begin{cases} J(\phi, \mathbf{z}) = \dfrac{\gamma}{2}\displaystyle\int_\Omega |\nabla \phi|^2\, d\mathbf{x} + \dfrac{\alpha\mu}{2}\int_\Omega (z_{11} + z_{22})^2\, d\mathbf{x} \\ \qquad + \dfrac{\beta\mu}{2}\displaystyle\int_\Omega [(2z_{12})^2 + (z_{22} - z_{11})^2]\, d\mathbf{x} + g j(\mathbf{z}) - (f, \phi). \end{cases}$$

It is thus clear that (5.1)-(5.2) can be expressed as the following problem.

$$\text{(5.3)}\qquad \begin{cases} \text{Find } \{\psi, \mathbf{s}\} \in W \text{ such that} \\ J(\psi, \mathbf{s}) \le J(\phi, \mathbf{z}) \quad \forall \{\phi, \mathbf{z}\} \in W. \end{cases}$$

In order to apply the general methods of Chapter 3 to this case, it is natural to introduce a supplementary variable $\mathbf{q} = \{q_i\}_{i=1}^2 \in (L^2(\Omega))^2$, related to $\mathbf{z}$ by the linear equations

$$\text{(5.4)}\qquad q_1 = 2z_{12}, \qquad q_2 = z_{22} - z_{11}.$$

It is these constraints (5.4) that we shall be treating via the introduction of an augmented Lagrangian function.

To formulate the problem in the notation of Chapter 3, we have that

$$
(5.5)\qquad \begin{cases} V=W, \\ H=(L^2(\Omega))^2, \\ B\in\mathscr{L}(V,H) \quad \text{defined by } B\{\phi,\mathbf{z}\}=\{2z_{12},\,z_{22}-z_{11}\}, \\ F(\mathbf{q})=g\displaystyle\int_\Omega |\mathbf{q}|\,d\mathbf{x}, \\ G(\phi,\mathbf{z})=\dfrac{\gamma}{2}\displaystyle\int_\Omega |\nabla\phi|^2\,d\mathbf{x}+\dfrac{\alpha\mu}{2}\displaystyle\int_\Omega (z_{11}+z_{22})^2\,d\mathbf{x} \\ \qquad\qquad +\dfrac{\beta\mu}{2}\displaystyle\int_\Omega [(2z_{12})^2+(z_{22}-z_{11})^2]\,d\mathbf{x}-(f,\phi). \end{cases}
$$

We then define, for $r>0$, $\{\phi,\mathbf{z}\}\in V$, $\mathbf{q}\in H$, and $\boldsymbol{\mu}\in H$, the augmented Lagrangian function $\mathscr{L}_r:(V\times H)\times H\to\mathbb{R}$ by

$$
(5.6)\qquad \begin{cases} \mathscr{L}_r(\{\phi,\mathbf{z}\},\mathbf{q};\boldsymbol{\mu})=G(\phi,\mathbf{z})+F(\mathbf{q}) \\ +\displaystyle\int_\Omega (2z_{12}-q_1)\mu_1\,d\mathbf{x}+\displaystyle\int_\Omega (z_{22}-z_{11}-q_2)\mu_2\,d\mathbf{x} \\ +\dfrac{r}{2}\displaystyle\int_\Omega |2z_{12}-q_1|^2\,d\mathbf{x}+\dfrac{r}{2}\displaystyle\int_\Omega |z_{22}-z_{11}-q_2|^2\,d\mathbf{x}. \end{cases}
$$

The solution of Problem (5.1)–(5.2) is then reduced to seeking a saddle-point of $\mathscr{L}_r$ on $(V\times H)\times H$. We also could have considered in (5.6) the augmented Lagrangian function associated with

$$
\begin{cases} F(\mathbf{q})=\dfrac{\beta\mu}{2}\displaystyle\int_\Omega |q|^2\,d\mathbf{x}+g\displaystyle\int_\Omega |\mathbf{q}|\,d\mathbf{x}, \\ G(\phi,\mathbf{z})=\dfrac{\gamma}{2}\displaystyle\int_\Omega |\nabla\phi|^2\,d\mathbf{x}+\dfrac{\alpha\mu}{2}\displaystyle\int_\Omega (z_{11}+z_{22})^2\,d\mathbf{x}-(f,\phi). \end{cases}
$$

In the following sections we will solve problems that correspond to the minimization of $\mathscr{L}_r$ on V, $\boldsymbol{\mu}$ and $\mathbf{q}$ being fixed. This minimization leads to solving a linear mixed problem in $\phi,\mathbf{z}$. Our earlier remarks pertaining to the space discretization and the use of numerical integration still apply; the solution of the fully discrete problems by variants of the algorithms described in the next section is straightforward.

5.3. Application of ALG1 to the solution of Problem (5.1)–(5.2). In view of § 5.2, it is natural to solve Problem (5.1)–(5.2) by using ALG1 of Chapter 3, § 4.4. We then obtain the following algorithm.

ALGORITHM (5.7)–(5.9).

$$\boldsymbol{\lambda}^0 \in H = (L^2(\Omega))^2 \text{ given,} \tag{5.7}$$

then, for $n \geq 0$, $\boldsymbol{\lambda}^n \in H$ being known, compute $\{\psi^n, \mathbf{s}^n\} \in V$ and $\mathbf{p}^n \in H$, and then $\boldsymbol{\lambda}^{n+1}$ by

$$\begin{cases} \{\psi^n, \mathbf{s}^n\} \in V, \quad \mathbf{p}^n \in H, \\ \mathcal{L}_r(\{\psi^n, \mathbf{s}^n\}, \mathbf{p}^n; \boldsymbol{\lambda}^n) \leq \mathcal{L}_r(\{\phi, \mathbf{z}\}, \mathbf{q}; \boldsymbol{\lambda}^n) \quad \forall \{\phi, \mathbf{z}\} \in V, \quad \mathbf{q} \in H, \end{cases} \tag{5.8}$$

$$\boldsymbol{\lambda}^{n+1} = \boldsymbol{\lambda}^n + \rho(B\{\psi^n, \mathbf{s}^n\} - \mathbf{p}^n). \tag{5.9}$$

We have the following theorem.

THEOREM 5.1. *Suppose that $\mathcal{L}_r$ has a saddle-point $\{\{\psi, \mathbf{s}\}, \mathbf{p}; \boldsymbol{\lambda}\}$ over $(V \times H) \times H$; then, if*

$$0 < \rho < 2r, \tag{5.10}$$

we have, for all $\boldsymbol{\lambda}^0 \in H$,

$$\lim_{n \to +\infty} \{\psi^n, \mathbf{s}^n\} = \{\psi, \mathbf{s}\} \quad \textit{strongly in } H_0^1(\Omega) \times (L^2(\Omega))^4, \tag{5.11}$$

$$\lim_{n \to +\infty} \mathbf{p}^n = \mathbf{p} \quad \textit{strongly in } (L^2(\Omega))^2, \tag{5.12}$$

$$\lim_{n \to +\infty} \boldsymbol{\lambda}^n = \boldsymbol{\lambda}^* \quad \textit{weakly in } (L^2(\Omega))^2, \tag{5.13}$$

where $\boldsymbol{\lambda}^*$ is such that $\{\{\psi, \mathbf{s}\}, \mathbf{p}; \boldsymbol{\lambda}^*\}$ is a saddle-point of $\mathcal{L}_r$ over $(V \times H) \times H$.

Proof. It is a consequence of the convergence results established in Chapter 3, § 4.6 (see also Fortin and Glowinski [1983, Chap. 3, § 4]). □

It is clear that, once again, the essential difficulty with this approach lies in the fact that system (5.8) has to be solved at each iteration; in view of the structure of $\mathcal{L}_r$, this problem can be solved by a block over-relaxation method like the one described in Chapter 3, § 4.5. As far as the choice of ρ is concerned, numerical experiments indicate once again that the optimal value lies close to $\rho = r$.

5.4. Application of ALG2 and ALG3 to the solution of Problem (5.1)–(5.2). As mentioned in Chapter 3, it may be interesting to solve Problem (5.1)–(5.2) by variants of ALG1. The first of these variants is ALG2, which has already been studied in detail in Chapter 3, § 4, and used extensively in Chapters 4 and 5. In this case, we obtain the following algorithm.

ALGORITHM (5.14)–(5.17).

$$\{\psi^0, \mathbf{s}^0\} \in V, \quad \boldsymbol{\lambda}^1 \in H \text{ are given;} \tag{5.14}$$

then, for $n \geq 1$, $\{\psi^{n-1}, \mathbf{s}^{n-1}\} \in V$ and $\boldsymbol{\lambda}^n \in H$ being known, we compute $\mathbf{p}^n$, $\{\psi^n, \mathbf{s}^n\}$, and $\boldsymbol{\lambda}^{n+1}$ successively by

$$\begin{cases} \mathbf{p}^n \in H, \\ \mathcal{L}_r(\{\psi^{n-1}, \mathbf{s}^{n-1}\}, \mathbf{p}^n; \boldsymbol{\lambda}^n) \leq \mathcal{L}_r(\{\psi^{n-1}, \mathbf{s}^{n-1}\}, \mathbf{q}; \boldsymbol{\lambda}^n) \quad \forall \mathbf{q} \in H, \end{cases} \tag{5.15}$$

(5.16) $$\begin{cases} \{\psi^n, \mathbf{s}^n\} \in V, \\ \mathcal{L}_r(\{\psi^n, \mathbf{s}^n\}, \mathbf{p}^n; \boldsymbol{\lambda}^n) \leq \mathcal{L}_r(\{\phi, \mathbf{z}\}, \mathbf{p}^n; \boldsymbol{\lambda}^n) \quad \forall \{\phi, \mathbf{z}\} \in V, \end{cases}$$

(5.17) $$\boldsymbol{\lambda}^{n+1} = \boldsymbol{\lambda}^n + \rho(B\{\psi^n, \mathbf{s}^n\} - \mathbf{p}^n).$$

It follows from Chapter 3, § 4.6, that the convergence results of Theorem 5.1 still hold if, instead of (5.10), we have

(5.18) $$0 < \rho < \frac{1+\sqrt{5}}{2} r.$$

We shall see that the solution of Problems (5.15) and (5.16) do not introduce any real (new) difficulties. Problem (5.16) is in fact a biharmonic problem

(5.19) $$\begin{cases} \text{Find } \{\psi^n, \mathbf{s}^n\} \in W, \text{ such that, } \forall \{\phi, \mathbf{z}\} \in W, \\ \gamma \int_\Omega \nabla\psi^n \cdot \nabla\phi \, d\mathbf{x} + \alpha\mu \int_\Omega (s_{11}^n + s_{22}^n)(z_{11} + z_{22}) \, d\mathbf{x} \\ + (\beta\mu + r) \int_\Omega [(2s_{12}^n)(2z_{12}) + (s_{22}^n - s_{11}^n)(z_{22} - z_{11})] \, d\mathbf{x} = (f, \phi) \\ - \int_\Omega (\lambda_1^n - rp_1^n)(2z_{12}) \, d\mathbf{x} - \int_\Omega (\lambda_2^n - rp_2^n)(z_{22} - z_{11}) \, d\mathbf{x}. \end{cases}$$

Problem (5.19) does have a unique solution which is easy to compute. In the case of problem (5.15), it can easily be shown that it has a unique solution given explicitly by

(5.20) $$\mathbf{p}^n = \frac{1}{r}(|\mathbf{X}^n| - g)^+ \frac{\mathbf{X}^n}{|\mathbf{X}^n|},$$

where $\mathbf{X}^n = \{X_1^n, X_2^n\}$ is defined by

(5.21) $$X_1^n = \lambda_1^n + 2rs_{12}^{n-1} \quad \text{and}$$

(5.22) $$X_2^n = \lambda_2^n + r(s_{22}^{n-1} - s_{11}^{n-1}),$$

where $|\mathbf{X}| = \sqrt{X_1^2 + X_2^2}$ and $f^+ = \sup(f, 0)$.

Once again, the optimal choice for ρ appears to be close to $\rho = r$; the optimal choice for r is a much more complicated problem, since this optimal value appears to depend on μ and g.

The second variant of ALG1 we wish to explore in the context of Problem (5.1)-(5.2) is ALG3 (see Chap. 3, § 4). We obtain the following algorithm.

ALGORITHM (5.23)-(5.27).

(5.23) $$\{\psi^0, \mathbf{s}^0\} \in V, \qquad \boldsymbol{\lambda}^1 \in H \text{ are given;}$$

then, for $n \geq 1$, $\{\psi^{n-1}, \mathbf{s}^{n-1}\} \in V$ and $\boldsymbol{\lambda}^n \in H$ being known, we compute $\mathbf{p}^n$, $\boldsymbol{\lambda}^{n+1/2}$, $\{\psi^n, \mathbf{s}^n\}$, and $\boldsymbol{\lambda}^{n+1}$ successively, by

$$\begin{cases} \mathbf{p}^n \in H, \\ \mathcal{L}_r(\{\psi^{n-1}, \mathbf{s}^{n-1}\}, \mathbf{p}^n; \boldsymbol{\lambda}^n) \leq \mathcal{L}_r(\{\psi^{n-1}, \mathbf{s}^{n-1}\}, \mathbf{q}; \boldsymbol{\lambda}^n) \quad \forall \mathbf{q} \in H, \end{cases} \tag{5.24}$$

$$\boldsymbol{\lambda}^{n+1/2} = \boldsymbol{\lambda}^n + \rho(B\{\psi^{n-1}, \mathbf{s}^{n-1}\} - \mathbf{p}^n), \tag{5.25}$$

$$\begin{cases} \{\psi^n, \mathbf{s}^n\} \in V, \\ \mathcal{L}_r(\{\psi^n, \mathbf{s}^n\}, \mathbf{p}^n; \boldsymbol{\lambda}^{n+1/2}) \leq \mathcal{L}_r(\{\phi, \mathbf{z}\}, \mathbf{p}^n; \boldsymbol{\lambda}^{n+1/2}) \quad \forall \{\phi, \mathbf{z}\} \in V, \end{cases} \tag{5.26}$$

$$\boldsymbol{\lambda}^{n+1} = \boldsymbol{\lambda}^{n+1/2} + \rho(B\{\psi^n, \mathbf{s}^n\} - \mathbf{p}^n). \tag{5.27}$$

From numerical experiments, Algorithm (5.23)–(5.27) appears to be much faster when the value of ρ lies between $r/2$ and $r/\sqrt{2}$ than when ρ is close to r; such a choice agrees with the analysis done in Chapter 3, § 5.6, for a simpler model problem.

In § 6, we describe the results of a number of numerical experiments in which finite-dimensional variants of Algorithms (5.14)–(5.17) and (5.23)–(5.27) were applied to the solution of Problems (2.8) and (3.1).

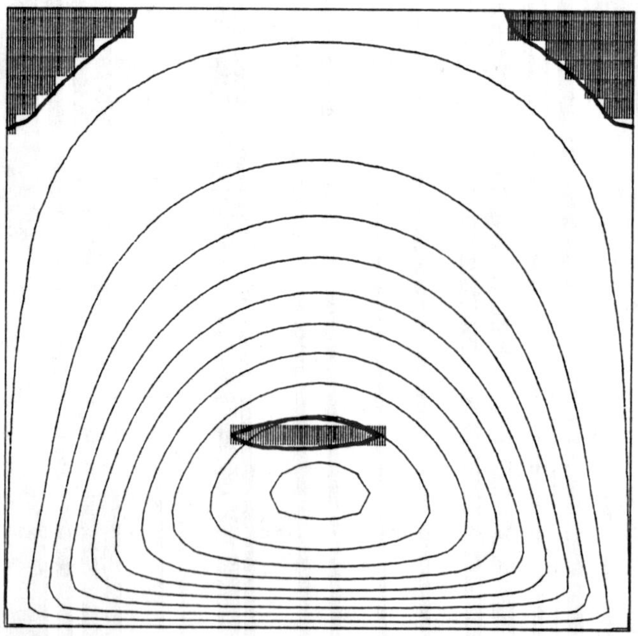

FIG. 6.1. *Bingham fluid flow, steady-state case* ($g = 0.5$) (*Figs.* 6.1–6.5 *courtesy of D. Begis*).

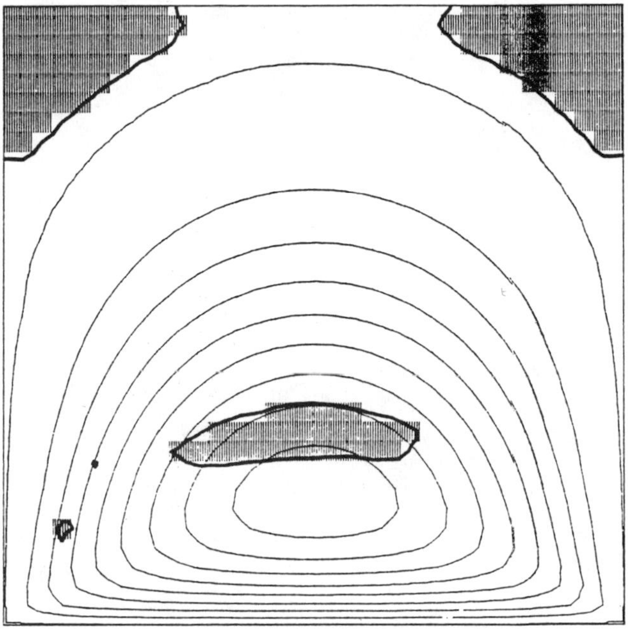

FIG. 6.2. *Bingham fluid flow, steady-state case* $(g=1)$.

6. Numerical experiments. In this section, we describe numerical results obtained by Begis [1979] using the mixed finite-element approximation described in § 3 and a finite-dimensional variant of Algorithm (5.14)–(5.17).

6.1. Formulation of the test problem. Let $\Omega=(0,1)\times(0,1)$. We consider a family of Bingham flows for which (using the notation of § 1) we have

$$\mu=1, \tag{6.1}$$

$$\mathbf{f}=\mathbf{0}, \tag{6.2}$$

$$\begin{cases} \mathbf{u}|_\Gamma=\mathbf{b}=\{b_1,b_2\} \quad \text{with } b_2=0 \quad \text{on } \Gamma,\\ b_1(0,x_2)=b_1(1,x_2)=0 \quad \forall x_2\in(0,1),\\ b_1(x_1,1)=0 \quad \forall x_1\in(0,1) \quad \text{and} \quad b_1(x_1,0)=1 \quad \forall x_1\in(0,1). \end{cases} \tag{6.3}$$

We are thus dealing with problems that can be classed as flows in a cavity with a sliding wall; we note that $\int_\Omega \mathbf{b}\cdot\mathbf{n}\,d\Gamma=0$ ($\mathbf{n}$: unit vector of the outward normal at Γ), but that $\mathbf{b}\notin H^{1/2}(\Gamma)\times H^{1/2}(\Gamma)$ (we have $\mathbf{b}\in H^s(\Gamma)\times H^s(\Gamma)$ for all $s<\frac{1}{2}$).

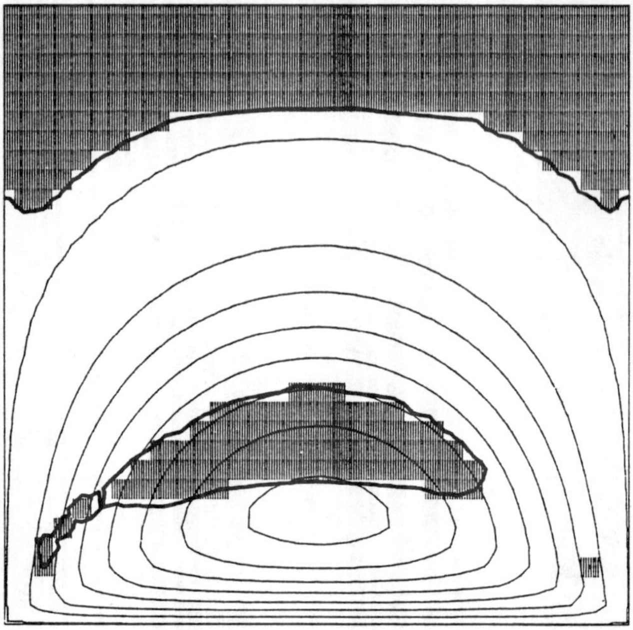

FIG. 6.3. *Bingham fluid flow, steady-state case* ($g = 2.5$).

6.2. Numerical results.[1]

6.2.1. Steady-state cases. We show in Figs. 6.1–6.5, for various values of g, the rigid regions (those where $\mathbf{u} = 0$ are shown hatched) and the viscoplastic regions (in white), as well as the streamlines (these are the equipotentials of ψ); we observe that the zones of rigidity increase with g (the asymmetries that can be seen are due to the asymmetries of the triangulation employed).

6.2.2. Unsteady cases. For various values of g, we have considered the case in which the material filling the cavity Ω is initially at rest (so $\mathbf{u}(\mathbf{x}, 0) = \mathbf{0}\ \forall \mathbf{x} \in \Omega$ that is $\psi(\mathbf{x}, 0) = 0\ \forall \mathbf{x} \in \Omega$) and is then set into motion by the sliding (defined by (6.3)) of the lower wall of Ω. When the steady state has been reached (to within a certain precision), the motion of the lower wall is stopped so that the return of the initial state $\mathbf{u} = \mathbf{0}$ (i.e., $\psi = 0$) in Ω can be observed.

In Fig. 6.6, we show, for various values of g (including $g = 0$), the behavior of

$$t \to \int_\Omega |\psi(\mathbf{x}, t)|^2 \, d\mathbf{x};$$

[1] Details concerning the practical implementation of the finite-element approximations and the iterative algorithms discussed in the previous sections may be found in Begis [1979].

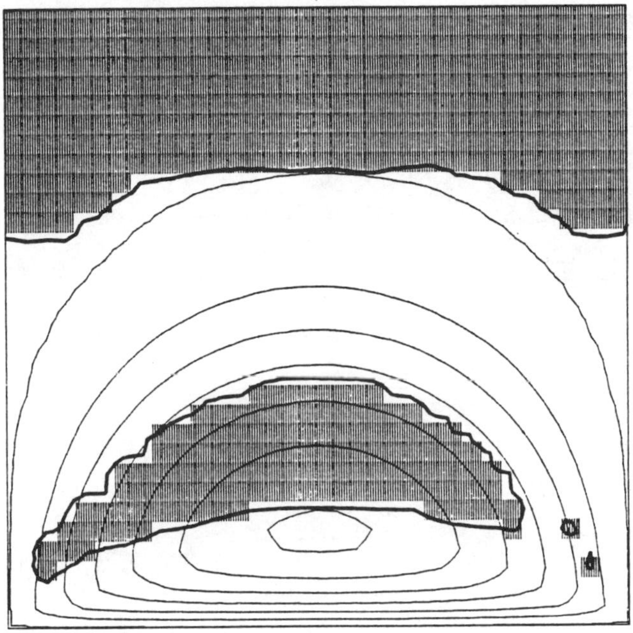

FIG. 6.4. *Bingham fluid flow, steady-state case* $(g = 5)$.

it may be noted that, for $g > 0$, the rigid state is reached in a finite time, which grows progressively smaller as g becomes larger. This agrees with physical intuition and can be justified theoretically.

Methods for the numerical simulation of the two-dimensional flow of Bingham fluids may be found in Fortin [1975] and Begis [1973], these being based on different principles (including the direct use of the velocity formulation of § 1.2). Numerous numerical results that are in agreement with those presented here are also given. See also Glowinski, Lions, and Tremolières [1983, Chap. 6].

7. Further comments. Since the Bingham flow problems discussed in this chapter could have been solved by the methods described in Chapter 4, using the velocity formulation directly, the reader might wonder about the necessity of the present chapter. In fact, the main motivation for Chapter 6 was to show that (at least for two-dimensional flows) stream function formulations can yield quite accurate numerical results. Another motivation was to show that the augmented Lagrangian techniques of Chapter 3 can successfully be applied to the solution of nonlinear fourth-order (in space) partial differential equations or inequalities. In the above example, the nonlinearity was associated to the

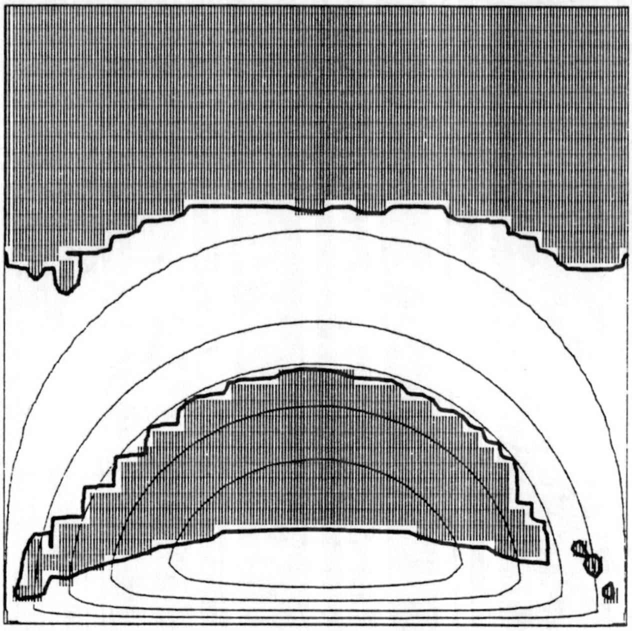

FIG. 6.5. *Bingham fluid flow, steady-state case* $(g = 10)$.

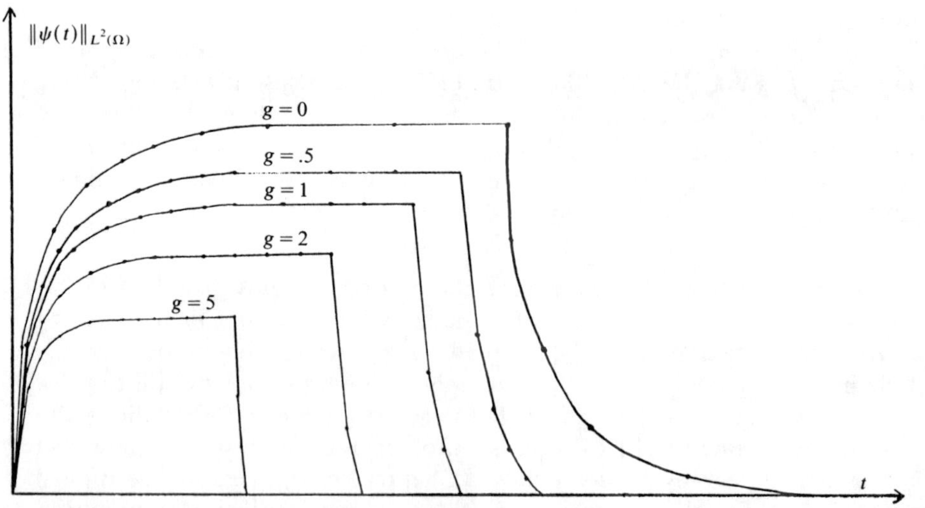

FIG. 6.6. *Bingham fluid flow, unsteady case.*

nondifferentiable function

$$j(\phi)=\int_\Omega \left[\left(2\frac{\partial^2\phi}{\partial x_1\,\partial x_2}\right)^2+\left(\frac{\partial^2\phi}{\partial x_2^2}-\frac{\partial^2\phi}{\partial x_1^2}\right)^2\right]^{1/2} d\mathbf{x}.$$

Indeed, in Glowinski, Marini, and Vidrascu [1984], the techniques of Chapter 3 were also applied to the solution of the following fourth-order variational problem.

$$\begin{cases}\text{Find } u\in K \text{ such that}\\ J(u)\le J(v) \quad \forall v\in K,\end{cases}$$

where

$$J(v)=\frac{1}{2}\int_\Omega |\Delta v|^2\,d\mathbf{x}-\int_\Omega fv\,d\mathbf{x},$$

and

$$K=\{v\mid v\in H_0^2(\Omega),\ \alpha\le\Delta v(\mathbf{x})\le\beta \text{ a.e.}\}.$$

Some preliminary results on this subject can also be found in Glowinski, Lions, and Tremolières [1981, App. 4].

In Chapter 8, we will apply the techniques of Chapter 3 to another class of nonlinear, fourth-order problems, namely, the elastostatic and elastodynamic analysis of flexible pipelines in the presence of large displacements. In these problems (originating from the off-shore petroleum industry), the pipelines are viewed as (long) nonlinear rods.

Chapter 7

Finite Elasticity

1. Classical formulations.

1.1. The physical problem. As seen in Chapter 2, § 5, equilibrium problems in finite elasticity consist of determining the final equilibrium position $\boldsymbol{\varphi}(\mathbf{x}) = \mathbf{x}+\mathbf{u}(\mathbf{x})$ of any particle $\mathbf{x}$ of an elastic body that occupies a known domain Ω in its reference configuration and that is subjected to large deformations due to the application of given external loads and imposed displacements (Chap. 2, Fig. 5.2). Such problems arise in the study of bodies that can undergo large deformations (because they are made of soft materials, or because they are assembled in slender structures, or because of buckling phenomena, or because of geometric singularities like cracks, etc.) but whose physical state depends mainly on the present deformations and not on their history. These situations lead to many interesting nonlinear phenomena such as buckling (Antman and Kenney [1981]), cavitation (Ball [1982]), or change of phases in metallurgy (James [1985]).

Since, for such problems, the considered body is elastic, the stresses at any point $\mathbf{x}$ of the body depend only on the deformation gradient $\mathbf{F}$. To be consistent with our previous notation, we label any particle $\mathbf{x}$ of the body by its position in its reference configuration and, when needed, we relate both the displacement field $\mathbf{u}$ and the particles $\mathbf{x}$ to the fixed orthonormal basis $(\mathbf{e}_i)$ chosen once and for all in the three-dimensional Euclidian space. To further simplify the mathematical formulation of our problem with almost no loss of generality, we suppose that the considered body is made of a hyperelastic material (see Chap. 1, § 5.2). Therefore, at any particle $\mathbf{x}$, the first Piola–Kirchhoff stress tensor $\mathbf{t}$ is given by the constitutive law

$$\mathbf{t} = \rho \frac{\partial \mathcal{W}}{\partial \mathbf{F}}(\mathbf{x}, \mathbf{F}) \qquad \text{(compressible case)}, \tag{1.1}$$

$$\mathbf{t} = \rho \frac{\partial \mathcal{W}}{\partial \mathbf{F}}(\mathbf{x}, \mathbf{F}) - p(\mathbf{x}) \frac{\partial \det(\mathbf{F})}{\partial \mathbf{F}}, \qquad \det \mathbf{F} = 1 \qquad \text{(incompressible case)}. \tag{1.2}$$

Recall that the first Piola–Kirchhoff stress tensor $\mathbf{t}$ characterizes those forces in the present configuration that are applied by one part of the body onto the remainder of the body through an elementary surface that is defined in the reference configuration. Moreover, ρ denotes the mass density in the reference configuration; $\mathbf{F}$ the deformation gradient ($\mathbf{F}=\mathbf{Id}+\partial\mathbf{u}/\partial\mathbf{x}$); p the hydrostatic pressure, which is a multiplier of the incompressibility constraint; and $\mathcal{W}$ the specific free-energy potential which, in hyperelasticity, is a function of $\mathbf{x}$ and $\mathbf{F}$ only. For example, for a Mooney–Rivlin incompressible material, we have

$$\rho\mathcal{W}(\mathbf{x},\mathbf{F})=C_{10}(\mathbf{x})(|\mathbf{F}|^2-3)+C_{01}(\mathbf{x})(|\mathrm{adj}\,\mathbf{F}|^2-3).$$

In this context, the body is subjected to body forces of intensity $\mathbf{f}$ per unit volume in the reference configuration and to surface tractions $\mathbf{g}$ measured per unit area in the reference configuration, prescribed on a portion Γ_2 of the boundary Γ of Ω. The displacement takes on prescribed values $\mathbf{u}_1$ on the complementary part Γ_1 of Γ_2 in Γ, and stresses and external loads are related through the virtual work theorem.

1.2. Weak formulations. In the compressible case, the elimination of the stress tensor $\mathbf{t}$ between the virtual work theorem and the constitutive law (1.1) characterizes the displacement field $\mathbf{u}$ at equilibrium as a solution of the variational problem

$$\boxed{\int_\Omega \rho\frac{\partial\mathcal{W}}{\partial\mathbf{F}}(\mathbf{x},\mathbf{Id}+\nabla\mathbf{u})\cdot\nabla\mathbf{v}\,d\mathbf{x}=\int_\Omega \mathbf{f}\cdot\mathbf{v}\,d\mathbf{x}+\int_{\Gamma_2}\mathbf{g}\cdot\mathbf{v}\,da \qquad \forall\mathbf{v}\in V,\quad \mathbf{u}\in K.} \tag{1.3}$$

This formulation, introduced in § 5.2 of Chapter 2, specifies the set K of kinematically admissible displacement fields and the space V of virtual displacements as

$$K=\{\mathbf{v}\in\mathbf{W}^{1,s}(\Omega),\,\mathbf{v}=\mathbf{u}_1 \text{ on } \Gamma_1,\,\det(\mathbf{Id}+\nabla\mathbf{v})>0 \text{ on } \Omega\}, \tag{1.4}$$

$$V=\{\mathbf{v}\in\mathbf{W}^{1,s}(\Omega),\,\mathbf{v}=\mathbf{0} \text{ on } \Gamma_1\}, \tag{1.5}$$

with s such that the integrals of (1.3) are defined for any $\mathbf{u}$ in K and $\mathbf{v}$ in V. As stated in Remark 5.1 of Chapter 2, the definition of K can be slightly modified in view of Ball's existence theory, but such a change would not affect our numerical strategy.

In the incompressible case, the elimination of the stress tensor $\mathbf{t}$ between the virtual work theorem and the constitutive law (1.2), together with the formal characterization of the term $p\,\partial/\partial\mathbf{F}(\det\mathbf{F})$ as an element $\mathbf{A}$ orthogonal to the surface $Y=\{\mathbf{H}\in\mathbb{R}^9, \det\mathbf{H}=1\}$, leads to the following variational formulation of equilibrium problems in finite elasticity

$$(1.6)\quad \begin{cases} \displaystyle\int_\Omega \left\{\rho\frac{\partial \mathcal{W}}{\partial \mathbf{F}}(\mathbf{Id}+\nabla\mathbf{u})+\mathbf{A}\right\}\cdot\nabla\mathbf{v}\,dx \\ \displaystyle\qquad =\int_\Omega \mathbf{f}\cdot\mathbf{v}\,d\mathbf{x}+\int_{\Gamma_2}\mathbf{g}\cdot\mathbf{v}\,da \qquad \forall\mathbf{v}\in V,\quad \mathbf{u}\in K, \\ \displaystyle\int_\Omega \mathbf{A}\cdot\mathbf{G}\,d\mathbf{x}=0 \qquad \forall\mathbf{G}\in dY(\mathbf{Id}+\nabla\mathbf{u}),\quad \mathbf{A}\in(L^{s^*}(\Omega))^{N\times N}. \end{cases}$$

In (1.6), as in Chapter 2, V, K, and $dY(\cdot)$ are defined by

$$(1.7)\qquad V=\{\mathbf{v}\in\mathbf{W}^{1,s}(\Omega),\mathbf{v}=\mathbf{0}\text{ on }\Gamma_1\},\qquad ss^*=s+s^*,$$

$$(1.8)\qquad K=\{\mathbf{v}\in\mathbf{W}^{1,s}(\Omega),\mathbf{v}=\mathbf{u}_1\text{ on }\Gamma_1,\det(\mathbf{Id}+\nabla\mathbf{v})=1\text{ on }\Omega\},$$

$$(1.9)\qquad dY(\mathbf{F})=\left\{\mathbf{G}\in(L^s(\Omega))^{N\times N},\frac{\partial\det}{\partial\mathbf{F}}(\mathbf{F})\cdot\mathbf{G}=0\text{ a.e. on }\Omega\right\}.$$

Remark 1.1. Formulations (1.3) and (1.6) are both quite general, because they allow any type of hyperelastic constitutive law, because the reference configuration Ω may be of arbitrary shape and does not have to be stress-free, and finally because the external loads **f** and **g** may well depend on the displacement field **u**. Unfortunately, no general existence theory is available for such formulations. The existence theory presented in Chapter 2 deals with minimization problems that are a bit more restrictive and are equivalent to (1.3) and (1.6) only in a formal sense.

Remark 1.2. As written, (1.3) and (1.6) correspond to genuine three-dimensional situations ($N=3$, $\Omega\subset\mathbb{R}^3$). They can also be used to model plane strains situations (situations that are invariant along x_3) simply by setting $N=2$, by defining $\Omega\subset\mathbb{R}^2$ as the section of the reference configuration by the plane $x_3=0$, and by considering only functions $\mathbf{v}(x_1,x_2)$ and $\mathbf{u}(x_1,x_2)$ defined over Ω with values in $\mathbb{R}^2$ to represent, respectively, the virtual and the real in-plane displacements of the particle $\{x_1,x_2,0\}$ of the body. Plane stresses situations can be reduced to plane strains problems by an adequate change of the energy potential $\mathcal{W}$.

2. Augmented Lagrangian formulation.

2.1. Formal derivation. As seen in Chapter 2, for dead loading (**f** and **g** independent of **u**), problems (1.3) and (1.6) formally correspond to the problem of minimizing the total potential energy $J(\cdot)$ over the set K of kinematically admissible fields, that is,

$$\text{find } \mathbf{u}\in K \text{ such that } J(\mathbf{u})\le J(\mathbf{v}) \quad \forall \mathbf{v}\in K, \tag{2.1}$$

where $J(\cdot)$ defined by

$$J(\mathbf{v})=\int_\Omega \rho\mathcal{W}(\mathbf{x}, \mathbf{Id}+\nabla\mathbf{v})\,d\mathbf{x}-\int_\Omega \mathbf{f}\cdot\mathbf{v}\,d\mathbf{x}-\int_{\Gamma_2}\mathbf{g}\cdot\mathbf{v}\,da, \tag{2.2}$$

and K given by (1.4) and (1.8), respectively.

Choosing $\mathbf{F}=\mathbf{Id}+\nabla\mathbf{u}$ as the local variable, which appears to be a very natural choice, this minimization problem is clearly of the form

$$(P)\qquad \mathcal{F}(B(\mathbf{u}-\mathbf{u}_1))+\mathcal{G}(\mathbf{u}-\mathbf{u}_1)\le \mathcal{F}(B\mathbf{v})+\mathcal{G}(\mathbf{v}) \qquad \forall \mathbf{v}\in V,\quad \mathbf{u}-\mathbf{u}_1\in V,$$

under the notation

$$V=\{\mathbf{v}\in \mathbf{W}^{1,s}(\Omega),\ \mathbf{v}=\mathbf{0} \text{ on } \Gamma_1\}, \tag{2.3}$$

$$B\mathbf{v}=\mathbf{Id}+\nabla(\mathbf{v}+\mathbf{u}_1), \tag{2.4}$$

$$\mathcal{G}(\mathbf{v})=\int_\Omega \rho\mathcal{W}_1(\mathbf{x}, \mathbf{Id}+\nabla(\mathbf{v}+\mathbf{u}_1))\,d\mathbf{x}-\int_\Omega \mathbf{f}\cdot(\mathbf{v}+\mathbf{u}_1)\,d\mathbf{x}-\int_{\Gamma_2}\mathbf{g}\cdot(\mathbf{v}+\mathbf{u}_1)\,da, \tag{2.5}$$

$$\mathcal{F}(\mathbf{G})=\int_\Omega \rho\mathcal{W}_2(\mathbf{x}, \mathbf{G})\,d\mathbf{x}+I_Y(\mathbf{G}). \tag{2.6}$$

Above, we have decomposed the specific free-energy potential $\mathcal{W}$ into

$$\mathcal{W}(\mathbf{x}, \mathbf{F})=\mathcal{W}_1(\mathbf{x}, \mathbf{F})+\mathcal{W}_2(\mathbf{x}, \mathbf{F}), \tag{2.7}$$

with $\mathcal{W}_1(\mathbf{x},\cdot)$ being an adequate simple function of $\mathbf{F}$, and we have introduced the indicator function I_Y defined by

$$I_Y(\mathbf{G})=0 \quad \text{if } \mathbf{G}\in Y, \qquad I_Y(\mathbf{G})=+\infty \quad \text{if not}, \tag{2.8}$$

where

$$Y=\{\mathbf{G}\in (L^s(\Omega))^{N\times N},\ \det \mathbf{G}>0 \text{ a.e. in } \Omega\} \tag{2.9}$$

for the compressible case and

$$Y=\{\mathbf{G}\in (L^s(\Omega))^{N\times N},\ \det \mathbf{G}=1 \text{ a.e. in } \Omega\} \tag{2.10}$$

for the incompressible case.

The reader will recall that problems similar to (P) were studied extensively in Chapter 3, § 4, where, in view of their numerical solution, they were associated with the saddle-point problem,

$$
(2.11)\qquad \begin{cases} \text{find } \{\mathbf{u}-\mathbf{u}_1, \mathbf{F}; \boldsymbol{\lambda}\} \in (V\times H)\times H \text{ such that} \\ \overline{\mathscr{L}}_r(\mathbf{u}-\mathbf{u}_1, \mathbf{F}; \boldsymbol{\mu}) \le \overline{\mathscr{L}}_r(\mathbf{u}-\mathbf{u}_1, \mathbf{F}; \boldsymbol{\lambda}) \le \overline{\mathscr{L}}_r(\mathbf{v}, \mathbf{G}; \boldsymbol{\lambda}) \\ \hfill \forall \{\mathbf{v}, \mathbf{G}; \boldsymbol{\mu}\} \in (V\times H)\times H, \end{cases}
$$

corresponding to the augmented Lagrangian $\overline{\mathscr{L}}_r$ defined by

$$
(2.12)\qquad \overline{\mathscr{L}}_r(\mathbf{v}, \mathbf{G}; \boldsymbol{\mu}) = \mathscr{F}(\mathbf{G}) + \mathscr{G}(\mathbf{v}) + \frac{r}{2}|B\mathbf{v}-\mathbf{G}|^2 + (\boldsymbol{\mu}, B\mathbf{v}-\mathbf{G}),
$$

and to the Hilbert space H (here, $(L^2(\Omega))^{N\times N}$) endowed with the scalar product $(\cdot,\cdot)$ and the associated norm $|\cdot|$.

It is thus very tempting to replace the variational formulations (1.3) and (1.6) of equilibrium problems in finite elasticity with the saddle-point problem (2.11) to be solved numerically by one of the algorithms (ALG1–ALG4) introduced in Chapter 3. However, before this can be done, it is necessary to modify Problem (2.11). Indeed, as written, (2.11) applies only to conservative loadings. More critically, because $\mathscr{F}(\cdot)$ is not convex, (2.11) may not be equivalent to the original problems (1.3) and (1.6). In the next section, we will therefore reformulate (2.11) in the light of these considerations.

2.2. Final formulation. To handle any type of loading and to recover equivalence with (1.3) and (1.6), we simply replace the saddle-point problem (2.11) by the associated formal Euler–Lagrange equations, i.e.,

$$
\begin{cases} \dfrac{\partial \overline{\mathscr{L}}_r}{\partial \mathbf{v}}(\mathbf{u}-\mathbf{u}_1, \mathbf{F}; \boldsymbol{\lambda}) = 0 \quad \text{in } V^*, \\[2mm] \dfrac{\partial \overline{\mathscr{L}}_r}{\partial \mathbf{G}}(\mathbf{u}-\mathbf{u}_1, \mathbf{F}; \boldsymbol{\lambda}) = 0 \quad \text{in } H^*, \\[2mm] \dfrac{\partial \overline{\mathscr{L}}_r}{\partial \boldsymbol{\mu}}(\mathbf{u}-\mathbf{u}_1, \mathbf{F}; \boldsymbol{\lambda}) = 0 \quad \text{in } H^*. \end{cases}
$$

However, because $\overline{\mathscr{L}}_r$ is not differentiable, we introduce our final augmented Lagrangian

$$
(2.13)\qquad \begin{cases} \mathscr{L}_r(\mathbf{v}, \mathbf{G}; \boldsymbol{\mu}) = \displaystyle\int_\Omega \rho \mathscr{W}_1(\mathbf{x}, \mathbf{Id}+\nabla\mathbf{v})\, d\mathbf{x} + \int_\Omega \rho \mathscr{W}_2(\mathbf{x}, \mathbf{G})\, d\mathbf{x} \\ \qquad\qquad + \dfrac{r}{2}\displaystyle\int_\Omega \eta(\mathbf{x})|\mathbf{G}-\mathbf{Id}-\nabla\mathbf{v}|^2\, d\mathbf{x} \\ \qquad\qquad + \displaystyle\int_\Omega \eta(\mathbf{x})\boldsymbol{\mu}\cdot(\mathbf{Id}+\nabla\mathbf{v}-\mathbf{G})\, d\mathbf{x}, \end{cases}
$$

and consider the following problem.

$$
(2.14)\quad \left\{
\begin{array}{l}
\text{Find } \{\mathbf{u}-\mathbf{u}_1, \mathbf{F}; \boldsymbol{\lambda}\} \in V \times Y \times (L^{s^*}(\Omega))^{N\times N} \text{ such that} \\[2mm]
\text{(i)}\quad \dfrac{\partial \mathcal{L}_r}{\partial \mathbf{v}}(\mathbf{u}, \mathbf{F}; \boldsymbol{\lambda}) \cdot \mathbf{v} = \displaystyle\int_\Omega \mathbf{f} \cdot \mathbf{v}\, d\mathbf{x} + \int_{\Gamma_2} \mathbf{g} \cdot \mathbf{v}\, da \quad \forall \mathbf{v} \in V, \\[2mm]
\text{(ii)}\quad \dfrac{\partial \mathcal{L}_r}{\partial \mathbf{G}}(\mathbf{u}, \mathbf{F}; \boldsymbol{\lambda}) \cdot \mathbf{G} = 0 \quad \forall \mathbf{G} \in dY(\mathbf{F}), \\[2mm]
\text{(iii)}\quad \dfrac{\partial \mathcal{L}_r}{\partial \boldsymbol{\mu}}(\mathbf{u}, \mathbf{F}; \boldsymbol{\lambda}) \cdot \boldsymbol{\mu} = 0 \quad \forall \boldsymbol{\mu} \in (L^{s^*}(\Omega))^{N\times N}.
\end{array}
\right.
$$

In (2.13) and (2.14), $\mathcal{W}_1$ and $\mathcal{W}_2$ are given by (2.7); V by (2.3); Y by (2.9); and $dY(\mathbf{F})$ by $(L^s(\Omega))^{N\times N}$ for the compressible case; and Y by (2.10) and $dY(\mathbf{F})$ by (1.9) for the incompressible case. Moreover, r denotes an arbitrary positive constant and $\eta(\mathbf{x})$ an arbitrary positive function bounded away from zero. As a further help to the interpretation of (2.14), we give the following explicit form of the derivatives of $\mathcal{L}_r$ at $\{\mathbf{u}, \mathbf{F}; \boldsymbol{\lambda}\}$, which follows from its definition (2.13).

$$
(2.15)\quad \left\{
\begin{array}{l}
\dfrac{\partial \mathcal{L}_r}{\partial \mathbf{v}}(\mathbf{u}, \mathbf{F}; \boldsymbol{\lambda}) \cdot \mathbf{v} = \displaystyle\int_\Omega \left\{ \rho \frac{\partial \mathcal{W}_1}{\partial \mathbf{F}}(\mathbf{x}, \mathbf{Id} + \nabla \mathbf{u}) \right. \\[2mm]
\qquad\qquad\qquad \left. + \eta(\mathbf{x})(r(\mathbf{Id} + \nabla \mathbf{u} - \mathbf{F}) + \boldsymbol{\lambda}) \right\} \cdot \nabla \mathbf{v}\, d\mathbf{x},
\end{array}
\right.
$$

$$
(2.16)\quad \frac{\partial \mathcal{L}_r}{\partial \mathbf{G}}(\mathbf{u}, \mathbf{F}; \boldsymbol{\lambda}) \cdot \mathbf{G} = \int_\Omega \left\{ \rho \frac{\partial \mathcal{W}_2}{\partial \mathbf{F}}(\mathbf{x}, \mathbf{F}) - \eta(\mathbf{x})(r(\mathbf{Id} + \nabla \mathbf{u} - \mathbf{F}) + \boldsymbol{\lambda}) \right\} \cdot \mathbf{G}\, d\mathbf{x},
$$

$$
(2.17)\quad \frac{\partial \mathcal{L}_r}{\partial \boldsymbol{\mu}}(\mathbf{u}, \mathbf{F}; \boldsymbol{\lambda}) \cdot \boldsymbol{\mu} = \int_\Omega \eta(\mathbf{x}) \boldsymbol{\mu} \cdot (\mathbf{Id} + \nabla \mathbf{u} - \mathbf{F})\, d\mathbf{x}.
$$

Remark 2.1. We will see in the next section that the choice of r, η, or $\mathcal{W}_1$ does not affect the values of the solutions of (2.14) and that, in this respect, this choice can be arbitrary. However, their selection does affect the convergence properties of the algorithms used, and, for this reason, a proper strategy should be followed for choosing r, η, and $\mathcal{W}_1$.

2.3. Equivalence theorem. We have the following theorem.

THEOREM 2.1. *The variational formulations* (1.3) *and* (1.6) *of equilibrium problems in finite elasticity are equivalent to the Lagrangian formulation* (2.14).

Proof. We will consider only the incompressible case, since the compressible one is quite similar.

First, let $\{\mathbf{u}, \mathbf{A}\}$ be a solution of (1.6). If we require that

$$\mathbf{F} = \mathbf{Id} + \nabla \mathbf{u}, \tag{2.18}$$

$$\boldsymbol{\lambda} = \left(\mathbf{A} + \rho \frac{\partial \mathcal{W}_2}{\partial \mathbf{F}}(\mathbf{x}, \mathbf{F}) \right) \Big/ \eta, \tag{2.19}$$

it follows that $\mathbf{F} \in Y$, because $\mathbf{u} \in K$. From the first relation of (1.6), we thus directly obtain

$$\frac{\partial \mathcal{L}_r}{\partial \mathbf{v}}(\mathbf{u}, \mathbf{F}; \boldsymbol{\lambda}) \cdot \mathbf{v} = \int_\Omega \mathbf{f} \cdot \mathbf{v} \, d\mathbf{x} + \int_{\Gamma_2} \mathbf{g} \cdot \mathbf{v} \, da \quad \forall \mathbf{v} \in V.$$

On the other hand, from the second relation of (1.6), we satisfy that

$$\frac{\partial \mathcal{L}_r}{\partial \mathbf{G}}(\mathbf{u}, \mathbf{F}; \boldsymbol{\lambda}) \cdot \mathbf{G} = 0 \quad \forall \mathbf{G} \in dY(\mathbf{F}).$$

Finally, from (2.18), we can verify that

$$\frac{\partial \mathcal{L}_r}{\partial \boldsymbol{\mu}}(\mathbf{u}, \mathbf{F}; \boldsymbol{\lambda}) \cdot \boldsymbol{\mu} = 0 \quad \forall \boldsymbol{\mu} \in (L^{s^*}(\Omega))^{N \times N}.$$

In summary, therefore, $\{\mathbf{u} - \mathbf{u}_1, \mathbf{F}; \boldsymbol{\lambda}\}$ is a solution of (2.14).

Conversely, let $\{\mathbf{u} - \mathbf{u}_1, \mathbf{F}; \boldsymbol{\lambda}\}$ be a solution of (2.14). Since η is strictly positive, (2.14)(iii) can hold only if we have

$$\mathbf{F} = \mathbf{Id} + \nabla \mathbf{u} \quad \text{a.e. in } \Omega, \tag{2.20}$$

which implies, since $\mathbf{F} \in Y$, that $\mathbf{u}$ belongs to K. Moreover, if we denote

$$\mathbf{A} = \eta \boldsymbol{\lambda} - \rho \frac{\partial \mathcal{W}_2}{\partial \mathbf{F}}(\mathbf{x}, \mathbf{F}), \tag{2.21}$$

it follows from (2.20) that (1.6) is directly satisfied by (2.14)(i) and (2.14)(ii). Thus, $\{\mathbf{u}, \mathbf{A}\}$ is a solution of (1.6), which completes our proof. □

Remark 2.2. In the incompressible case, it follows from the constitutive law (1.2), the formal identification $\mathbf{A} = -p \, \partial \det / \partial \mathbf{F}(\mathbf{F})$, and the identity (2.21) that the first Piola–Kirchhoff stress tensor $\mathbf{t}$ is given from the solution $\{\mathbf{u} - \mathbf{u}_1, \mathbf{F}; \boldsymbol{\lambda}\}$ of (2.14) by the formula

$$\mathbf{t} = \rho \frac{\partial \mathcal{W}_1}{\partial \mathbf{x}}(\mathbf{x}, \mathbf{F}) + \eta \boldsymbol{\lambda}. \tag{2.22}$$

Similarly, in the compressible case, we have

$$\mathbf{t} = \rho \frac{\partial \mathcal{W}_1}{\partial \mathbf{x}}(\mathbf{x}, \mathbf{F}) + \eta \boldsymbol{\lambda} = \rho \frac{\partial \mathcal{W}_1}{\partial \mathbf{x}}(\mathbf{x}, \mathbf{F}) + \rho \frac{\partial \mathcal{W}_2}{\partial \mathbf{x}}(\mathbf{x}, \mathbf{F}). \tag{2.23}$$

3. Finite-element discretization.

3.1. The discrete augmented Lagrangian formulation. In view of its numerical solution, the Lagrangian problem (2.14) must first be approximated by a problem with a finite number of unknowns. This approximation is classically realized in the finite-element method by replacing the spaces V and H on which (2.14) is set with the finite-dimensional subspaces V_h and H_h. Here, introducing

$$V_h \subset V, \qquad \dim V_h < +\infty, \tag{3.1}$$

$$H_h \subset (L^\infty(\Omega))^{N\times N}, \qquad \dim H_h < +\infty, \tag{3.2}$$

$$\begin{cases} \nabla_h : \mathbf{W}^{1,s}(\Omega) \to H_h, \\ \nabla_h \mathbf{v} = L^2 \text{ projection of } \nabla\mathbf{v} \text{ on } H_h, \end{cases} \tag{3.3}$$

$$\begin{cases} \mathscr{L}_r^h(\mathbf{v}, \mathbf{G}; \boldsymbol{\mu}) = \displaystyle\int_\Omega \rho \mathscr{W}_1(\mathbf{x}, \mathbf{Id}+\nabla\mathbf{v})\, d\mathbf{x} + \int_\Omega \rho\mathscr{W}_2(\mathbf{x}, \mathbf{G})\, d\mathbf{x} \\ \qquad + \dfrac{r}{2}\displaystyle\int_\Omega \eta(\mathbf{x})|\mathbf{G}-\mathbf{Id}-\nabla_h\mathbf{v}|^2\, d\mathbf{x} \\ \qquad + \displaystyle\int_\Omega \eta(\mathbf{x})\boldsymbol{\mu}\cdot(\mathbf{Id}+\nabla_h\mathbf{v}-\mathbf{G})\, d\mathbf{x}, \end{cases} \tag{3.4}$$

we obtain the following finite-element approximation of the Lagrangian problem (2.14).

$$\begin{cases} \text{Find } \{\mathbf{u}_h - \mathbf{u}_1, \mathbf{F}_h; \boldsymbol{\lambda}_h\} \in V_h \times (Y \cap H_h) \times H_h \text{ such that} \\ \text{(i)} \quad \dfrac{\partial \mathscr{L}_r^h}{\partial \mathbf{v}}(\mathbf{u}_h, \mathbf{F}_h; \boldsymbol{\lambda}_h)\cdot \mathbf{v}_h = \displaystyle\int_\Omega \mathbf{f}\cdot\mathbf{v}_h\, d\mathbf{x} + \int_{\Gamma_2} \mathbf{g}\cdot\mathbf{v}_h\, da \quad \forall \mathbf{v}_h \in V_h, \\ \text{(ii)} \quad \dfrac{\partial \mathscr{L}_r^h}{\partial \mathbf{G}}(\mathbf{u}_h, \mathbf{F}_h; \boldsymbol{\lambda}_h)\cdot \mathbf{G}_h = 0 \quad \forall \mathbf{G}_h \in dY(\mathbf{F}_h)\cap H_h, \\ \text{(iii)} \quad \dfrac{\partial \mathscr{L}_r^h}{\partial \boldsymbol{\mu}}(\mathbf{u}_h, \mathbf{F}_h; \boldsymbol{\lambda}_h)\cdot \boldsymbol{\mu}_h = 0 \quad \forall \boldsymbol{\mu}_h \in H_h. \end{cases} \tag{3.5}$$

In (3.5) we have kept the definitions of (2.14) for the functions $\mathscr{W}_1$ and $\mathscr{W}_2$, for the sets Y and $dY(\mathbf{F})$, for the constant r, and for the positive weight η.

3.2. A first choice of discrete spaces. As our first choice, we construct the approximate spaces V_h and H_h using quadrilateral (resp., hexahedral if $\Omega \subset \mathbb{R}^3$)

isoparametric finite elements, the displacements of V being interpolated at each vertex, and the matrices of H being interpolated at the center of each element. More precisely, we suppose that Ω is a polygonal (resp., polyhedral) domain of $\mathbb{R}^2$ (resp., $\mathbb{R}^3$) which can be decomposed into a finite number N_h of quadrilaterals (resp., hexahedrals) Ω_l such that the following conditions are satisfied.

(i) $\bar{\Omega} = \bigcup_{l=1,N_h} \bar{\Omega}_l$.

(ii) The diameter of any Ω_l is bounded by h.

(iii) Any Ω_l contains a ball of radius αh, α being given once and for all and independent of h.

(iv) Two different elements Ω_l have either nothing, a vertex, an edge, or a face in common.

By extension, we will still refer to such a partition $\mathcal{T}_h$ of Ω as a regular "triangulation" of Ω. Once $\mathcal{T}_h$ is given (see Fig. 3.1), the spaces H_h and V_h are simply

(3.6) $$H_h = \{\mathbf{G}_h : \Omega \to \mathbb{R}^{N\times N}, \mathbf{G}_h|_{\Omega_l} = \text{constant } \forall l = 1, \cdots, N_h\}.$$

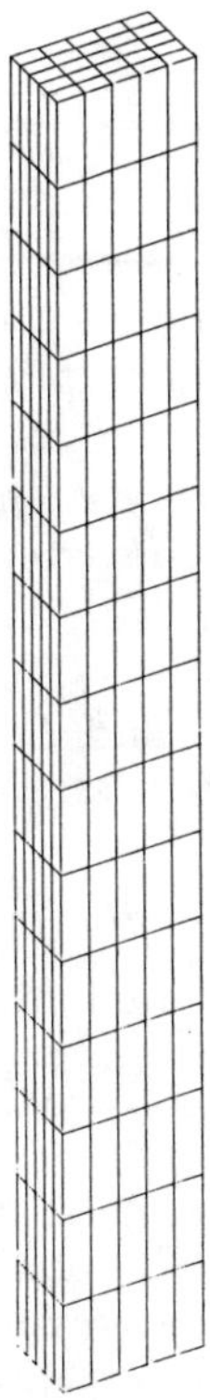

FIG. 3.1. *A three-dimensional triangulation.*

$$(3.7)\qquad \begin{cases} V_h=\{\mathbf{v}_h:\bar{\Omega}\to\mathbb{R}^N, \mathbf{v}_h \text{ continuous}, \mathbf{v}_h=\mathbf{0} \text{ on } \Gamma_1, \\ \mathbf{v}_h|_{\Omega_l}=\mathbf{v}_l\circ\boldsymbol{\varphi}_l^{-1}\mathbf{v}_l\in[Q_1(\hat{\Omega})]^N\ \forall l=1,\cdots,N_h\}. \end{cases}$$

In (3.7), $\boldsymbol{\varphi}_l$ is the mapping from the reference element $\hat{\Omega}=(-1,+1)^N$ into Ω_l defined by

$$\boldsymbol{\varphi}_l(\hat{\mathbf{x}})=\sum_{\alpha=1}^{2^N}\mathbf{x}^{\alpha l}\hat{\varphi}^{\alpha}(\hat{\mathbf{x}}),$$

where $\mathbf{x}^{\alpha l}$ denotes the coordinates of the α-vertex of the element Ω_l and where $\hat{\varphi}^{\alpha}$ denotes the function of $Q_1(\hat{\Omega})$ with values 1 at vertex α and 0 at all other vertices. The space $Q_1(\hat{\Omega})$ is the usual space of N-linear polynomials defined over $\hat{\Omega}$, that is,

$$Q_1(\hat{\Omega})=\left\{\hat{q},\ \hat{q}(\hat{\mathbf{x}})=a_0+\sum_{i=1}^{N}a_i\hat{x}_i+\sum_{i\neq j}a_{ij}\hat{x}_i\hat{x}_j+\sum_{i\neq j\neq k}a_{ijk}\hat{x}_i\hat{x}_j\hat{x}_k\right\}.$$

The mapping $\boldsymbol{\nabla}_h$ from V_h into H_h is then given by

$$(3.8)\qquad \boldsymbol{\nabla}_h(\mathbf{v})|_{\Omega_l}=\frac{1}{\text{meas }(\Omega_l)}\int_{\Omega_l}\boldsymbol{\nabla}\mathbf{v}\,d\mathbf{x}.$$

3.3. Another choice of discrete spaces. The above choice of finite elements leads to local numerical unstabilities in certain, specific situations—the computed displacement field $\mathbf{u}_h$ and multiplier $\boldsymbol{\lambda}_h$ may oscillate next to Γ_1 for irregular loadings. For $\Omega\subset\mathbb{R}^2$, these oscillations can be suppressed either by considering triangulations of the type shown in Fig. 3.2 or by choosing asymmetric, triangular finite elements (Ruas [1981]).

In the latter case, the domain Ω is divided into triangles (this triangulation being regular in the sense discussed in § 3.2), the displacements are interpolated at each vertex and at one midside, and the matrices of H are interpolated at the center of each triangle (Fig. 3.3). More precisely, we have

$$(3.9)\qquad H_h=\{\mathbf{G}_h:\Omega\to\mathbb{R}^{N\times N}, \mathbf{G}_h|_{\Omega_l}=\text{const } \forall l=1,\cdots,N_h\},$$

$$(3.10)\qquad \begin{cases} V_h=\{\mathbf{v}_h:\bar{\Omega}\to\mathbb{R}^N, \mathbf{v}_h \text{ continuous}, \mathbf{v}_h=\mathbf{0} \text{ on } \Gamma_1, \mathbf{v}_h|_{\Omega_l} \text{ is} \\ \quad \text{a second-order polynomial whose restriction to two sides} \\ \quad \text{of } \Omega_l \text{ is a first-order polynomial } \forall l=1,\cdots,N_h\}, \end{cases}$$

the mapping $\boldsymbol{\nabla}_h$ still being defined by (3.8).

Remark 3.1. A priori, many different choices exist for V_h and H_h. Nevertheless, we will see in § 3.5 that V_h and H_h cannot be chosen independently and must satisfy a compatibility condition, which restricts the choice to few finite elements.

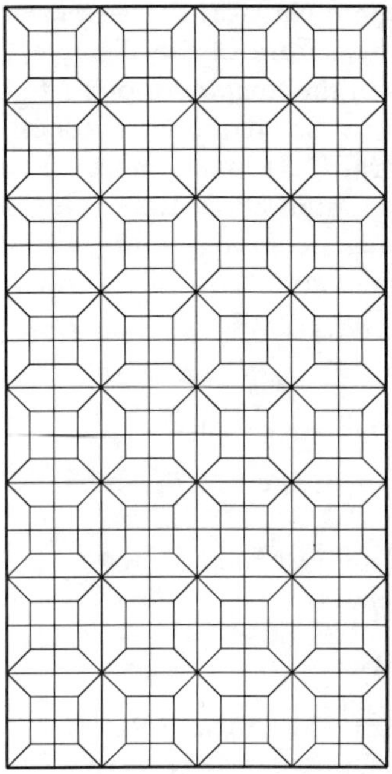

FIG. 3.2. *A two-dimensional triangulation of special type.*

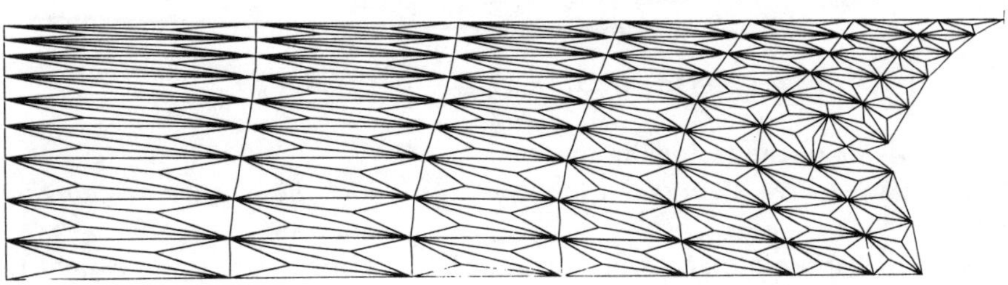

(a)

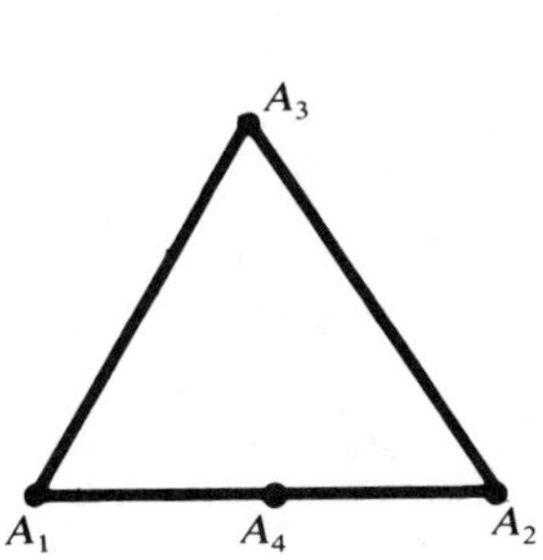

(b)

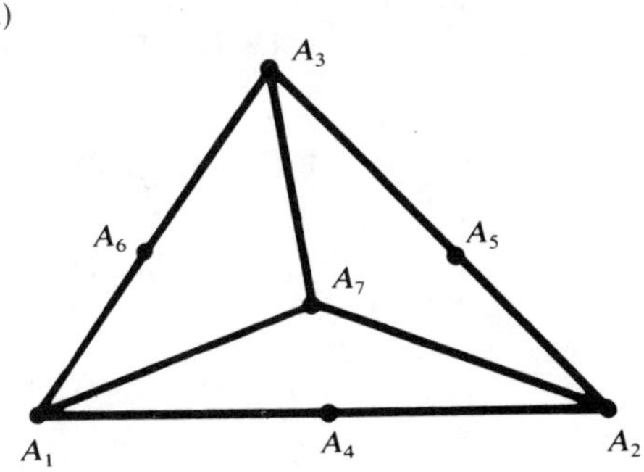

(c)

FIG. 3.3. (a) *A complete triangulation of the Ruas type*; (b) *the basic element*; (c) *three elements together.*

3.4. Relationship to more classical finite-element formulations. So that we can further investigate the nature of the discrete Lagrangian problem (3.5), we will restrict ourselves in the remainder of § 3 to the case of incompressible materials in which we have the spaces H_h made of piecewise constants. This will enable us to transform (3.5) into a more classical problem whose study will bring us useful information about the choice of V_h and H_h.

More precisely, assuming incompressibility, we introduce a space V_h satisfying (3.1), a regular "triangulation" $\mathcal{T}_h$ of Ω, and the spaces

$$P_h = \{q_h \in L^\infty(\Omega),\ q_h|_{\Omega_l} = \text{const}\ \forall l = 1, \cdots, N_h\}, \tag{3.11}$$

$$H_h = (P_h)^{N\times N}. \tag{3.12}$$

We then consider the following problem.

$$\left\{\begin{aligned} &\text{Find } (\mathbf{u}_h - \mathbf{u}_1) \in V_h \text{ and } p_h \in P_h \text{ such that}\\ &\text{(i)}\ \int_\Omega \rho \frac{\partial \mathcal{W}_1}{\partial \mathbf{F}}(\mathbf{x}, \mathbf{Id} + \nabla \mathbf{u}_h) \cdot \nabla \mathbf{v}_h\, d\mathbf{x}\\ &\quad + \int_\Omega \rho \frac{\partial \mathcal{W}_2}{\partial \mathbf{F}}(\mathbf{x}, \mathbf{Id} + \nabla_h \mathbf{u}_h) \cdot \nabla_h \mathbf{v}_h\, d\mathbf{x}\\ &\quad - \int_\Omega p_h \frac{\partial \det}{\partial \mathbf{F}}(\mathbf{Id} + \nabla_h \mathbf{u}_h) \cdot \nabla_h \mathbf{v}_h\, d\mathbf{x}\\ &\quad - \int_\Omega \mathbf{f} \cdot \mathbf{v}_h\, d\mathbf{x} - \int_{\Gamma_2} \mathbf{g} \cdot \mathbf{v}_h\, da = 0 \quad \forall \mathbf{v}_h \in V_h;\\ &\text{(ii)}\ \int_\Omega q_h(\det(\mathbf{Id} + \nabla_h \mathbf{u}_h) - 1)\, d\mathbf{x} = 0 \quad \forall q_h \in P_h. \end{aligned}\right. \tag{3.13}$$

This leads us to the following theorem.

THEOREM 3.1. *The variational problem* (3.13) *and the discrete Lagrangian problem* (3.5) *are equivalent.*

Proof. First observe that, as in Theorem 2.1, Problem (3.5) is equivalent to

$$\left\{\begin{aligned} &\int_\Omega \rho \frac{\partial \mathcal{W}_1}{\partial \mathbf{F}}(\mathbf{x}, \mathbf{Id} + \nabla \mathbf{u}_h) \cdot \nabla \mathbf{v}_h\, d\mathbf{x}\\ &\qquad + \int_\Omega \left\{\rho \frac{\partial \mathcal{W}_2}{\partial \mathbf{F}}(\mathbf{x}, \mathbf{Id} + \nabla_h \mathbf{u}_h) + \mathbf{A}_h\right\} \cdot \nabla_h \mathbf{v}_h\, d\mathbf{x}\\ &\qquad = \int_\Omega \mathbf{f} \cdot \mathbf{v}_h\, d\mathbf{x} + \int_{\Gamma_2} \mathbf{g} \cdot \mathbf{v}_h\, da \qquad \forall \mathbf{v}_h \in V_h, \quad \mathbf{u}_h \in K_h, \end{aligned}\right. \tag{3.14}$$

$$\int_\Omega \mathbf{A}_h \cdot \mathbf{G}_h\, d\mathbf{x} = 0 \qquad \forall \mathbf{G}_h \in H_h \cap dY(\mathbf{Id} + \nabla_h \mathbf{u}_h), \quad \mathbf{A}_h \in H_h, \tag{3.15}$$

with

$$\mathbf{F}_h = \mathbf{Id} + \nabla_h \mathbf{u}_h, \qquad \boldsymbol{\lambda}_h = \left(\rho \frac{\partial \mathcal{W}_2}{\partial \mathbf{F}}(\mathbf{x}, \mathbf{F}_h) + \mathbf{A}_h\right) \Big/ \eta,$$

and

$$K_h=\{\mathbf{v}_h, \mathbf{v}_h-\mathbf{u}_1\in V_h, (\mathbf{Id}+\nabla_h\mathbf{v}_h)\in Y\}.$$

Then, let $\{\mathbf{u}_h, \mathbf{A}_h\}$ be a solution of (3.14)–(3.15). By definition of K_h and Y, $\det(\mathbf{Id}+\nabla_h\mathbf{u}_h)$ is equal to 1 almost everywhere in Ω, and thus

$$\int_\Omega q(\det(\mathbf{Id}+\nabla_h\mathbf{u}_h)-1)\,d\mathbf{x}=0\quad \forall q\in L^1(\Omega). \tag{3.16}$$

In particular, (3.16) holds for any $q=q_h$ in P_h, which directly yields (3.13)(ii). On the other hand, writing (3.15) with $\mathbf{G}_h=\mathbf{0}$ everywhere except on an arbitrary element Ω_l, and considering the fact that $\mathbf{A}_h$, $\mathbf{G}_h$, and $\mathbf{Id}+\nabla_h\mathbf{u}_h$ belong to H_h and are thus constant over Ω_l, we get

$$\text{meas}\,(\Omega_l)\mathbf{A}_h|_{\Omega_l}\cdot\mathbf{G}=0\quad \forall\mathbf{G}\in\ker(B_l),$$

where B_l is the linear application defined from $\mathbb{R}^{N\times N}$ into $\mathbb{R}$ by

$$B_l\mathbf{G}=\frac{\partial\det}{\partial\mathbf{F}}(\mathbf{Id}+\nabla_h\mathbf{u}_h|_{\Omega_l})\cdot\mathbf{G}.$$

In other words, the restriction of $\mathbf{A}_h$ to Ω_l belongs to the orthogonal of $\ker(B_l)$ in $\mathbb{R}^{N\times N}$, that is, since we are in finite dimension, to the image of B_l^t. Therefore, there exists p_l in $\mathbb{R}$ such that

$$\mathbf{A}_h|_{\Omega_l}=-B_l^t p_l=-p_l\frac{\partial\det}{\partial\mathbf{F}}(\mathbf{Id}+\nabla_h\mathbf{u}_h|_{\Omega_l}). \tag{3.17}$$

Writing (3.17) successively for any $l=1$ to N_h, and introducing the function p_h of P_h defined by

$$p_h|_{\Omega_l}=p_l\quad \forall l=1, N_h,$$

we finally obtain

$$\mathbf{A}_h=-p_h\frac{\partial\det}{\partial\mathbf{F}}(\mathbf{Id}+\nabla_h\mathbf{u}_h). \tag{3.18}$$

Substituting (3.18) into (3.14) yields (3.13)(i). The pair $\{\mathbf{u}_h, p_h\}$ is therefore a solution of (3.13).

Conversely, let $\{\mathbf{u}_h, p_h\}$ be a solution of (3.13). Writing (3.13)(ii) with $q_h=0$ everywhere except on an arbitrary element Ω_l, and considering the fact that q_h and $(\mathbf{Id}+\nabla_h\mathbf{u}_h)$ are constant over Ω_l, we get

$$\text{meas}\,(\Omega_l)(\det(\mathbf{Id}+\nabla_h\mathbf{u}_h|_{\Omega_l})-1)q=0\quad \forall q\in\mathbb{R},$$

that is,

$$\det(\mathbf{Id}+\nabla_h\mathbf{u}_h)|_{\Omega_l}=\det(\mathbf{Id}+\nabla_h\mathbf{u}_h|_{\Omega_l})=1.$$

Thus, $\mathbf{Id}+\boldsymbol{\nabla}_h\mathbf{u}_h$ belongs to Y, and $\mathbf{u}_h$ belongs to K_h. Introducing $\mathbf{A}_h = -p_h\,(\partial \det/\partial \mathbf{F})(\mathbf{Id}+\boldsymbol{\nabla}_h\mathbf{u}_h)$, we can easily verify (3.14)–(3.15) from (3.13)(i) and the definition of $\mathbf{A}_h$. Therefore, $\{\mathbf{u}_h, \mathbf{A}_h\}$ is a solution of (3.14)–(3.15), and our proof is complete. □

3.5. Linearization and compatibility condition. Before solving the fully discrete, nonlinear problem (3.5) or (3.13), it is useful to study its linearization around a solution. This linearized problem consists of finding the rate of variation $\{\dot{\mathbf{u}}_h, \dot{p}_h\}$ of the known solution $\{\mathbf{u}_h, p_h\}$ corresponding to a rate of variation $\{\dot{\mathbf{f}}, \dot{\mathbf{g}}\}$ of the external loads. Writing Problem (3.13) with obvious notation as

$$\mathcal{F}(\mathbf{u}_h, p_h, \mathbf{f}, \mathbf{g}) = \mathbf{0} \quad \text{in } V_h^* \times P_h^*,$$

$\{\dot{\mathbf{u}}_h, \dot{p}_h\}$ is determined by solving

$$\frac{\partial \mathcal{F}}{\partial \mathbf{u}}\dot{\mathbf{u}}_h + \frac{\partial \mathcal{F}}{\partial p}\dot{p}_h + \frac{\partial \mathcal{F}}{\partial \mathbf{f}}\dot{\mathbf{f}} + \frac{\partial \mathcal{F}}{\partial \mathbf{g}}\dot{\mathbf{g}} = \mathbf{0} \quad \text{in } V_h^* \times P_h^*. \tag{3.19}$$

When all computations are complete, the linearized problem is finally

$$\boxed{\begin{cases} a(\dot{\mathbf{u}}_h, \mathbf{v}_h) + b(\dot{p}_h, \mathbf{v}_h) = L(\mathbf{v}_h) & \forall \mathbf{v}_h \in V_h, \quad \dot{\mathbf{u}}_h \in V_h, \\ b(q_h, \dot{\mathbf{u}}_h) = 0 \quad \forall q_h \in P_h, \quad \dot{p}_h \in P_h, \end{cases}} \tag{3.20}$$

where the bilinear forms $a(\cdot,\cdot)$ and $b(\cdot,\cdot)$ and the linear form $L(\cdot)$ are given, respectively, by

$$\begin{cases} a(\mathbf{w}_h, \mathbf{v}_h) = \displaystyle\int_\Omega \rho\left(\frac{\partial^2 \mathcal{W}_1}{\partial \mathbf{F}^2}(\mathbf{x}, \mathbf{Id}+\boldsymbol{\nabla}\mathbf{u}_h)\cdot\boldsymbol{\nabla}\mathbf{w}_h\right)\cdot\boldsymbol{\nabla}\mathbf{v}_h\,d\mathbf{x} \\ \qquad + \displaystyle\int_\Omega \rho\left(\frac{\partial^2 \mathcal{W}_2}{\partial \mathbf{F}^2}(\mathbf{x}, \mathbf{Id}+\boldsymbol{\nabla}_h\mathbf{u}_h)\cdot\boldsymbol{\nabla}_h\mathbf{w}_h\right)\cdot\boldsymbol{\nabla}_h\mathbf{v}_h\,d\mathbf{x} \\ \qquad + \displaystyle\int_\Omega p_h\left(\frac{\partial^2 \det}{\partial \mathbf{F}^2}(\mathbf{Id}+\boldsymbol{\nabla}_h\mathbf{u}_h)\cdot\boldsymbol{\nabla}_h\mathbf{w}_h\right)\cdot\boldsymbol{\nabla}_h\mathbf{v}_h\,d\mathbf{x}; \end{cases} \tag{3.21}$$

$$b(q_h, \mathbf{v}_h) = -\int_\Omega q_h \frac{\partial \det}{\partial \mathbf{F}}(\mathbf{Id}+\boldsymbol{\nabla}_h\mathbf{u}_h)\cdot\boldsymbol{\nabla}_h\mathbf{v}_h\,d\mathbf{x}; \tag{3.22}$$

$$L(\mathbf{v}_h) = \int_\Omega \dot{\mathbf{f}}\cdot\mathbf{v}_h\,d\mathbf{x} + \int_{\Gamma_2}\dot{\mathbf{g}}\cdot\mathbf{v}_h\,da. \tag{3.23}$$

As written, (3.20) is a mixed variational problem. Such problems have been extensively studied by Brezzi [1974] and Babuska [1973], among others. For these problems to be well posed, $a(\cdot,\cdot)$ must be coercive on $\ker \mathcal{B}$ and $\mathcal{B}$ must be onto, with $\mathcal{B}$ from V_h into P_h^* defined by

$$\langle \mathcal{B}(\mathbf{v}_h), q_h\rangle = b(q_h, \mathbf{v}_h) \quad \forall q_h \in P_h, \quad \mathbf{v}_h \in V_h.$$

It is also known that $\mathcal{B}$ is onto if and only if the spaces V_h and P_h satisfy the following fundamental compatibility condition

$$\inf_{\substack{q_h \in P_h \\ \|q_h\|=1}} \sup_{\substack{\mathbf{v}_h \in V_h \\ \|\mathbf{v}_h\|=1}} \{b(q_h, \mathbf{v}_h)\} > 0,$$

that is, replacing $b(\cdot, \cdot)$ by its definition,

$$(3.24)\qquad \boxed{\inf_{\substack{q_h \in P_h \\ \|q_h\|=1}} \sup_{\substack{\mathbf{v}_h \in V_h \\ \|\mathbf{v}_h\|=1}} \left\{ \int_\Omega q_h \frac{\partial \det}{\partial \mathbf{F}} (\mathbf{Id} + \nabla_h \mathbf{u}_h) \cdot \nabla_h \mathbf{v}_h \, d\mathbf{x} \right\} = C_0(h) > 0.}$$

In summary, for the linearized problem to be well posed, which is certainly, in general, a minimal requirement to impose on the discrete problem, the spaces V_h and P_h and the desired solution $\mathbf{u}_h$ must satisfy the foregoing compatibility condition.

Remark 3.2. Although condition (3.24) appears to be important and realistic, it is very difficult to verify in practice. Nonetheless, this forbids P_h to be too large; otherwise, (3.24) would be obviously violated.

Remark 3.3. If we introduce the function $\boldsymbol{\varphi}_h(\mathbf{x}) = \mathbf{u}_h(\mathbf{x}) + \mathbf{x}$ and suppose that $\nabla_h = \nabla$, then, from the identity $\partial \det/\partial \mathbf{F}(\mathbf{F}) \cdot \mathbf{G} = (\text{adj } \mathbf{F})^t \cdot \mathbf{G} = \text{tr}\,(\mathbf{G}\mathbf{F}^{-1}) \det(\mathbf{F})$, (3.24) can be expressed as

$$(3.25)\qquad \inf_{\substack{q_h \in P_h \\ \|q_h\|=1}} \sup_{\substack{\mathbf{v}_h \in V_h \\ \|\mathbf{v}_h\|=1}} \int_{\boldsymbol{\varphi}_h(\Omega)} (q_h \circ \boldsymbol{\varphi}_h^{-1}) \, \text{div}\,(\mathbf{v}_h \circ \boldsymbol{\varphi}_h^{-1}) \, d\bar{\mathbf{x}} > 0.$$

In other words, (3.24) is a linear compatibility condition associated with the divergence operator on the deformed configuration.

Remark 3.4. For a given pair of spaces V_h and P_h, (3.24) may or may not be satisfied depending on the value of the discrete solution $\mathbf{u}_h$.

Remark 3.5. Condition (3.24), which appears here as a necessary condition to be satisfied by V_h and P_h, is in fact sufficient to guarantee the existence of solutions for the full discrete problem and, for a given $\mathbf{u}_h$, the uniqueness of the associated pressure field p_h (Le Tallec [1981]).

In view of the foregoing remarks, a practical strategy for choosing spaces V_h and P_h that is likely to satisfy (3.24), or, in other words, that is likely to lead to a reasonable discrete problem, is the following.

Choose V_h and P_h to satisfy (3.24) for $\mathbf{u}_h = \mathbf{0}$. In other words, from (3.25), choose a pair of finite-element spaces that are well adapted to the mixed finite-element discretization of the Stokes problem. Such pairs, among which we find the elements of § 3.2 and § 3.3, are well known, and can be found, for example, in Girault and Raviart [1986]. Then verify that, at the computed solution $\mathbf{u}_h$, the spaces $\boldsymbol{\varphi}_h(V_h)$ and $\boldsymbol{\varphi}_h(P_h)$ keep the same structure, so that (3.25), that is, (3.24), is likely to hold.

3.6. Existence and convergence results. For completeness, we give a brief survey of existence and convergence results for the incompressible discrete problem (3.13) in the case of dead loading, with $\mathscr{W}_1 = \mathscr{W}$ and $\nabla_h = \nabla$. The interested reader can find more details in Le Tallec [1982].

First, let us recall our notation. As in Chapter 2, we have that

$$J(\mathbf{v}) = \int_\Omega \rho \mathscr{W}(\mathbf{x}, \mathbf{Id} + \nabla \mathbf{v})\, d\mathbf{x} - \int_\Omega \mathbf{f} \cdot \mathbf{v}\, d\mathbf{x} - \int_{\Gamma_2} \mathbf{g} \cdot \mathbf{v}\, da; \tag{3.26}$$

$$\begin{cases} K = \{\mathbf{v} \in \mathbf{W}^{1,s}(\Omega), \mathbf{v} = \mathbf{u}_1 \text{ on } \Gamma_1, \operatorname{adj}(\mathbf{Id} + \nabla \mathbf{v}) \in (L^q(\Omega))^{N \times N}, \\ \qquad \det(\mathbf{Id} + \nabla \mathbf{v}) = 1 \text{ a.e. in } \Omega\}; \end{cases} \tag{3.27}$$

$$K_h = \left\{ \mathbf{v}_h, \mathbf{v}_h - \mathbf{u}_1 \in V_h, \int_\Omega q_h \{\det(\mathbf{Id} + \nabla \mathbf{v}_h) - 1\}\, d\mathbf{x} = 0\ \forall q_h \in P_h \right\}. \tag{3.28}$$

Moreover, we suppose that the following assumptions hold.

(i) J satisfies the following conditions:

$$\mathscr{W}(\mathbf{x}, \mathbf{F}) = a_1 |\mathbf{F}|^s + a_2 |\operatorname{adj} \mathbf{F}|^q + \mathscr{G}_0(\mathbf{x}, \mathbf{F}, \operatorname{adj} \mathbf{F}), \qquad a_i > 0, \tag{3.29}$$

$$\mathscr{G}_0(\mathbf{x}, \cdot, \cdot) \text{ is convex on } \mathbb{R}^9 \times \mathbb{R}^9, \tag{3.30}$$

$$\mathscr{G}_0(\cdot, \mathbf{F}, \mathbf{G}) \text{ is measurable on } \Omega \quad \forall \{\mathbf{F}, \mathbf{G}\} \in \mathbb{R}^9 \times \mathbb{R}^9, \tag{3.31}$$

$$\mathscr{G}_0(\mathbf{x}, \mathbf{F}, \mathbf{G}) \geq a(\mathbf{x}) \qquad \forall \{\mathbf{F}, \mathbf{G}\} \in \mathbb{R}^9 \times \mathbb{R}^9, \quad a \in L^1(\Omega), \tag{3.32}$$

$$J \text{ is strongly continuous on } \mathbf{W}^{1,r}(\Omega). \tag{3.33}$$

(ii) There exists a continuous solution $\mathbf{u}$ and an element $\mathbf{v}_1$ in K such that

$$\mathbf{u} \in \mathbf{W}^{1,Nr}(\Omega), \qquad J(\mathbf{u}) \leq J(\mathbf{v}) \quad \forall \mathbf{v} \in K, \tag{3.34}$$

$$\mathbf{v}_1 \in K \cap K_h \quad \forall h, \quad \text{and } J(\mathbf{v}_1) < +\infty. \tag{3.35}$$

(iii) The spaces V_h and P_h satisfy the compatibility condition (3.24) uniformly around $\mathbf{u}$, that is,

$$\begin{cases} \text{there exist } \varepsilon_0 > 0 \text{ and } C_0(h) > 0 \text{ such that, for any } \mathbf{w}_h \text{ verifying} \\ \qquad \mathbf{w}_h \in V_h + \mathbf{u}_1, \qquad \|\mathbf{w}_h - \mathbf{u}\|_{1,r} \leq \varepsilon_0, \\ \qquad \displaystyle\int_\Omega q_h (\det(\mathbf{Id} + \nabla \mathbf{w}_h) - 1)\, d\mathbf{x} \leq \varepsilon_0 \|q_h\|_{0,r^*} \quad \forall q_h \in P_h, \\ \text{we have} \\ \qquad \displaystyle\inf_{\substack{q_h \in P_h \\ \|q_h\|_{0,r^*} = 1}} \sup_{\substack{\mathbf{v}_h \in V_h \\ \|\mathbf{v}_h\|_{1,r} = 1}} \left\{ \int_\Omega q_h \frac{\partial \det}{\partial \mathbf{F}} (\mathbf{Id} + \nabla \mathbf{w}_h) \cdot \nabla \mathbf{v}_h\, d\mathbf{x} \right\} \geq C_0(h). \end{cases} \tag{3.36}$$

(iv) V_h and P_h are such that

$$\lim_{h \to 0} \left\{ C_0(h)^{-2} h^{(1-N)/r} \inf_{\mathbf{w}_h \in V_h + \mathbf{u}_1} (\|\mathbf{u} - \mathbf{w}_h\|_{1,Nr}) \right\} = 0, \tag{3.37}$$

(3.38) $\mathscr{D}(\Omega)$ is included in the closure of $\bigcup_h P_h$ for the $L^\infty(\Omega)$ norm.

Under the assumptions (3.29)–(3.38), it is then possible to prove the following.

(i) The existence of discrete solutions. There exists a solution $\{\mathbf{u}_h, p_h\}$ to the discrete problem, where $\mathbf{u}_h$ realizes the minimum of J on K_h and where p_h is unique once $\mathbf{u}_h$ is determined.

(ii) The consistency of the approximation. For any h, there exists $\mathbf{w}_h$ in K_h such that

$$\|\mathbf{u}-\mathbf{w}_h\|_{1,r} \leq C(\mathbf{u}) C_0(h)^{-1} \inf_{\mathbf{z}_h \in V_h + \mathbf{u}_1} \|\mathbf{u}-\mathbf{z}_h\|_{1,Nr}.$$

(iii) The convergence of the discrete solution. Any sequence $(\mathbf{u}_h)_h$ of global minimizers of J on K_h decomposes into subsequences, each of which converge strongly for the K topology toward stable solutions of the continuous problem (i.e., minimizers of J on K).

The proof of (i)–(iii) is rather lengthy. The existence of $\mathbf{u}_h$ is a consequence of the Weierstrass theorem. The existence and uniqueness of p_h follow from the closed range theorem (see the proof of Theorem 2.2 in Chap. 2). For any h, $\mathbf{w}_h$ is constructed by solving the equation $\mathbf{w}_h \in K_h$ by Newton's method. The weak convergence of $(\mathbf{u}_h)_h$ follows from the uniform boundedness of $(\mathbf{u}_h)_h$, the weak lower semi-continuity of J, and the weak continuity of the applications adj $(\nabla \cdot)$ and det $(\nabla \cdot)$. Finally, the strong convergence of $(\mathbf{u}_h)_h$ is implied by the weak convergence of $(\mathbf{u}_h)_h$ and by the convergence of the real sequence $(J(\mathbf{u}_h))_h$.

4. Iterative numerical solution of the augmented Lagrangian formulations.

4.1. Basic iterative method. In the previous sections, we have introduced and analyzed augmented Lagrangian formulations of equilibrium problems in finite elasticity. These formulations turn out to be equivalent to the original variational formulation (1.3) or (1.6), at least before discretization by the finite element method. Their major interest is that, as written, they can be solved numerically by one of the algorithms (ALG1–ALG4) introduced in Chapter 3. For problems in finite elasticity, the algorithm that we have used in practice and that appears to be the most stable is ALG1. Combined with block relaxation techniques and applied to the discrete problem (3.5), this algorithm is as follows.

ALGORITHM (4.1)–(4.6).

(4.1) $$\boldsymbol{\lambda}^1 \in H_h \text{ and } \mathbf{F}^0 \in Y \cap H_h \text{ given;}$$

then, for $n \geq 1$, $\boldsymbol{\lambda}^n$ being known, determine $\mathbf{u}^n$, $\mathbf{F}^n$, and $\boldsymbol{\lambda}^{n+1}$ by setting

(4.2) $$\mathbf{F}^{n,0} = \mathbf{F}^{n-1},$$

and by solving sequentially, for $1 \leq k \leq k_{\max}$,

$$(4.3)\qquad \begin{cases} \dfrac{\partial \mathcal{L}_r^h}{\partial \mathbf{v}}(\mathbf{u}^{n,k}, \mathbf{F}^{n,k-1}; \boldsymbol{\lambda}^n) \cdot \mathbf{v} = \displaystyle\int_\Omega \mathbf{f} \cdot \mathbf{v}\, d\mathbf{x} + \int_{\Gamma_2} \mathbf{g} \cdot \mathbf{v}\, da \\ \forall \mathbf{v} \in V_h, \quad (\mathbf{u}^{n,k} - \mathbf{u}_1) \in V_h, \end{cases}$$

$$(4.4)\qquad \frac{\partial \mathcal{L}_r^h}{\partial \mathbf{G}}(\mathbf{u}^{n,k}, \mathbf{F}^{n,k}; \boldsymbol{\lambda}^n) \cdot \mathbf{G} = 0 \quad \forall \mathbf{G} \in H_h \cap dY(\mathbf{F}^{n,k}), \quad \mathbf{F}^{n,k} \in H_h \cap Y,$$

and, finally, by setting

$$(4.5)\qquad \mathbf{u}^n = \mathbf{u}^{n,k_{\max}}, \qquad \mathbf{F}^n = \mathbf{F}^{n,k_{\max}},$$

$$(4.6)\qquad \boldsymbol{\lambda}^{n+1} = \boldsymbol{\lambda}^n + r(\mathbf{Id} + \boldsymbol{\nabla}_h \mathbf{u}^n - \mathbf{F}^n).$$

Algorithm (4.1)-(4.6) is very simple; it reduces the solution of (3.5) mainly to a sequence of problems ((4.3)) formulated in displacements (to be studied in § 4.2) and of problems ((4.4)) formulated in deformation gradients (to be studied in § 5). Observe, in addition, that the good values of $k_{\max}$ appear to be between 1 and 5 and that the algorithm is stopped in practice as soon as we have

$$\frac{\|\mathbf{F}^n - \mathbf{Id} - \boldsymbol{\nabla}_h \mathbf{u}^n\|_{0,2}}{\|\mathbf{F}^n\|_{0,2}} + \frac{\|\mathbf{u}^n - \mathbf{u}^{n-1}\|_{0,2}}{\|\mathbf{u}^n\|_{0,2}} \le \varepsilon\ (\approx 10^{-4}).$$

4.2. Problem (4.3) formulated in displacements. From the definition of the augmented Lagrangian $\mathcal{L}_r^h$, Problem (4.3) can be expressed as follows.

$$(4.7)\qquad \begin{cases} \text{Find } (\mathbf{u} - \mathbf{u}_1) \in V_h \text{ such that} \\ \displaystyle\int_\Omega \rho \frac{\partial \mathcal{W}_1}{\partial \mathbf{F}}(\mathbf{x}, \mathbf{Id} + \boldsymbol{\nabla}\mathbf{u}) \cdot \boldsymbol{\nabla}\mathbf{v}\, d\mathbf{x} + \int_\Omega \eta(\mathbf{x})(r(\mathbf{Id} + \boldsymbol{\nabla}_h \mathbf{u} - \mathbf{F}) + \boldsymbol{\lambda}) \cdot \boldsymbol{\nabla}_h \mathbf{v}\, d\mathbf{x} \\ \qquad = \displaystyle\int_\Omega \mathbf{f} \cdot \mathbf{v}\, d\mathbf{x} + \int_{\Gamma_2} \mathbf{g} \cdot \mathbf{v}\, da \quad \forall \mathbf{v} \in V_h, \end{cases}$$

where, for simplicity, we have dropped the subscript h and the superscripts n and k from all variables. If we did choose $\mathcal{W}_1$ as a convex function of $\mathbf{F}$, then, except for the possible dependence of $\mathbf{f}$ and $\mathbf{g}$ on $\mathbf{u}$, (4.7) corresponds to the variational formulation of an unconstrained convex minimization problem on V_h, which can be solved by one of the many numerical techniques that exist for such problems (see Polak [1971] and Glowinski [1984]).

In fact, however, Problem (4.3) in displacements can be simplified further. If we choose $\mathcal{W}_1$ as a quadratic function of $\mathbf{F}$ (see Remark 2.1), and if we approximate the external loads $\mathbf{f}(\mathbf{u})$ and $\mathbf{g}(\mathbf{u})$ by their values $\mathbf{f}(\mathbf{u}^{n,k-1})$ and $\mathbf{g}(\mathbf{u}^{n,k-1})$ at the previous iterate, then (4.3) reduces to the following linear system.

$$
(4.8)\qquad \begin{cases} \text{Find } \mathbf{u} = \mathbf{u}_1 + \sum_{j=1}^{\dim V_h} u^j \boldsymbol{\varphi}_j \text{ such that} \\ \sum_{j=1}^{\dim V_h} a_{ij} u^j = b_i \quad \forall 1 \le i \le \dim V_h. \end{cases}
$$

In (4.8), $(\boldsymbol{\varphi}_i)_{i=1,\dim V_h}$ is a known basis of V_h, and the coefficients a_{ij} and b_i are given by

$$
(4.9)\qquad a_{ij} = \int_\Omega \left\{ \rho \left(\frac{\partial^2 \mathscr{W}_1}{\partial \mathbf{F}^2} \cdot \nabla \boldsymbol{\varphi}_i \right) \cdot \nabla \boldsymbol{\varphi}_j + r\eta(\mathbf{x}) \nabla_h \boldsymbol{\varphi}_i \cdot \nabla_h \boldsymbol{\varphi}_j \right\} d\mathbf{x},
$$

$$
(4.10)\qquad \begin{aligned} b_i = &-\int_\Omega \left\{ \rho \left(\frac{\partial^2 \mathscr{W}_1}{\partial \mathbf{F}^2} \cdot (\mathbf{Id} + \nabla \mathbf{u}_1) \right) \cdot \nabla \boldsymbol{\varphi}_i \right. \\ &\left. + \eta(\mathbf{x})(r(\mathbf{Id} + \nabla_h \mathbf{u}_1 - \mathbf{F}) + \boldsymbol{\lambda}) \cdot \nabla_h \boldsymbol{\varphi}_i \right\} d\mathbf{x}, \\ &+ \int_\Omega \mathbf{f}(\mathbf{u}^{n,k-1}) \cdot \boldsymbol{\varphi}_i \, d\mathbf{x} + \int_{\Gamma_2} \mathbf{g}(\mathbf{u}^{n,k-1}) \cdot \boldsymbol{\varphi}_i \, da. \end{aligned}
$$

By construction, the linear system (4.8) is associated with a sparse, symmetric, positive-definite matrix that does not change during the iterations. When this matrix is computed and factorized, the solution of (4.8) becomes a standard, cheap, and stable operation.

In our numerical experiments, we solved Problem (4.3) in displacements by solving the associated linear systems (4.8) using either a standard Cholesky method or an incomplete Cholesky conjugate-gradient (ICCG) method. This last method, developed by Meijerink and van der Vorst [1977] and Ajiz and Jennings [1984], multiplies the linear system (4.8) by the inverse of an incomplete Cholesky factorization of the matrix $\mathscr{A}$, and solves the resulting system by a conjugate-gradient method (Chap. 3, § 2.4.1). This saves both computer storage and running time when dealing with large systems ($\dim V_h \ge 1000$).

5. Solution of local problems formulated in deformation gradients.

5.1. Formulation of the problem and preliminary lemma. We now turn to the study of the most specific step of Algorithm (4.1)–(4.6), that is, the solution of Problem (4.4) formulated in deformation gradients. From the definition of $\mathscr{L}_r^h$, (4.4) can be expressed as follows.

$$
(5.1)\qquad \begin{cases} \text{Find } \mathbf{F} \in H_h \cap Y \text{ such that} \\ \int_\Omega \left\{ \rho \dfrac{\partial \mathscr{W}_2}{\partial \mathbf{F}}(\mathbf{x}, \mathbf{F}) - \eta(\mathbf{x})(r(\mathbf{Id} + \nabla_h \mathbf{u} - \mathbf{F}) + \boldsymbol{\lambda}) \right\} \cdot \mathbf{G} \, d\mathbf{x} = 0 \\ \qquad\qquad\qquad\qquad\qquad\qquad \forall \mathbf{G} \in H_h \cap dY(\mathbf{F}), \end{cases}
$$

where, for simplicity, we have dropped the subscript h and the superscripts n and k from all variables. Recall that H_h is a given, finite-dimensional approximation of $(L^\infty(\Omega))^{N\times N}$, that $\nabla_h\mathbf{v}$ denotes the L^2 projection of $\nabla\mathbf{v}$ on H_h, that $\mathscr{W}_2(\mathbf{x},\cdot)$ represents that part of the specific free energy not taken into account by the problem in displacements, that r and $\eta(\mathbf{x})$ are given positive constants, and that, at this step, the values of $\mathbf{u}$ and $\boldsymbol{\lambda}$ are known. Moreover, in the compressible case, we have

$$Y\cap H_h=\{\mathbf{G}\in H_h,\ \det\mathbf{G}>0\}, \tag{5.2}$$

$$dY(\mathbf{F})\cap H_h=H_h, \tag{5.3}$$

and, in the incompressible case, we have

$$Y\cap H_h=\{\mathbf{G}\in H_h,\ \det\mathbf{G}=1\text{ in }\Omega\}, \tag{5.4}$$

$$dY(\mathbf{F})\cap H_h=\left\{\mathbf{G}\in H_h,\ \frac{\partial\det}{\partial\mathbf{F}}(\mathbf{F})\cdot\mathbf{G}=0\right\}. \tag{5.5}$$

To study Problem (5.1), we also recall that the singular values $G_1\ge G_2\cdots\ge G_N$ of real $N\times N$ matrix $\mathbf{G}$ are the square roots of the eigenvalues of $\mathbf{G}\mathbf{G}^\top$, and that a real function W defined on the space $\mathbb{R}^{N\times N}$ of real $N\times N$ matrices is said to be isotropic if and only if it is a symmetric function of the singular values of its argument. With this definition, equivalent to the one given in equation (6.5) of Chapter 1, we can now prove the following lemma.

LEMMA 5.1. *Let $\mathscr{W}:\mathbb{R}^{N\times N}\to\mathbb{R}$ be differentiable and isotropic. Then, for any choice $\{\mathbf{D},\mathbf{G},\mathbf{Q},\mathbf{R}\}$ of $N\times N$ matrices with $\mathbf{D}$ diagonal and $\mathbf{Q}$ and $\mathbf{R}$ orthogonal, we have the following.*

$$\frac{\partial\mathscr{W}}{\partial\mathbf{F}}(\mathbf{D})\text{ is diagonal,} \tag{5.6}$$

$$\frac{\partial\mathscr{W}}{\partial\mathbf{F}}(\mathbf{QGR})=\mathbf{Q}\frac{\partial\mathscr{W}}{\partial\mathbf{F}}(\mathbf{G})\mathbf{R}. \tag{5.7}$$

Proof. To obtain (5.6), we simply observe that, for $\mathbf{D}$ diagonal,

$$\det((\mathbf{D}+t\mathbf{H})(\mathbf{D}+t\mathbf{H})^t-\mu\mathbf{Id})=\det(\mathbf{D}\mathbf{D}^t-\mu\mathbf{Id})$$
$$+2t\sum_{i=1}^N H_{ii}D_{ii}\prod_{j\ne i}(D_{jj}^2-\mu)+o(t)\qquad\forall t,\mu\in\mathbb{R},\quad\mathbf{H}\in\mathbb{R}^{N\times N}$$

In other words, if $\mathbf{H}$ has no nonzero components on the diagonal, then, at the first order of t, the singular values of $\mathbf{D}$ and $(\mathbf{D}+t\mathbf{H})$ are identical, which implies, since $\mathscr{W}$ is isotropic, that

$$\mathscr{W}(\mathbf{D}+t\mathbf{H})=\mathscr{W}(\mathbf{D})+o(t)\qquad\forall\mathbf{H}\in\mathbb{R}^{N\times N}\text{ where }H_{ii}=0. \tag{5.8}$$

From (5.8), we then deduce that

$$\frac{\partial\mathscr{W}}{\partial\mathbf{F}}(\mathbf{D})\cdot\mathbf{H}=\lim_{t\to0}\frac{1}{t}(\mathscr{W}(\mathbf{D}+t\mathbf{H})-\mathscr{W}(\mathbf{D}))=0\qquad\forall\mathbf{H}\in\mathbb{R}^{N\times N}\text{ where }H_{ii}=0,$$

which yields

$$\frac{\partial \mathscr{W}}{\partial \mathbf{F}}(\mathbf{D})_{ij}=0 \quad \forall i \neq j,$$

that is, (5.6) precisely.

Similarly, to obtain (5.7), we observe that, for orthogonal matrices **Q** and **R**, we have

$$\begin{aligned}\det((\mathbf{QGR})(\mathbf{QGR})^{t}-\mu\mathbf{Id}) &= \det(\mathbf{QGG}^{t}\mathbf{Q}^{t}-\mu\mathbf{Id})\\ &= \det(\mathbf{GG}^{t}-\mu\mathbf{Id}) \quad \forall \mathbf{G}\in\mathbb{R}^{N\times N}.\end{aligned}$$

In other words, **G** and **QGR** have the same singular values, and thus

$$\mathscr{W}(\mathbf{QGR})=\mathscr{W}(\mathbf{G}) \quad \forall \mathbf{G}\in\mathbb{R}^{N\times N}. \tag{5.9}$$

Applying (5.9) to **G** and $\mathbf{G}+t\mathbf{Q}^{t}\mathbf{HR}^{t}$, we then obtain

$$\begin{aligned}\frac{\partial \mathscr{W}}{\partial \mathbf{F}}(\mathbf{QGR})\cdot\mathbf{H} &= \lim_{t\to 0}\frac{1}{t}(\mathscr{W}(\mathbf{QGR}+t\mathbf{H})-\mathscr{W}(\mathbf{QGR}))\\ &= \lim_{t\to 0}\frac{1}{t}(\mathscr{W}(\mathbf{G}+t\mathbf{Q}^{t}\mathbf{HR}^{t})-\mathscr{W}(\mathbf{G}))\\ &= \frac{\partial \mathscr{W}}{\partial \mathbf{F}}(\mathbf{G})\cdot\mathbf{Q}^{t}\mathbf{HR}^{t}\\ &= \mathbf{Q}\frac{\partial \mathscr{W}}{\partial \mathbf{F}}(\mathbf{G})\mathbf{R}\cdot\mathbf{H} \quad \forall \mathbf{H}\in\mathbb{R}^{N\times N},\end{aligned}$$

which is (5.7), and our proof is complete. □

5.2. Solution procedure. With Lemma 5.1, Problem (5.1) reduces to the solution, in parallel, of N_h nonlinear equations set on $\mathbb{R}$, $\mathbb{R}^2$, or $\mathbb{R}^3$. Indeed, let us define, in the compressible case,

(5.10) $M=N=3$ (for three-dimensional problems, where $\Omega\subset\mathbb{R}^3$),

$\quad\quad\quad\;\; =2$ (for plane strains problems, where $\Omega\subset\mathbb{R}^2$),

$$\begin{cases}\mathbf{T}:\mathbb{R}^{\mathrm{M}}\to\mathbb{R}^{\mathrm{N\times N}}, \quad \text{where}\\ \mathbf{T}(t_1,\cdots,t_M)=\begin{bmatrix} t_1 & 0 & 0\\ 0 & \cdot & 0\\ 0 & 0 & t_M\end{bmatrix},\end{cases} \tag{5.11}$$

and, in the incompressible case,

$$M=N-1, \tag{5.12}$$

$$\mathbf{T}:\mathbb{R}^M \to \mathbb{R}^{N\times N}, \quad \text{where}$$

(5.13)
$$\mathbf{T}(t_1, \cdots, t_M) = \begin{bmatrix} t_1 & 0 & 0 \\ 0 & t_M & 0 \\ 0 & 0 & 1/(t_1 \cdots t_M) \end{bmatrix}.$$

We then have the following theorem.

THEOREM 5.2. *If $\mathscr{W}_2$ is isotropic and if H_h is a space of piecewise constant functions defined by*

(5.14)
$$H_h = \{\mathbf{G}:\Omega \to \mathbb{R}^{N\times N}, \mathbf{G}|_{\Omega_l} = \text{const } \forall l = 1, \cdots, N_h\},$$

then a solution $\mathbf{F}$ of (5.1) can be obtained by the following sequence of computations.

For $l = 1, \cdots, N_h$,

(5.15) *compute* $\mathbf{A}_l = \eta(r(\mathbf{Id} + \boldsymbol{\nabla}_h \mathbf{u}) + \boldsymbol{\lambda})|_{\Omega_l}$;

(5.16) *diagonalize* $\mathbf{A}_l$ *into* $\mathbf{A}_l = \mathbf{Q}_l \mathbf{D}_l \mathbf{R}_l$, *with* $\mathbf{Q}_l$ *and* $\mathbf{R}_l$ *orthogonal,* $(\mathbf{D}_l)_{11} \geq \cdots \geq (\mathbf{D}_l)_{NN}$, $\det \mathbf{D}_l = \det \mathbf{A}_l$;

(5.17) *solve* $\dfrac{\partial J_l}{\partial t}(t_{il}) = 0$ *in* $\mathbb{R}^M$;

(5.18) *set* $\mathbf{F}|_{\Omega_l} = \mathbf{Q}_l \mathbf{T}(t_{il}) \mathbf{R}_l$.

In (5.17), the function J_l is defined by

(5.19)
$$J_l(t_i) = \rho \mathscr{W}_2(\mathbf{x}, \mathbf{T}(t_i))|_{\Omega_l} + \frac{r}{2}\eta|\mathbf{T}(t_i)|^2 - \mathbf{D}_l \cdot \mathbf{T}(t_i).$$

Proof. We will consider only the incompressible case, since the compressible one is less complex. Therefore, let $\mathbf{F}$ be given by (5.18) with $\mathbf{T}$ defined as in (5.13). We have that

$$\det(\mathbf{F})|_{\Omega_l} = \det(\mathbf{Q}_l)\det(\mathbf{T})\det(\mathbf{R}_l) = \det(\mathbf{T})\det(\mathbf{A}_l)/\det(\mathbf{D}_l) = 1,$$

and, thus, $\mathbf{F}$ belongs to $H_h \cap Y$.

Then, let us compute $H_h \cap dY(\mathbf{F})$. By definition, this is

(5.20)
$$H_h \cap dY(\mathbf{F}) = \left\{\mathbf{G} \in H_h, \frac{\partial \det}{\partial \mathbf{F}}(\mathbf{F}|_{\Omega_l}) \cdot \mathbf{G}|_{\Omega_l} = 0 \ \forall l = 1, \cdots, N_h\right\}.$$

However, $\mathbf{F}$ being constructed by (5.18) and the function det $(\cdot)$ being satisfying, and therefore (5.7), we have

$$\frac{\partial \det}{\partial \mathbf{F}}(\mathbf{F}|_{\Omega_l}) = \mathbf{Q}_l \frac{\partial \det}{\partial \mathbf{F}}(\mathbf{T}(t_{il}))\mathbf{R}_l = \mathbf{Q}_l(\operatorname{adj}(\mathbf{T}(t_{il})))^t\mathbf{R}_l.$$

It therefore follows that (5.20) can be expressed as

$$H_h \cap dY(\mathbf{F}) = \{\mathbf{G} \in H_h, [\operatorname{adj}(\mathbf{T}(t_{il}))]^t \cdot (\mathbf{Q}_l^t\mathbf{G}|_{\Omega_l}\mathbf{R}_l^t) = 0 \ \forall l = 1, \cdots, N_h\},$$

which, by construction of $\mathbf{T}$, finally yields

$$H_h \cap dY(\mathbf{F}) = \left\{\mathbf{G} \in H_h, \sum_{i=1}^{M} (\mathbf{Q}_l^t\mathbf{G}|_{\Omega_l}\mathbf{R}_l^t)_{ii}/t_{il} + (\mathbf{Q}_l^t\mathbf{G}|_{\Omega_l}\mathbf{R}_l^t)_{NN} \prod_{i=1}^{M} t_{il} = 0 \ \forall l = 1, \cdots, N_h\right\}. \tag{5.21}$$

We are now ready to compute the quantity

$$q = \int_\Omega \left\{\rho \frac{\partial \mathcal{W}_2}{\partial \mathbf{F}}(\mathbf{x}, \mathbf{F}) + r\eta\mathbf{F} - \eta[r(\mathbf{Id} + \nabla_h\mathbf{u}) + \boldsymbol{\lambda}]\right\} \cdot \mathbf{G}\, d\mathbf{x},$$

where $\mathbf{F}$ is given by (5.18) and $\mathbf{G}$ is arbitrary in $H_h \cap dY(\mathbf{F})$. Since all functions appearing in the above integral are constant over Ω_l, we have

$$q = \sum_{l=1}^{N_h} \operatorname{meas}(\Omega_l)\left\{\rho \frac{\partial \mathcal{W}_2}{\partial \mathbf{F}}(\mathbf{x}, \mathbf{F})|_{\Omega_l} + r\eta\mathbf{F}|_{\Omega_l} - \mathbf{A}_l\right\} \cdot \mathbf{G}|_{\Omega_l}, \tag{5.22}$$

with $\mathbf{A}_l$ given by (5.15). From (5.16) and (5.18), and applying Lemma 5.1 to $\mathcal{W}_2(\mathbf{x}, \cdot)|_{\Omega_l}$, (5.22) can be expressed as

$$q = \sum_{l=1}^{N_h} \operatorname{meas}(\Omega_l)\mathbf{Q}_l\left\{\rho \frac{\partial \mathcal{W}_2}{\partial \mathbf{F}}(\mathbf{x}, \mathbf{T}(t_{il}))|_{\Omega_l} + r\eta\mathbf{T}(t_{il}) - \mathbf{D}_l\right\}\mathbf{R}_l \cdot \mathbf{G}|_{\Omega_l}. \tag{5.23}$$

However, by construction, $\mathbf{T}$ and $\mathbf{D}_l$ are diagonal, and $(\partial\mathcal{W}_2/\partial\mathbf{F})(\mathbf{x}, \mathbf{T})$ is diagonal from Lemma 5.1; thus, (5.23) takes the explicit form

$$q = \sum_{l=1}^{N_h} \operatorname{meas}(\Omega_l) \sum_{i=1}^{N} \left\{\rho \frac{\partial \mathcal{W}_2}{\partial \mathbf{F}}(\mathbf{x}, \mathbf{T})|_{\Omega_l} + r\eta\mathbf{T} - \mathbf{D}_l\right\}_{ii} (\mathbf{Q}_l^t\mathbf{G}|_{\Omega_l}\mathbf{R}_l^t)_{ii}.$$

From the characterization (5.21) of $H_h \cap dY(\mathbf{F})$, this implies

$$\begin{aligned} q &= \sum_{l=1}^{N_h} \operatorname{meas}(\Omega_l)\Bigg(\sum_{i=1}^{M} (\mathbf{Q}_l^t\mathbf{G}|_{\Omega_l}\mathbf{R}_l^t)_{ii}\Bigg\{\left\{\rho \frac{\partial \mathcal{W}_2}{\partial \mathbf{F}}(\mathbf{x}, \mathbf{T})|_{\Omega_l} + r\eta\mathbf{T} - \mathbf{D}_l\right\}_{ii} \\ &\qquad - \left\{\rho \frac{\partial \mathcal{W}_2}{\partial \mathbf{F}}(\mathbf{x}, \mathbf{T})|_{\Omega_l} + r\eta\mathbf{T} - \mathbf{D}_l\right\}_{NN} t_{il}^{-1} \prod_{j=1}^{M} (t_{jl})^{-1}\Bigg\}\Bigg) \\ &= \sum_{l=1}^{N_h} \operatorname{meas}(\Omega_l)\left(\sum_{i=1}^{M} (\mathbf{Q}_l^t\mathbf{G}|_{\Omega_l}\mathbf{R}_l^t)_{ii} \frac{\partial J_l}{\partial t_i}(t_{1l}, \cdots, t_{Ml})\right). \end{aligned}$$

Therefore, from (5.17), $q=0$, which means that $\mathbf{F}\in Y\cap H_h$ is a solution of (5.1) and our proof is completed. □

5.3. Further remarks. The following remarks illustrate both the feasibility and the performance of the solution procedure described in Theorem 5.2.

Remark 5.1. Since $N=2$ or 3, the diagonalization of $\mathbf{A}_l$ in $\mathbb{R}^{N\times N}$ can be achieved by a direct method which, in the general three-dimensional case, proceeds as follows.

(i) Computation of $\mathbf{A}_l\mathbf{A}_l^t$;

(ii) tridiagonalization of $\mathbf{A}_l\mathbf{A}_l^t$;

(iii) computation of the eigenvalues $\mu_1\geq\mu_2\geq\mu_3$ of the tridiagonal matrix by computation of the roots of the associated characteristic polynomial (by Cardan's formulas, for example);

(iv) computation of the corresponding normalized eigenvectors $(\mathbf{g}_j)$ by solving $\mathbf{A}_l\mathbf{A}_l^t\mathbf{g}_j=\mu_j\mathbf{g}_j$, $|\mathbf{g}_j|^2=1$;

(v) computation of $(\mathbf{D}_l)_{11}=\sqrt{\mu_1}$, $(\mathbf{D}_l)_{22}=\sqrt{\mu_2}$, $(\mathbf{D}_l)_{33}=\sqrt{\mu_3}\operatorname{sgn}(\det\mathbf{A}_l)$;

(vi) computation of $(\mathbf{Q}_l)_{ij}=(\mathbf{g}_i)_j$;

(vii) computation of $\mathbf{R}_l=\mathbf{D}_l^{-1}\mathbf{Q}_l^t\mathbf{A}_l$.

Remark 5.2. The nonlinear equation (5.17) always has a solution corresponding to the absolute minimum of J_l over $\mathbb{R}^M$. Indeed, (5.17) consists of finding a critical point of the "potential" energy J_l over the set of admissible diagonal matrices, which is parameterized on $\mathbb{R}^M$ by the map $\mathbf{T}$. By construction, J_l is coercive and continuous on this set, and thus attains its minimum. This minimal point is a critical point of J_l and thus corresponds to a solution of (5.17).

Remark 5.3. The nonlinear equation (5.17) in $\mathbb{R}^M$ is solved numerically by Newton's method with line search, the initial guess being the solution (t_{il}) at the previous resolution of (4.4). In that respect, it is interesting to choose r sufficiently large in order to guarantee the local convexity of J_l around the computed solution. Indeed, there will then be local uniqueness of the solution, local convergence of Newton's method, and, thus, consistency to Algorithm (4.1)-(4.6), which in the same neighborhood will always pick the same solution of (4.4).

Remark 5.4. The solution procedure of Theorem 5.2 respects and uses at its maximum the isotropy and, if relevant, the incompressibility of the considered material. Indeed, it reduces the problem in deformation gradients to local problems (5.17) whose only unknowns are the independent singular values (t_{il}) of $\mathbf{F}$ at the exclusion of any rotational component of $\mathbf{F}$.

6. Numerical results.

6.1. Implementation of Algorithm (4.1)–(4.6). In all our numerical tests, we implemented Algorithm (4.1)-(4.6) in the case of quadratic potentials $\mathcal{W}_1$, of isotropic potentials $\mathcal{W}_2$, and of spaces H_h made of piecewise constant functions. In view of (4.8) and of Theorem 5.2, this algorithm is very easy to code, as indicated by the flow chart shown in Fig. 6.1.

Now, for a given problem, the practical choice of r, η, and $\mathscr{W}_1$ is not so clear. Due to the lack of convexity of the original problems (1.3) and (1.6), there are no theoretical results on the convergence of this algorithm that could help us with this choice. The only numerical evidence is that Algorithm (4.1)–(4.6) diverges if r is too small and converges very slowly if r is too large.

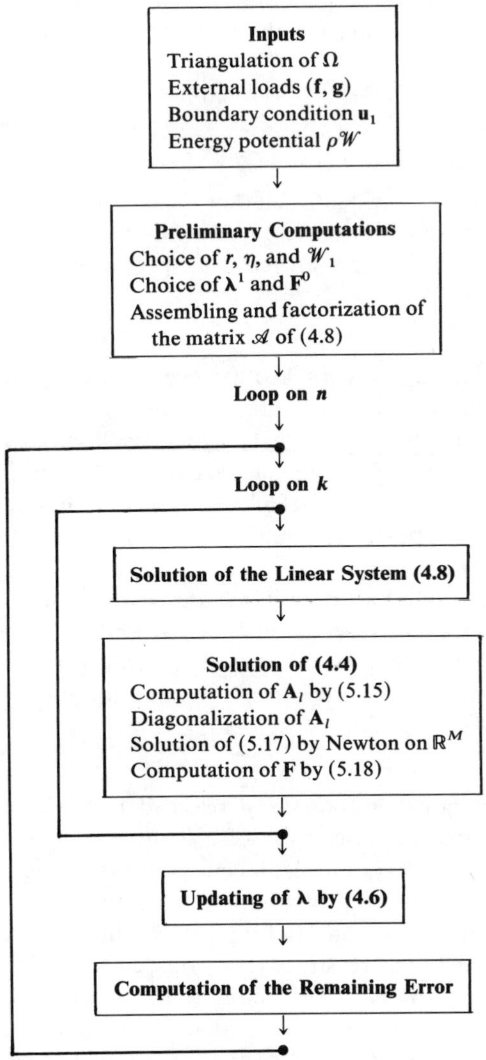

FIG. 6.1. *Computer flow chart of Algorithm* (4.1)–(4.6).

help us with this choice. The only numerical evidence is that Algorithm (4.1)–(4.6) diverges if r is too small and converges very slowly if r is too large.

For heterogeneous materials of Ogden type, whose specific energy potential $\mathscr{W}$ is given by

$$(6.1)\quad \begin{cases} \rho\mathscr{W}(\mathbf{x},\mathbf{F}) = C_{10}(\mathbf{x})(|\mathbf{F}|^2-3)+C_{01}(\mathbf{x})(|\mathrm{adj}\,\mathbf{F}|^2-3) \\ \qquad\qquad + a(\mathbf{x})[(\det \mathbf{F})^2-1]-(2a(\mathbf{x})+2C_{10}(\mathbf{x})+4C_{01}(\mathbf{x}))\ln(\det \mathbf{F}), \end{cases}$$

and which reduce to incompressible Mooney–Rivlin materials when $a(\mathbf{x}) = +\infty$, the strategy that we have used with good success has consisted of setting

$$(6.2)\qquad \rho\mathscr{W}_1(\mathbf{x},\mathbf{F}) = C_{10}(\mathbf{x})(|\mathbf{F}|^2-3),$$

$$(6.3)\qquad \eta(\mathbf{x}) = 2C_{10}(\mathbf{x}),$$

and of choosing r between 2 and 20 such that local convexity of J_l was roughly achieved in the nonlinear equation (5.17). In this range, the choice of r was usually not critical but could, nevertheless, if properly done, accelerate the convergence by a factor of 2.

6.2. Example. Stretching of a cracked rectangular bar. We consider a thick rectangular slab of Mooney–Rivlin material with a nonpropagating crack in its middle that is subjected to vertical stretching forces applied at its extremities.

The initial configuration of the lower part of the bar and of the crack is indicated in Fig. 6.2a. This bar is stretched under the action of the external loads, and its equilibrium position, computed under the plane strains assumption, is shown in Fig. 6.2b.

This solution was obtained after 20 iterations of Algorithm (4.1)–(4.6) with $k_{\max} = 1$; H_h and V_h, respectively, given by (3.6) and (3.7); $\mathscr{W}_1$ and η given by (6.2) and (6.3); and $r = 4$. The computed stresses at the boundary match the applied tension with a 10^{-4} precision. The computational time was 3.2 sec on a CDC 6400.

6.3. Example. Combined inflation and extension of a circular cylindrical tube. We consider a circular tube made of a Mooney–Rivlin material that is inflated by imposing a fixed radial displacement on the inner surface Γ_{r1} and by leaving the outer surface free of tractions. An analytical solution of this problem is given in Chadwick and Haddon [1972] under the assumption that both extremities of the tube are stress-free and remain horizontal. We have approximated these conditions by restricting the axial displacement to zero at the mid-cross-section Γ_{z1} of the body and by leaving the upper extremity traction-free. The resulting reference configuration of the upper half of the tube is described in Fig. 6.3.

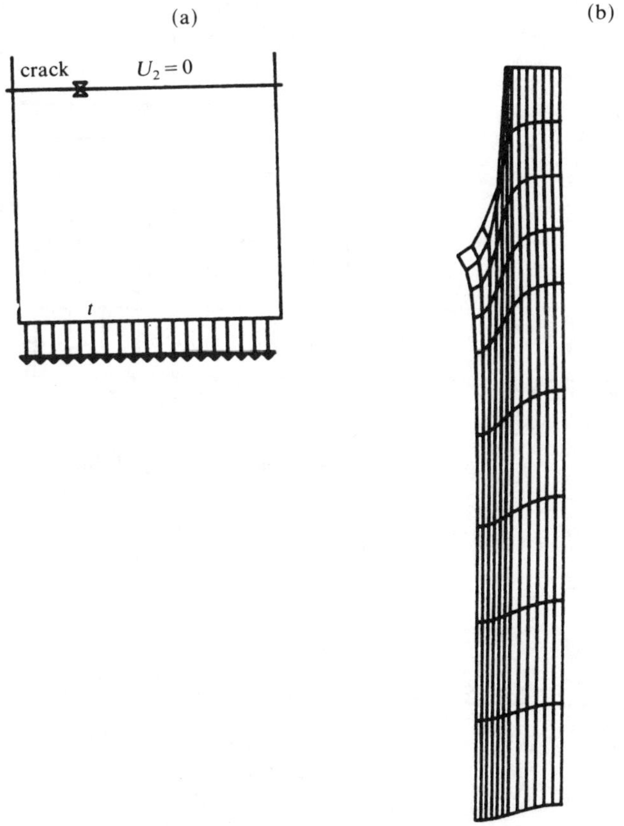

FIG. 6.2. *Stretching of a cracked, rectangular bar* (*Example* 6.2). *Cl* = 1.0 psi; *traction* = 6.0 psi; $H = 1.75$ in, 4.44 cm; $L = 1.95$ in, 4.95 cm; *crack* = 0.50 in, 1.27 cm; *energy* = 2.212 ft-lb, 2.999 J; $u_{\max} = 3.77$ in, 9.57 cm. (a) *The initial configuration of the crack and the lower part of the bar.* (b) *The equilibrium position after the bar is stretched under the action of external loads.*

The reference configuration and the loading being axisymmetric, we restricted ourselves to the calculation of axisymmetric positions of equilibrium. With this assumption, the definition of the space V of test functions becomes

$$(6.4)\qquad \begin{cases} V=\{\mathbf{v}\in \mathbf{W}^{1,s}(\Omega), \mathbf{v}(\mathbf{x})=v_r(r,z)\mathbf{e}_r(\theta)+v_z(r,z)\mathbf{e}_z, \\ \qquad v_r=0 \text{ on } \Gamma_{r1},\ v_z=0 \text{ on } \Gamma_{z1}\} \end{cases}$$

where $\{r, z, \theta\}$ and $\{\mathbf{e}_r, \mathbf{e}_z, \mathbf{e}_\theta\}$ denote, respectively, the polar coordinates of $\mathbf{x}$ and the associated local basis. The approximation of V and $(L^s(\Omega))^{3\times 3}$ is then achieved by considering a regular "triangulation" of the meridian section ω of Ω into quadrilaterals ω_l ($\{\omega_l\}_{l=1}^{N_h}$, with $\omega = \bigcup_{l=1}^{N_h} \omega_l$; see Fig. 6.3) and by defining, under the notations of § 3.2,

$$(6.5)\qquad \Omega_l = \{\mathbf{x}, \{r, z\} \in \omega_l\},$$

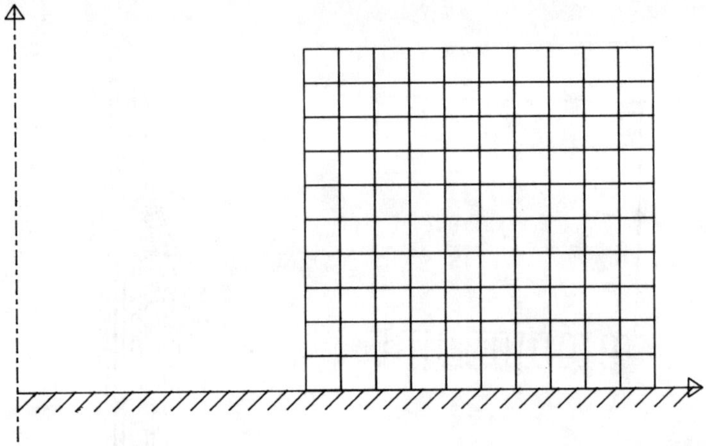

FIG. 6.3. *Reference configuration of a circular tube made of Mooney–Rivlin material* (*Example* 6.3).

$$H_h = \{\mathbf{G}_h : \Omega \to \mathbb{R}^{3\times 3}, \mathbf{G}_h|_{\Omega_l} = \text{const } \forall l = 1, \cdots, N_h\}, \tag{6.6}$$

$$\begin{cases} V_h = \{\mathbf{v}_h : \Omega \to \mathbb{R}^3, \mathbf{v}_h \text{ continuous}, \mathbf{v}_h \cdot \mathbf{e}_r = 0 \text{ on } \Gamma_{r1}, \\ \qquad \mathbf{v}_h \cdot \mathbf{e}_z = 0 \text{ on } \Gamma_{1z}, \mathbf{v}_h|_{\Omega_l} = v_{lr} \circ \boldsymbol{\varphi}_l^{-1}(r, z)\mathbf{e}_r \\ \qquad + v_{lz} \circ \boldsymbol{\varphi}_l^{-1}(r, z)\mathbf{e}_z, \{v_{lr}, v_{lz}\} \in [Q_1(\hat{\omega})]^2 \, \forall l = 1, \cdots, N_h\}. \end{cases} \tag{6.7}$$

Solving the discrete problem (3.5) by Algorithm (4.1)–(4.6) with V_h and H_h given by (6.6)–(6.7) and $k_{\max} = 3$, $r = 8$, $C_{10} = 0.4375$ MPa, $C_{01} = 0.0635$ MPa, and $\eta = 1$ MPa, leads to the numerical results described in Table 6.1 and Fig. 6.4 under the following notation.

$$\text{Data}\begin{cases} N = (\text{outer radius/inner radius}) \text{ in the reference configuration}, \\ Q = \text{final inner radius/initial inner radius}. \end{cases}$$

TABLE 6.1
Comparison of analytical and numerical results for combined inflation and extension of a circular cylindrical tube (*Example* 6.3).

N	1.4	1.4	1.8	1.8	2.2	2.2	2.2
Q	1.2	1.6	1.6	2.0	1.6	2.0	2.2
EXTV							
Anal.	0.9460	0.8583	0.8991	0.8432	0.9252	0.8794	0.8578
Num.	0.9460	0.8582	0.8995	0.8434	0.9261	0.8801	0.8584
EXTH							
Anal.	1.1191	1.3700	1.2486	1.4334	1.1774	1.3146	1.3879
Num.	1.1192	1.3701	1.2489	1.4339	1.1778	1.3154	1.3882

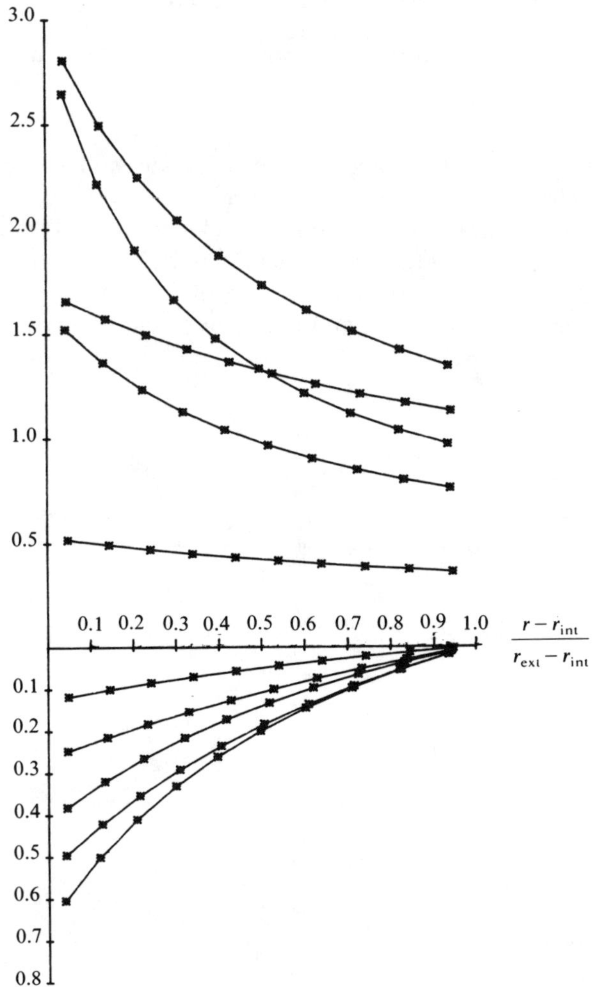

FIG. 6.4. *Analytical and numerical values of stresses in Example* 6.3. $C1 = 0.8750$ psi; $C2 = 0.1250$ psi; *height* = 1.2 in, 3.04 cm.

$$\text{Results}\begin{cases} EXTV = \text{final height/initial height}, \\ EXTH = \text{final outer radius/initial outer radius}, \\ \sigma_{\theta\theta} = \text{Cauchy stresses along } \mathbf{e}_\theta \text{ in the mid-cross-section } \Gamma_{z1}, \\ \sigma_{zz} = \text{Cauchy stresses along } \mathbf{e}_z \text{ in the mid-cross-section } \Gamma_{z1}. \end{cases}$$

Further details on this computation can be found in Glowinski and Le Tallec [1982].

6.4. Example. Post-buckling solution of a three-dimensional beam. This example illustrates the capability of Algorithm (4.1)–(4.6) to compute stable post-buckling equilibrium positions of elastic bodies even in a three-dimensional situation. It considers a $0.2 \times 0.2 \times 2$ beam that is compressed along its axis and subjected to a pressure of 10^{-4} MPa on one of its lateral faces. The beam is made of a compressible hyperelastic material whose energy

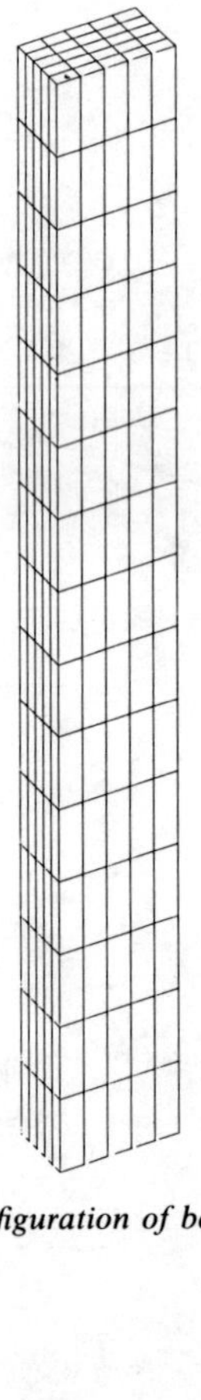

FIG. 6.5. *Initial configuration of beam* (*Example* 6.4).

potential is given by (6.1) with $C_{10}=0.5$ MPa, $C_{01}=0.125$ MPa, and $a=25$ MPa; the compression is achieved by an imposed displacement of 0.2 m of its upper extremity, the lower one remaining fixed.

For symmetry reasons, we only compute the upper half of the beam, using the spaces H_h and V_h defined in (3.6) and (3.7) with $N=3$ (Fig. 6.5). Algorithm (4.1)–(4.6) obtained two solutions for this problem, one unstable and symmetric (Fig. 6.6) and characterized by small lateral displacements, and another stable and characterized by large lateral displacements (Fig. 6.7).

For this example, it is very interesting to monitor the quantity

$$\log [\|\mathbf{Id}+\nabla_h \mathbf{u}^n - \mathbf{F}^n\|_{0,2}^2]$$

during the iterations (4.1)–(4.6) (Fig. 6.8). This quantity, which measures the lack of convergence of our algorithm, first decreases to a minimum that corresponds to the unstable, symmetric solution and then automatically diverges during few steps, finally converging to zero when the stable, buckled solution is approached. Parallel to this graph, we show in Fig. 6.9 the evolution of the potential energy during the iterations of the algorithm.

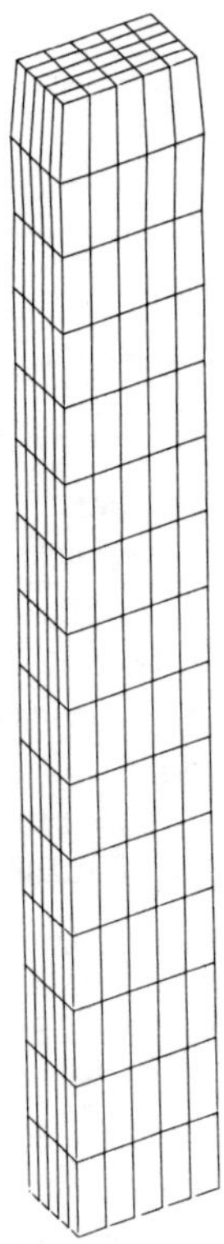

FIG. 6.6. *Symmetric solution of Example* 6.4.

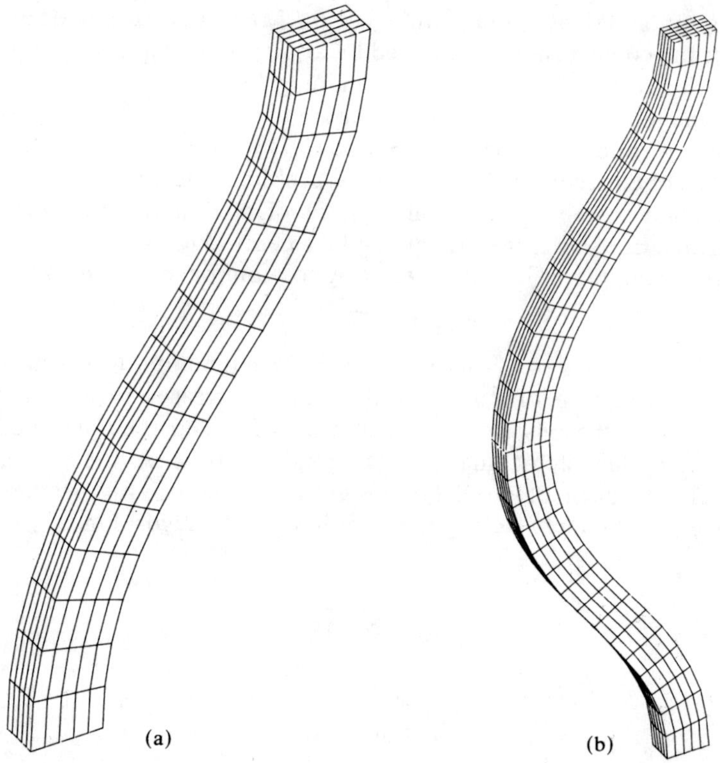

FIG. 6.7. *The buckled solution of Example* 6.4. (a) *The upper half of the beam.* (b) *The whole, deformed beam.*

7. Equilibrium problems with contact.

7.1. The physical problem. To further illustrate augmented Lagrangian methods in finite elasticity, we will briefly describe how these methods apply to situations with contact constraints. The corresponding physical problem is the same as in § 1.1, but now the boundary Γ of Ω contains a third part Γ_c (Fig. 7.1) where, due to the presence of a plane rigid obstacle in the neighborhood of the considered body, the displacement $\mathbf{u} \cdot \mathbf{n}$ of the body perpendicular to the obstacle cannot exceed a given value, that is,

$$\mathbf{u}(\mathbf{x}) \cdot \mathbf{n} \leq e_{\max}(\mathbf{x}) \quad \forall \mathbf{x} \in \Gamma_c. \tag{7.1}$$

To impose this constraint, the obstacle exerts a reaction force on Γ_c which, in the case of a contact without friction, is of the form

$$\mathbf{g}_c = g_c \mathbf{n}, \qquad g_c \leq 0, \qquad g_c(\mathbf{u} \cdot \mathbf{n} - e_{\max}) = 0. \tag{7.2}$$

7.2. Variational formulation. The variational formulation of problems with contact is obtained as in the contact-free case but, in addition, one must take

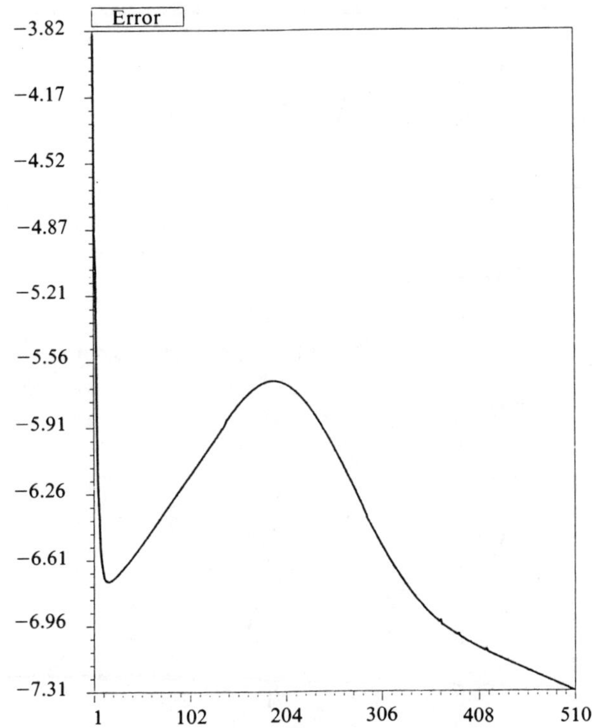

FIG. 6.8. *The error curve for Example* 6.4.

into account the reaction $\mathbf{g}_c$ in the virtual work theorem and impose the kinematic constraint (7.1) on the real displacement field. Then, using the definitions and notation of § 1.2, the variational formulation of equilibrium problems in incompressible finite elasticity with frictionless contact becomes as follows.

$$(7.3)\quad \begin{cases} \displaystyle\int_\Omega \left\{\rho \frac{\partial \mathcal{W}}{\partial \mathbf{F}}(\mathbf{x}, \mathbf{Id}+\nabla\mathbf{u}) + \mathbf{A}\right\} \cdot \nabla \mathbf{v}\, d\mathbf{x} \\ \displaystyle\quad = \int_\Omega \mathbf{f}\cdot\mathbf{v}\, d\mathbf{x} + \int_{\Gamma_2} \mathbf{g}\cdot\mathbf{v}\, da + \int_{\Gamma_c} g_c \mathbf{v}\cdot\mathbf{n}\, da \qquad \forall \mathbf{v}\in V, \quad \mathbf{u}\in K, \end{cases}$$

$$(7.4)\quad \int_\Omega \mathbf{A}\cdot\mathbf{G}\, d\mathbf{x} = 0 \qquad \forall \mathbf{G} \in dY(\mathbf{Id}+\nabla\mathbf{u}), \quad \mathbf{A}\in (L^{s^*}(\Omega))^{N\times N},$$

$$(7.5)\quad \begin{array}{c} (\mathbf{u}\cdot\mathbf{n} - e_{\max}) \le 0 \quad \text{on } \Gamma_c, \qquad g_c \le 0 \quad \text{on } \Gamma_c, \\ g_c(\mathbf{u}\cdot\mathbf{n} - e_{\max}) = 0 \quad \text{on } \Gamma_c, \end{array}$$

with a similar formulation for the compressible case.

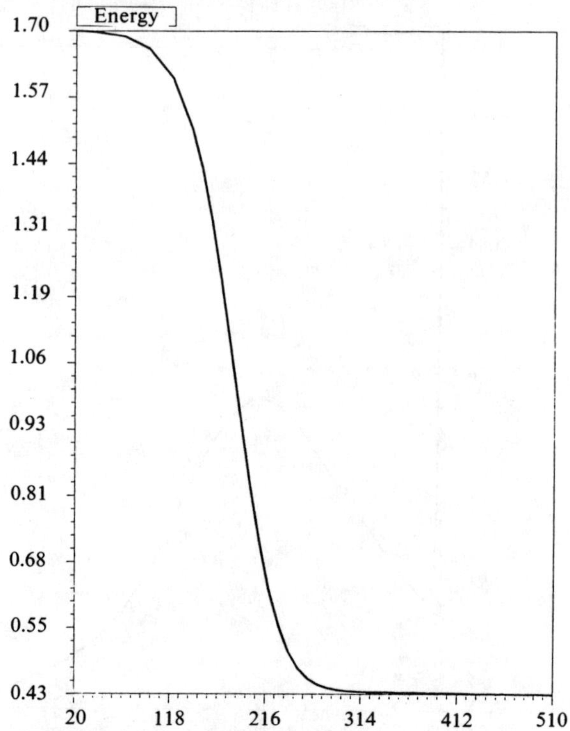

FIG. 6.9. *The energy curve in Example* 6.4.

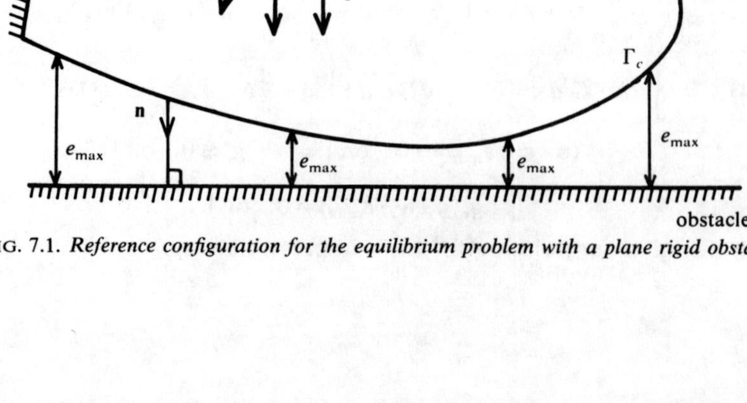

FIG. 7.1. *Reference configuration for the equilibrium problem with a plane rigid obstacle.*

Remark 7.1. Equation (7.5) is a particular case of the constitutive equation

$$-g_c \in \partial\left(\frac{1}{3\varepsilon}\sup(0, \mathbf{u}\cdot\mathbf{n}-e_{\max})^3\right) \tag{7.6}$$

corresponding to the choice $\varepsilon=0$. For computational purposes, this value $\varepsilon=0$ may not be optimal, and it is often more interesting to work with a strictly positive small value of ε.

7.3. Augmented Lagrangian formulation. The basic idea in the augmented Lagrangian formulation of the problem is to take as a local variable the pair

$$\{\mathbf{F}, d\}=\{\mathbf{Id}+\nabla\mathbf{u}, (\mathbf{u}\cdot\mathbf{n}-e_{\max})|_{\Gamma_c}\}. \tag{7.7}$$

Proceeding as in § 2, we introduce the augmented Lagrangian

$$\begin{cases} \mathcal{L}_r^c(\mathbf{v}, (\mathbf{G}, e); (\boldsymbol{\mu}, \mu_c)) \\ \quad = \displaystyle\int_\Omega \rho\mathcal{W}_1(\mathbf{x}, \mathbf{Id}+\nabla\mathbf{v})\, d\mathbf{x} + \int_\Omega \rho\mathcal{W}_2(\mathbf{x}, \mathbf{G})\, d\mathbf{x} \\ \quad \displaystyle +\frac{1}{3\varepsilon}\int_{\Gamma_c} \sup(0, e)^3\, da \\ \quad \displaystyle +\frac{r}{2}\left(\int_\Omega \eta(\mathbf{x})|\mathbf{Id}+\nabla\mathbf{v}-\mathbf{G}|^2\, d\mathbf{x} + \int_{\Gamma_c} \eta_c(\mathbf{x})(\mathbf{v}\cdot\mathbf{n}-e_{\max}-e)^2\, da\right) \\ \quad \displaystyle +\left(\int_\Omega \eta(\mathbf{x})\boldsymbol{\mu}\cdot(\mathbf{Id}+\nabla\mathbf{v}-\mathbf{G})\, d\mathbf{x} + \int_{\Gamma_c} \eta_c(\mathbf{x})\mu_c(\mathbf{v}\cdot\mathbf{n}-e_{\max}-e)\, da\right) \end{cases} \tag{7.8}$$

and consider the following problem.

Find $\{\mathbf{u}-\mathbf{u}_1, (\mathbf{F}, d); (\boldsymbol{\lambda}, \lambda_c)\} \in V\times(Y\times L^s(\Gamma_c))\times(L^{s^*}(\Omega))^{N\times N}\times L^{s^*}(\Gamma_c)$ such that

$$\begin{cases} \displaystyle\frac{\partial\mathcal{L}_r^c}{\partial\mathbf{v}}(\mathbf{u}, (\mathbf{F}, d); (\boldsymbol{\lambda}, \lambda_c))\cdot\mathbf{v} = \int_\Omega \mathbf{f}\cdot\mathbf{v}\, d\mathbf{x} + \int_{\Gamma_2} \mathbf{g}\cdot\mathbf{v}\, da \quad \forall\mathbf{v}\in V, \\ \displaystyle\frac{\partial\mathcal{L}_r^c}{\partial\mathbf{G}}(\mathbf{u}, (\mathbf{F}, d); (\boldsymbol{\lambda}, \lambda_c))\cdot\mathbf{G}=0 \quad \forall\mathbf{G}\in dY(\mathbf{F}), \\ \displaystyle\frac{\partial\mathcal{L}_r^c}{\partial e}(\mathbf{u}, (\mathbf{F}, d); (\boldsymbol{\lambda}, \lambda_c))\cdot e=0 \quad \forall e\in L^s(\Gamma_c), \\ \displaystyle\frac{\partial\mathcal{L}_r^c}{\partial\boldsymbol{\mu}}(\mathbf{u}, (\mathbf{F}, d); (\boldsymbol{\lambda}, \lambda_c))\cdot\boldsymbol{\mu}=0 \quad \forall\boldsymbol{\mu}\in(L^{s^*}(\Omega))^{N\times N}, \\ \displaystyle\frac{\partial\mathcal{L}_r^c}{\partial\mu_c}(\mathbf{u}, (\mathbf{F}, d); (\boldsymbol{\lambda}, \lambda_c))\cdot\mu_c=0 \quad \forall\mu_c\in L^{s^*}(\Gamma_c). \end{cases} \tag{7.9}$$

In (7.8)–(7.9), the notation is that of § 2, and η_c denotes an arbitrary positive function defined on Γ_c and bounded away from zero.

For the incompressible case, the equivalence between the original problems (7.3), (7.4), and (7.6) and the Lagrangian problem (7.9) is proved as in Theorem 2.1 through the identifications

$$\mathbf{F} = \mathbf{Id} + \nabla \mathbf{u}, \qquad d = (\mathbf{u} \cdot \mathbf{n} - e_{\max})|_{\Gamma_c},$$

$$\boldsymbol{\lambda} = \left(\mathbf{A} + \rho \frac{\partial \mathcal{W}_2}{\partial \mathbf{F}} (\mathbf{x}, \mathbf{F}) \right) \Big/ \eta, \qquad \lambda_c = -g_c / \eta_c.$$

Similarly, for the compressible case, equivalence is achieved through the identifications

$$\mathbf{F} = \mathbf{Id} + \nabla \mathbf{u}, \qquad d = (\mathbf{u} \cdot \mathbf{n} - e_{\max})|_{\Gamma_c},$$

$$\boldsymbol{\lambda} = \rho \frac{\partial \mathcal{W}_2}{\partial \mathbf{F}} (\mathbf{x}, \mathbf{F}) / \eta, \qquad \lambda_c = -g_c / \eta_c.$$

7.4. Finite-element discretization. To approximate the Lagrangian problem (7.9), the spaces V, $(L^s(\Omega))^{N \times N}$, and $L^s(\Gamma_c)$ must first be replaced by finite-dimensional subspaces V_h, H_h, and H_{ch}. For this purpose, we introduce regular "triangulations" of Ω and Γ_c and a space V_h such that

$$\Omega = \bigcup_{l=1}^{N_h} \Omega_l, \tag{7.10}$$

$$\Gamma_c = \bigcup_{l=1}^{N_{ch}} \Gamma_l, \tag{7.11}$$

$$V_h \subset V, \qquad \dim V_h < +\infty, \tag{7.12}$$

and we define

$$H_h = \{\mathbf{G}_h : \Omega \to \mathbb{R}^{N \times N}, \ \mathbf{G}_h|_{\Omega_l} = \text{const} \ \forall l = 1, \cdots, N_h\}, \tag{7.13}$$

$$H_{ch} = \{q_h : \Gamma_c \to \mathbb{R}, \ q_h|_{\Gamma_l} = \text{const} \ \forall l = 1, \cdots, N_{ch}\}, \tag{7.14}$$

$$\nabla_h : \mathbf{W}^{1,s}(\Omega) \to H_h, \qquad \nabla_h \mathbf{v}|_{\Omega_l} = \frac{1}{\text{meas}(\Omega_l)} \int_{\Omega_l} \nabla \mathbf{v} \, d\mathbf{x}, \tag{7.15}$$

$$s_h : L^s(\Gamma_c) \to H_{ch}, \qquad s_h(q)|_{\Gamma_l} = \frac{1}{\text{meas}(\Gamma_l)} \int_{\Gamma_l} q \, da. \tag{7.16}$$

For example, we can partition Ω into quadrilaterals (resp., hexahedrals if $N = 3$), use as the triangulation of Γ_c the trace on Γ_c of the triangulation of

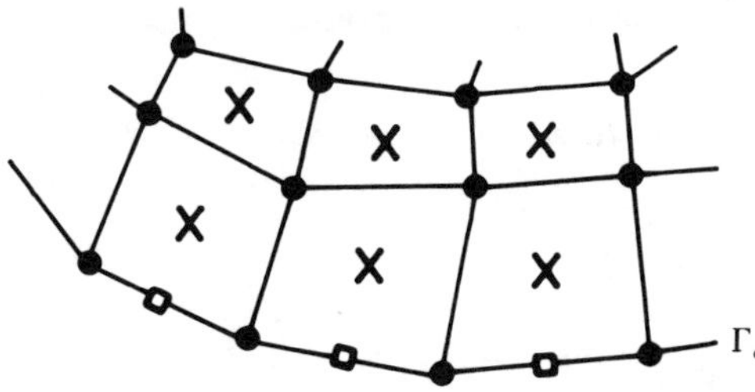

FIG. 7.2. *A first choice of triangulation. Key*: ● = *nodes of* V_h; × = *nodes of* H_h; □ = *nodes of* H_{ch}.

Ω, and construct V_h by (3.7) (Fig. 7.2). Alternatively, for $N=2$, we can partition Ω into triangles, use as the initial "triangulation" of Γ_c the trace on Γ_c of the triangulation of Ω, construct V_h by (3.10), and finally divide into two pieces any segment Γ_l where $v_h \in V_h$ is a second-order polynomial (Fig. 7.3). After discretization, the augmented Lagrangian (7.8) becomes

$$
(7.17)\quad \begin{cases}
\mathscr{L}_r^{ch}(\mathbf{v}_h, (\mathbf{G}_h, e_h); (\boldsymbol{\mu}_h, \mu_{ch})) = \displaystyle\int_\Omega \rho \mathscr{W}_1(\mathbf{x}, \mathbf{Id} + \nabla \mathbf{v}_h)\, d\mathbf{x} \\[2ex]
\quad + \displaystyle\int_\Omega \rho \mathscr{W}_2(\mathbf{x}, \mathbf{G}_h)\, d\mathbf{x} + \frac{1}{3\varepsilon}\int_{\Gamma_c} \sup(0, e_h)^3\, da \\[2ex]
\quad + \dfrac{r}{2}\Bigg(\displaystyle\int_\Omega \eta(\mathbf{x})|\mathbf{Id} + \nabla_h \mathbf{v}_h - \mathbf{G}_h|^2\, d\mathbf{x} \\[2ex]
\quad + \displaystyle\int_{\Gamma_c} \eta_c(\mathbf{x})(s_h(\mathbf{v}_h \cdot \boldsymbol{n}) - e_{\max} - e_h)^2\, da\Bigg) \\[2ex]
\quad + \Bigg(\displaystyle\int_\Omega \eta(\mathbf{x})\boldsymbol{\mu}_h \cdot (\mathbf{Id} + \nabla_h \mathbf{v}_h - \mathbf{G}_h)\, d\mathbf{x} \\[2ex]
\quad + \displaystyle\int_{\Gamma_c} \eta_c(\mathbf{x})\mu_{ch}(s_h(\mathbf{v}_h \cdot \mathbf{n}) - e_{\max} - e_h)\, da\Bigg)
\end{cases}
$$

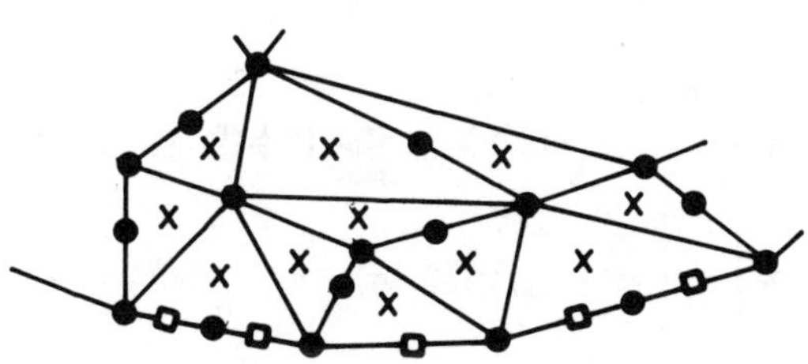

FIG. 7.3. *A second choice of triangulation. Key*: ● = *nodes of* V_h; × = *nodes of* H_h; □ = *nodes of* H_{ch}.

and is associated with the discrete Lagrangian problem

$$
(7.18)\quad \begin{cases}
\dfrac{\partial \mathcal{L}_r^{ch}}{\partial \mathbf{v}}(\mathbf{u}_h, (\mathbf{F}_h, d_h); (\boldsymbol{\lambda}_h, \lambda_{ch})) \cdot \mathbf{v}_h = \displaystyle\int_\Omega \mathbf{f} \cdot \mathbf{v}_h \, d\mathbf{x} + \int_{\Gamma_2} \mathbf{g} \cdot \mathbf{v}_h \, da \\
\qquad\qquad\qquad\qquad\qquad\qquad\qquad\qquad \forall \mathbf{v}_h \in V_h \quad (\mathbf{u}_h - \mathbf{u}_1) \in V_h, \\
\dfrac{\partial \mathcal{L}_r^{ch}}{\partial \mathbf{G}}(\mathbf{u}_h, (\mathbf{F}_h, d_h); (\boldsymbol{\lambda}_h, \lambda_{ch})) \cdot \mathbf{G}_h = 0 \\
\qquad\qquad\qquad\qquad\qquad\qquad \forall \mathbf{G}_h \in H_h \cap dY(\mathbf{F}_h), \quad \mathbf{F}_h \in H_h \cap Y, \\
\dfrac{\partial \mathcal{L}_r^{ch}}{\partial e}(\mathbf{u}_h, (\mathbf{F}_h, d_h); (\boldsymbol{\lambda}_h, \lambda_{ch})) \cdot e_h = 0 \quad \forall e_h \in H_{ch}, \quad d_h \in H_{ch}, \\
\dfrac{\partial \mathcal{L}_r^{ch}}{\partial \boldsymbol{\mu}}(\mathbf{u}_h, (\mathbf{F}_h, d_h); (\boldsymbol{\lambda}_h, \lambda_{ch})) \cdot \boldsymbol{\mu}_h = 0 \qquad \forall \boldsymbol{\mu}_h \in H_h, \quad \boldsymbol{\lambda}_h \in H_h, \\
\dfrac{\partial \mathcal{L}_r^{ch}}{\partial \mu_c}(\mathbf{u}_h, (\mathbf{F}_h, d_h); (\boldsymbol{\lambda}_h, \lambda_{ch})) \cdot \mu_{ch} = 0 \qquad \forall \mu_{ch} \in H_{ch}, \quad \lambda_{ch} \in H_{ch}.
\end{cases}
$$

7.5. Numerical algorithm. The discrete Lagrangian formulation (7.18) of frictionless contact problems in finite elasticity can again be solved by ALG1. Using the explicit form of the gradients of $\mathcal{L}_r^{ch}$, this algorithm is as follows.

ALGORITHM (7.19)–(7.26).

$$
(7.19)\qquad \boldsymbol{\lambda}^0 \in H_h, \quad \lambda_c^0 \in H_{ch}, \mathbf{F}^0 \in Y \cap H_h, \text{ and } d^0 \in H_{ch} \text{ given;}
$$

then, for $n \geq 0$, $\{\boldsymbol{\lambda}^n, \lambda_c^n\}$ being known, determine $\mathbf{u}^n$, $\{\mathbf{F}^n, d^n\}$, and $\{\boldsymbol{\lambda}^{n+1}, \lambda_c^{n+1}\}$ by setting

$$
(7.20)\qquad \mathbf{F}^{n,0} = \mathbf{F}^{n-1}, \qquad d^{n,0} = d^{n-1},
$$

and then by solving sequentially, for $1 \leq k \leq k_{\max}$,

$$
(7.21)\quad \begin{cases}
\displaystyle\int_\Omega \rho \frac{\partial \mathcal{W}_1}{\partial \mathbf{F}}(\mathbf{x}, \mathbf{Id} + \nabla \mathbf{u}^{n,k}) \cdot \nabla \mathbf{v} \, d\mathbf{x} \\
\qquad + \displaystyle\int_\Omega \eta(\mathbf{x})(r(\mathbf{Id} + \nabla_h \mathbf{u}^{n,k} - \mathbf{F}^{n,k-1}) + \boldsymbol{\lambda}^n) \cdot \nabla_h \mathbf{v}_h \, d\mathbf{x} \\
\qquad + \displaystyle\int_{\Gamma_c} \eta_c(\mathbf{x})(r(s_h(\mathbf{u}^{n,k} \cdot \mathbf{n}) - e_{\max} - d^{n,k-1}) + \lambda_c^n) \, s_h(\mathbf{v} \cdot \mathbf{n}) \, da \\
\qquad = \displaystyle\int_\Omega \mathbf{f} \cdot \mathbf{v} \, d\mathbf{x} + \int_{\Gamma_2} \mathbf{g} \cdot \mathbf{v} \, da \qquad \forall \mathbf{v} \in V_h, \quad (\mathbf{u}^{n,k} - \mathbf{u}_1) \in V_h,
\end{cases}
$$

$$\text{(7.22)}\quad \begin{cases} \displaystyle\int_\Omega \left\{\rho \frac{\partial \mathscr{W}_2}{\partial \mathbf{F}}(\mathbf{x}, \mathbf{F}^{n,k}) - \eta(\mathbf{x})(r(\mathbf{Id} + \nabla_h \mathbf{u}^{n,k} - \mathbf{F}^{n,k}) + \boldsymbol{\lambda}^n)\right\} \cdot \mathbf{G} = 0 \\ \qquad \forall \mathbf{G} \in H_h \cap dY(\mathbf{F}^{n,k}), \quad \mathbf{F}^{n,k} \in Y \cap H_h, \end{cases}$$

$$\text{(7.23)}\quad \begin{cases} \displaystyle\int_\Omega \{g_c - \eta_c(\mathbf{x})(r(s_h(\mathbf{u}^{n,k} \cdot \mathbf{n}) - e_{\max} - d^{n,k}) + \lambda_c^n)\} e\, da = 0 \\ \qquad \forall e \in H_{ch}, \quad d^{n,k} \in H_{ch}, \quad g_c \in \partial\left(\dfrac{1}{3\varepsilon} \sup(0, d^{n,k})^3\right), \end{cases}$$

and by setting

$$\text{(7.24)}\qquad \mathbf{u}^n = \mathbf{u}^{n,k_{\max}}, \quad \mathbf{F}^n = \mathbf{F}^{n,k_{\max}}, \quad d^n = d^{n,k_{\max}},$$

$$\text{(7.25)}\qquad \boldsymbol{\lambda}^{n+1} = \boldsymbol{\lambda}^n + r(\mathbf{Id} + \nabla_h \mathbf{u}^n - \mathbf{F}^n),$$

$$\text{(7.26)}\qquad \lambda_c^{n+1} = \lambda_c^n + r(s_h(\mathbf{u}^n \cdot \mathbf{n}) - e_{\max} - d^n).$$

Of all the steps (7.19)-(7.26), the only one that has not been studied is (7.23). However, this is also one of the easiest steps, since its solution is given simply by

$$\begin{cases} d^{n,k}|_{\Gamma_l} = a_l/r \quad \text{if } a_l \le 0, \\ \qquad = -\varepsilon r \eta_c/2 + (\varepsilon^2 r^2 \eta_c^2/4 + \varepsilon a_l \eta_c)^{1/2} \quad \text{if } a_l > 0, \end{cases}$$

with

$$a_l = (r[s_h(\mathbf{u}^{n,k} \cdot \mathbf{n}) - e_{\max}] + \lambda_c^n)|_{\Gamma_l}.$$

7.6. Numerical examples. We consider Mooney–Rivlin materials for which we have

$$\rho \mathscr{W}_1(\mathbf{x}, \mathbf{F}) = C_{10}(|\mathbf{F}|^2 - 3), \qquad \rho \mathscr{W}_2(\mathbf{x}, \mathbf{F}) = C_{01}(|\mathrm{adj}\, \mathbf{F}|^2 - 3).$$

The first example considers a cylinder, of radius unity, at rest on a flat plane (Fig. 7.4) and subjected to a vertical body force of 400 kN/m^3. The result was obtained after 40 iterations of Algorithm (7.19)-(7.26) with

$$C_{10} = 435\ \text{kPa}, \qquad C_{01} = 65\ \text{kPa},$$
$$r = 5, \qquad \eta = \eta_c = 1000.0, \qquad k_{\max} = 1.$$

Here 120 finite elements of the RUAS-type (§ 3.3) were used to approximate half the cylinder.

The second numerical example, shown in Fig. 7.5, considers a three-dimensional gutter of length 20.00 m, of thickness 0.20 m, of internal radius 1.00 m, hung over the solid plane $y = -3.00$ m, and subjected on its upper face to a vertical surface loading of 0.5 kPa. For symmetry reasons, only one fourth of the body was computed using 270 hexahedral finite elements (§ 3.2). Taking

$$C_{10} = 50\ \text{kPa}, \qquad C_{01} = 0\ \text{kPa}, \qquad r = 5,$$

this computation required 70 iterations of Algorithm (7.19)-(7.26).

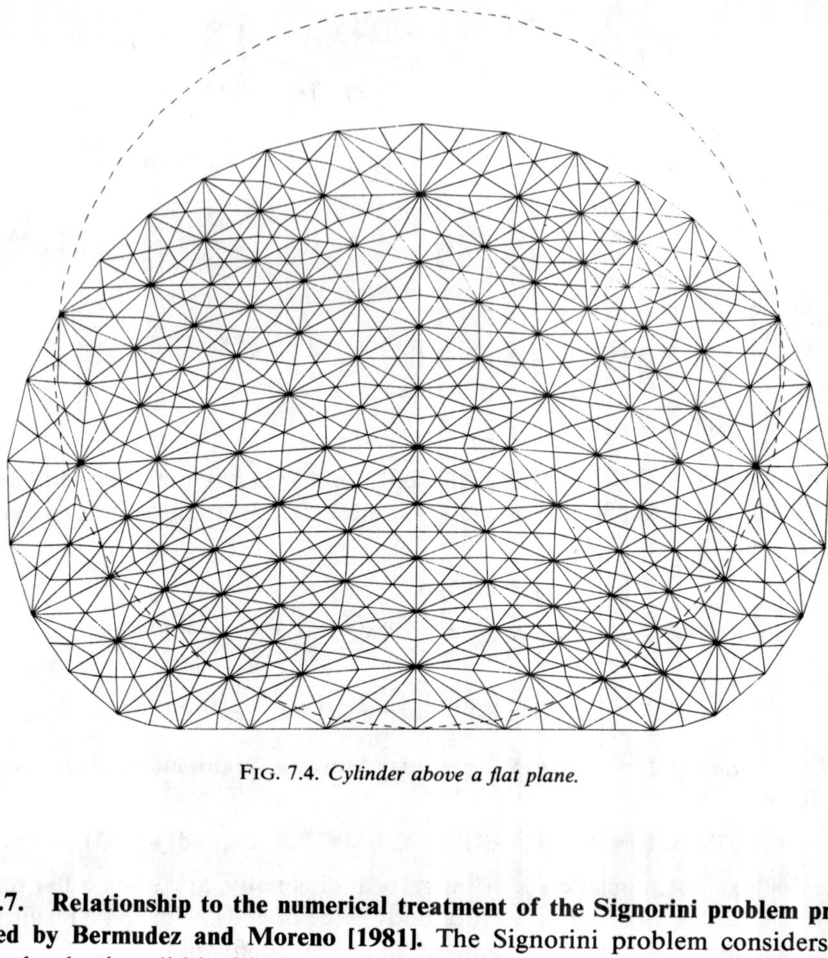

FIG. 7.4. *Cylinder above a flat plane.*

7.7. Relationship to the numerical treatment of the Signorini problem proposed by Bermudez and Moreno [1981]. The Signorini problem considers a linearly elastic solid in frictionless contact with a rigid support and takes the abstract form of Problem (P) within the notation

$$\mathcal{G}(\mathbf{v})=\frac{1}{8}\int_\Omega a_{ijkl}\left(\frac{\partial v_i}{\partial x_j}+\frac{\partial v_j}{\partial x_i}\right)\left(\frac{\partial v_k}{\partial x_l}+\frac{\partial v_l}{\partial x_k}\right)d\mathbf{x}-\int_\Omega \mathbf{f}\cdot\mathbf{v}\,d\mathbf{x}-\int_{\Gamma_2}\mathbf{g}\cdot\mathbf{v}\,da,$$

$$H=L^2(\Gamma),$$

$$\mathcal{F}(e)=\begin{cases}0 & \text{if } e\leq 0 \quad \text{on } \Gamma_c,\\ +\infty & \text{if not,}\end{cases}$$

$$B\mathbf{v}=\mathbf{v}\cdot\mathbf{n}|_{\Gamma_c}.$$

Writing this problem as

$$0\in\partial\mathcal{G}(\mathbf{u})+B^t\partial\mathcal{F}(B\mathbf{u}),$$

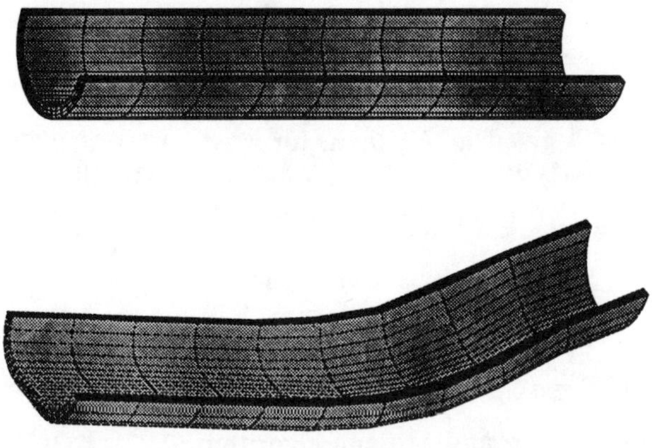

FIG. 7.5. *Three-dimensional gutter subjected to surface loading.*

and introducing the operator

$$H^{\omega}_{1/2\omega}=2\omega\left(\mathbf{Id}-\left(\frac{\omega\,\mathbf{Id}+\partial\mathcal{F}}{2\omega}\right)^{-1}\right),$$

which is the Yosida approximation of $\partial\mathcal{F}-\omega\,\mathbf{Id}$, Bermudez and Moreno [1981] have proposed the following algorithm for solving (P).

ALGORITHM (7.27)-(7.30).

$$q^0\in H \text{ given;} \tag{7.27}$$

then, for $n\geq 0$, q^n being known, determine q^{n+1} by solving sequentially:

$$0\in(\partial\mathcal{G}(\mathbf{u}^n)+\omega B^tB\mathbf{u}^n+B^tq^n),\qquad \mathbf{u}^n\in V, \tag{7.28}$$

$$q^{n+1/2}=H^{\omega}_{1/2\omega}(B\mathbf{u}^n+q^n/2\omega), \tag{7.29}$$

$$q^{n+1}=\rho q^{n+1/2}+(1-\rho)q^n. \tag{7.30}$$

If we set $r=\omega$ and $\lambda^n=+q^n+\omega d^{n-1}$, and if we introduce the augmented Lagrangian $\mathcal{L}_r$ defined in (2.12), the equations (7.28)-(7.30) can be rewritten as

$$\begin{aligned}
&\mathcal{L}_r(\mathbf{u}^n, d^{n-1}; \lambda^n)\leq\mathcal{L}_r(\mathbf{v}, d^{n-1}; \lambda^n)\qquad \forall\mathbf{v}\in V, \mathbf{u}^n\in V,\\
&\lambda^{n+1/2}=q^n+\omega B\mathbf{u}^n=\lambda^n+\omega(B\mathbf{u}^n-d^{n-1}),\\
&\omega B\mathbf{u}^n+\lambda^{n+1/2}=(q^n+2\omega B\mathbf{u}^n)\in[\omega p^{n+1/2}+\partial\mathcal{F}(p^{n+1/2})],\\
&q^{n+1/2}=\omega B\mathbf{u}^n+\lambda^{n+1/2}-2\omega p^{n+1/2},\\
&q^{n+1}=\rho q^{n+1/2}+(1-\rho)q^n.
\end{aligned}$$

Within the identification $d^n = p^{n+1/2}$, Algorithm (7.27)–(7.30) takes the following final form.

ALGORITHM (7.31)–(7.35).

(7.31) λ^0 and d^{-1} given in H; then, for $n \geq 0$, λ^n and d^{n-1} being known, determine $\mathbf{u}^n$, d^n, and λ^{n+1} by solving sequentially

$$\mathscr{L}_r(\mathbf{u}^n, d^{n-1}; \lambda^n) \leq \mathscr{L}_r(\mathbf{v}, d^{n-1}; \lambda^n) \quad \forall \mathbf{v} \in V, \quad \mathbf{u}^n \in V, \tag{7.32}$$

$$\lambda^{n+1/2} = \lambda^n + r(B\mathbf{u}^n - d^{n-1}), \tag{7.33}$$

$$\mathscr{L}_r(\mathbf{u}^n, d^n; \lambda^{n+1/2}) \leq \mathscr{L}_r(\mathbf{u}^n, e; \lambda^{n+1/2}) \qquad \forall e \in H, \quad d^n \in H, \tag{7.34}$$

$$\lambda^{n+1} = q^{n+1} + \omega d^n = \lambda^{n+1/2} + (2\rho - 1)r(B\mathbf{u}^n - d^n). \tag{7.35}$$

Therefore, the Bermudez and Moreno algorithm (7.27)–(7.30) appears finally as a particular case of ALG3 when applied to Problem (P). However, the use of Algorithm (7.31)–(7.35) to solve the large deformation problem (7.3)–(7.5) did not actually result in any speed up compared with the basic ALG1 algorithm (7.19)–(7.26).

Chapter 8

Large Displacement Calculations of Flexible Rods

1. Introduction and description of the physical problem. In this chapter we discuss the application of the techniques of Chapter 3 to the numerical modeling of *flexible and inextensible rods.* Related problems have been considered by many authors using both mathematical and computational approaches (see, for example, Antman and Kenney [1981] and Simo [1985] and the references therein).

An important motivation for this chapter is the study of the flexible pipelines used in off-shore oil production. Indeed, engineers are interested in the static and dynamic behaviors of these pipes as a result of the effects of streams, waves, and obstacles. Figure 1.1 explains some further notation associated with the problems under study.

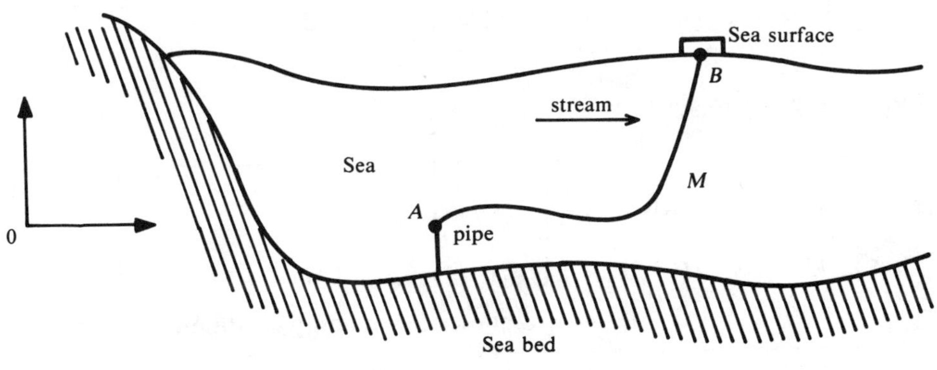

FIG. 1.1. *Physical configuration.*

We shall use the following notation for modeling such problems:

A, B: extremities of the rod;
s: curvilinear abscissa,
$s(A)=0$,
$s(B)=L$ ($L=$ length of the pipe);
$M(s)$: generic point of the rod whose present position is $\bar{\mathbf{x}}(s)$.

Furthermore, we suppose that the rods under consideration fulfill the following conditions, which not only simplify the problem but also provide realistic results on the behavior of such structures.

(i) The rods are inextensible;
(ii) they are of a small diameter with respect to their length L, so that the cross-sections of the rods can be assumed to remain undeformed and perpendicular to their center lines;
(iii) they are very flexible and therefore can handle large displacements while staying in an elastic regime.

Using this framework, we shall first discuss the mathematical modeling of the static problem without torsion in § 2, and then consider its finite-element approximation in § 3 and its augmented Lagrangian treatment in § 4. Then, in §§ 5 and 6, we will generalize this approach to handle nonconservative loading (stream, etc.), contact constraints, and torsional and dynamical effects.

2. Mathematical modeling of the torsion-free static problem.

2.1. Formulation of the problem. We begin with the general, energetic formulation (5.13) presented in Chapter 2 to model equilibrium problems in finite elasticity, using the assumptions of § 1, and neglecting torsional effects. The problem of finding the stable equilibrium positions of a transversely isotropic elastic rod under dead loading then reduces to the following minimization problem.

(2.1) Minimize the total potential energy $J(\cdot)$ over the set K of kinematically admissible configurations,

where J and K are given by

$$J(\mathbf{y})=\frac{1}{2}\int_0^L EI|\mathbf{y}''|^2\,ds-\int_0^L \mathbf{f}\cdot\mathbf{y}\,ds, \tag{2.2}$$

$$K=\{\mathbf{y}\,|\,\mathbf{y}\in H^2(0,L;\mathbb{R}^3),\ |\mathbf{y}'(s)|=1 \text{ on } [0,L], \text{ plus adequate boundary conditions}\}. \tag{2.3}$$

In (2.2) and (2.3), EI (>0) is the *flexural stiffness* of the rod, $\mathbf{f}$ the *lineic density of external loads*, and $\mathbf{y}'$ and $\mathbf{y}''$ the first and second derivatives of the function $\mathbf{y}$ with respect to the curvilinear abscissa s.

Model (2.1) is due to Euler (the Elastica problem). It can be derived directly from the three-dimensional model (5.13) of Chapter 2. The solutions $\bar{\mathbf{x}}$ of problem (2.1) represent the final positions of the center line of the rod.

Remark 2.1. The minimization problem (2.1) has to be understood in a local sense. We note that this problem is neither quadratic nor convex.

Remark 2.2. For a *rod subjected to its own weight* only, $\mathbf{f}$ reduces to $\rho\mathbf{g}$ where ρ denotes the *lineic density* of the rod and $\mathbf{g}$ the *gravity acceleration* vector. If the rod lies under water, to take into account the hydrostatic pressure we have to use ρ defined by $\rho=\rho_0-\sigma\rho_w$, where ρ_0 and σ are the *intrinsic lineic density* and the *cross-section* of the rod, respectively, and where ρ_w is the *volumic density* of the water.

Remark 2.3. More general situations than (2.1) will be discussed in §§ 5 and 6.

2.2. Mathematical analysis of the static problem. We suppose from now on that $(EI)\in L^\infty(0, L)$, with $0<(EI)_{\min}\leq(EI)(s)\leq(EI)_{\max}$ almost everywhere. The mathematical analysis of problems like (2.1) is by itself a fascinating subject that goes back to Euler; however, since our investigation is computationally oriented we shall limit our analysis to a very simple existence theorem and to some comments on nonuniqueness properties. For a very complete mathematical analysis of related problems we refer the reader to Antman and Rosenfeld [1978]. We suppose that the boundary conditions in (2.3) are given by either

$$\mathbf{y}(0)=\mathbf{x}_A, \qquad \mathbf{y}(L)=\mathbf{x}_B, \tag{2.4}$$

with $\mathbf{x}_A$, $\mathbf{x}_B$ given in $\mathbb{R}^3$, or by

$$\begin{cases}\mathbf{y}(0)=\mathbf{x}_A, & \mathbf{y}'(0)=\mathbf{x}'_A,\\ \mathbf{y}(L)=\mathbf{x}_B, & \mathbf{y}'(L)=\mathbf{x}'_B,\end{cases} \tag{2.5}$$

where, in (2.5), $\mathbf{x}_A$, $\mathbf{x}_B$, $\mathbf{x}'_A$, and $\mathbf{x}'_B$ are given in $\mathbb{R}^3$, with $|\mathbf{x}'_A|=|\mathbf{x}'_B|=1$.

We have the following theorem concerning existence properties.

THEOREM 2.1. *Suppose that* $|\mathbf{x}_B-\mathbf{x}_A|<L$, *that* (2.4) *or* (2.5) *hold, and that the linear function* $\mathbf{v}\to\int_0^L \mathbf{f}\cdot\mathbf{v}\,ds$ *is continuous on* $H^2(0, L; \mathbb{R}^3)$. *Then the minimization problem* (2.1) *has at least one solution.*

Proof. Since $|\mathbf{x}_B-\mathbf{x}_A|<L$, the set K defined by (2.3), with (2.4) or (2.5), is not empty. Let $\{\mathbf{x}_n\}_n$ be a minimizing sequence; then we clearly have some constant C such that

$$\|\mathbf{x}_n\|_{2,2}\leq C \qquad \forall n\geq 0. \tag{2.6}$$

It follows from (2.6) that $\mathbf{x}_n$ is bounded in $H^2(0, L; \mathbb{R}^3)$; then the compactness of the embedding of $H^2(0, L; \mathbb{R}^3)$ in $C^1([0, L]; \mathbb{R}^3)$ implies the existence of a

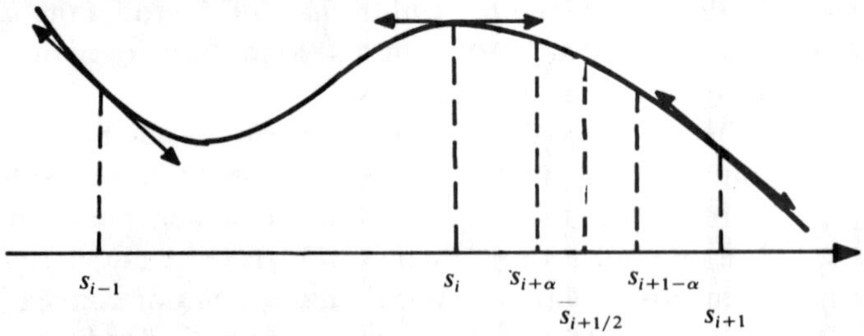

FIG. 2.1. *Hermite cubic approximation.*

subsequence—still denoted by $\{\mathbf{x}_n\}_n$—and of $\bar{\mathbf{x}} \in H^2(0, L; \mathbb{R}^3)$ such that

$$\mathbf{x}_n \to \bar{\mathbf{x}} \quad \text{strongly in } C^1([0, L]; \mathbb{R}^3), \tag{2.7}$$

$$\mathbf{x}_n'' \to \bar{\mathbf{x}}'' \quad \text{weakly in } L^2(0, L; \mathbb{R}^3). \tag{2.8}$$

Since $\mathbf{x}_n \in K$ for all n, it follows from (2.3), (2.4), (2.5), (2.7), and (2.8) that $\bar{\mathbf{x}} \in K$. It follows also from (2.7) and (2.8), and from the weak semi-continuity on $H^2(0, L; \mathbb{R}^3)$ of the (convex) function $J(\cdot)$ defined by (2.2), that

$$J(\bar{\mathbf{x}}) \le \liminf J(\mathbf{x}_n) = \inf_{\mathbf{y} \in K} J(\mathbf{y}). \tag{2.9}$$

Clearly, (2.9) and $\bar{\mathbf{x}} \in K$ imply that $\bar{\mathbf{x}}$ is a solution of (2.1) □

Remark 2.4. If (2.4) holds and if $|\mathbf{x}_B - \mathbf{x}_A| = L$, then (2.1) clearly has a unique solution, which is given by

$$\bar{\mathbf{x}}(s) = \mathbf{x}_A + \frac{s}{L}(\mathbf{x}_B - \mathbf{x}_A). \tag{2.10}$$

If $|\mathbf{x}_B - \mathbf{x}_A| = L$ and if K is defined from (2.5), then K is empty in general.

Remark 2.5 (On the nonuniqueness). If $EI = 0$ in $J(\cdot)$, if $\mathbf{f} = \rho \mathbf{g}$ with $\rho =$ constant, and if the boundary conditions are given by (2.4), then problem (2.1) has a unique solution corresponding to a catenoid curve (we suppose, of course, that $|\mathbf{x}_B - \mathbf{x}_A| \le L$). If $\mathbf{f} = \mathbf{0}$ and if the boundary conditions are given by (2.4), then (2.1) reduces to Euler's Elastica problem and there corresponds, to each solution of (2.1), the solution obtained by symmetry with respect to the line AB. Finally, if both EI and $\mathbf{f}$ are different from zero we also have nonuniqueness in general.

3. Finite-element approximation of the static problem.

3.1. Approximation of $H^2(0, L; \mathbb{R}^3)$. Since K is a subset of $H^2(0, L; \mathbb{R}^3)$, a most important step toward the numerical solution of (2.1) is to define a convenient approximation of $H^2(0, L; \mathbb{R}^3)$. For this purpose, let us introduce $\{s_i\}_{i=0}^n$ such that (see Fig. 2.1)

$$0 = s_0 < s_1 < \cdots < s_{i-1} < s_i < s_{i+1} < \cdots < s_N = L \tag{3.1}$$

and approximate $H^2(0, L; \mathbb{R}^3)$ by

$$(3.2) \qquad V_h = \{\mathbf{v}_h \in C^1([0, L]; \mathbb{R}^3), \mathbf{v}_{h|[s_i, s_{i+1}]} \in \mathbf{P}_3(s_i, s_{i+1}) \; \forall 0 \le i < N-1\}.$$

In (3.2) and below, $\mathbf{P}_k(s_i, s_{i+1})$ denotes the set of polynomials in s with coefficients in $\mathbb{R}^3$ and of degree less than or equal to k. We have that $V_h \subset H^2(0, L; \mathbb{R}^3)$ and $\dim V_h = 6(N+1)$ and, as usual, h denotes $\max_i |s_{i+1} - s_i|$. It is also clear that V_h corresponds to a finite-element approximation of the Hermite cubic type with degrees of freedom

$$\{\mathbf{v}_h(s_i)\}_{i=0}^N \quad \text{and} \quad \{\mathbf{v}'_h(s_i)\}_{i=0}^N.$$

3.2. Approximation of K. Since we are using a piecewise Hermite cubic approximation, we have no difficulty in approximating boundary conditions such as (2.4) and (2.5). Concerning the inextensibility condition $|\mathbf{y}'(s)| = 1$ for all $s \in [0, L]$, the obvious choice is to use

$$(3.3) \qquad |\mathbf{y}'_h(s_i)| = 1 \quad \forall 0 \le i \le N.$$

We have, however, observed that in situations involving strong curvature, (3.3) leads to inaccurate results. We have therefore found it more efficient to approximate the inextensibility condition by

$$(3.4) \qquad |\mathbf{y}'_h(s_{i+\alpha})| = |\mathbf{y}'_h(s_{i+1/2})| = |\mathbf{y}'_h(s_{i+1-\alpha})| = 1 \quad \forall 0 \le i < N,$$

where $s_{i+\alpha}$, $s_{i+1/2}$, and $s_{i+1-\alpha}$ are the three Gauss–Legendre points of the segment $[s_i, s_{i+1}]$.

Using (3.3) (resp., (3.4)) to approximate the inextensibility condition, we then approximate K by the set K_h of functions of V_h that satisfy the boundary conditions and the approximate inextensibility condition. It is obvious that, in any case, K_h is a closed subset of V_h.

3.3. Approximation of the minimization problem (2.1). From §§ 3.1, and 3.2, we approximate Problem (2.1) as follows.

(3.5) | Minimize the energy function $J(\cdot)$ over K_h. |

Concerning the existence of solutions of (3.5), we have the following variant of Theorem 2.1.

THEOREM 3.1. *Under the assumptions of Theorem* 2.1, *Problem* (3.5) *has at least a solution.*

Proof. Because K_h is not empty, we can introduce a minimizing sequence $\{\mathbf{x}_h^n\}_n$ of J on K_h. Then, because K_h is a closed subset of the finite-dimensional space V_h, and because J is continuous on V_h, if we can prove that $\{\mathbf{x}_h^n\}_n$ is bounded in V_h, the existence result will follow from elementary compactness arguments. Let us therefore prove the boundedness of $\{\mathbf{x}_h^n\}_n$.

Suppose for simplicity that K_h is defined from (3.3). We have that

$$\mathbf{y}'(s)=\mathbf{y}'(s_i)+\int_{s_i}^{s}\mathbf{y}_h''(\sigma)\,d\sigma,$$

which implies, combined with (3.3) and using Schwarz' inequality, that

$$|\mathbf{y}_h'(s)|\leq 1+h^{1/2}\|\mathbf{y}_h''\|_{0,2}\qquad \forall s\in[0,L],\quad \mathbf{y}_h\in K_h. \tag{3.6}$$

We have, similarly, that

$$\mathbf{y}(s)=\mathbf{y}(0)+\int_0^s\mathbf{y}'(\sigma)\,d\sigma,$$

which, from the boundary conditions, yields

$$|\mathbf{y}(s)|\leq|\mathbf{x}_A|+L(1+h^{1/2}\|\mathbf{y}_h''\|_{0,2})\qquad \forall s\in[0,L],\quad \mathbf{y}_h\in K_h. \tag{3.7}$$

On the other hand, we have by definition of J

$$\begin{cases} J(\mathbf{y}_h)\geq\dfrac{(EI)_{\min}}{2}\displaystyle\int_0^L|\mathbf{y}_h''|^2\,ds-\|\mathbf{f}\|(\|\mathbf{y}\|_{0,2}^2+\|\mathbf{y}'\|_{0,2}^2+\|\mathbf{y}''\|_{0,2}^2)^{1/2}\\ \forall\mathbf{y}_h\in V_h.\end{cases} \tag{3.8}$$

Using (3.6)–(3.8), it is then very easy to prove that, for the minimizing sequence $\{\mathbf{x}_h^n\}$, $\|\mathbf{x}_h^n\|_{2,2}$ is bounded. □

3.4. Convergence of the approximate solutions. The following lemma will play an important role in our proof of convergence.

LEMMA 3.2. *Suppose that, for all h, $\mathbf{x}_h\in V_h$ and satisfies* (3.3) *together with*

$$\lim_{h\to0}\{\mathbf{x}_h\}=\mathbf{x}\quad \textit{weakly in } H^2(0,L;\mathbb{R}^3). \tag{3.9}$$

Then $|\mathbf{x}'(s)|=1$ for all $s\in[0,L]$.

Proof. Assuming, for example, that K_h is defined from (3.3), we have, as in (3.6),

$$||\mathbf{y}_h'(s)|-1|\leq h^{1/2}\|\mathbf{y}_h''\|_{0,2}\quad \forall\mathbf{y}_h\in K_h.$$

Since we have from (3.9) that $\{\mathbf{x}_h\}\in K_h$ is bounded in $H^2(0,L;\mathbb{R}^3)$, this implies the existence of C such that

$$||\mathbf{x}_h'(s)|-1|\leq Ch^{1/2}\qquad \forall s\in[0,L],\quad \text{and } \forall h. \tag{3.10}$$

Moreover, (3.9) implies also the strong convergence of $\{\mathbf{x}_h\}$ toward $\mathbf{x}$ in $C^1([0,L];\mathbb{R}^3)$ and, hence, we have the desired result by going to the limit as h goes to zero in (3.10).

A similar method should be used for K_h defined from (3.4). □

We are now ready to prove the following theorem.

THEOREM 3.3. *Suppose that K_h is given from* (3.3) *and that $\bar{\mathbf{x}}$ is an isolated solution of* (2.1) *in the sense that there exists a neighborhood $\mathcal{N}$ of $\bar{\mathbf{x}}$ such that*

$$J(\bar{\mathbf{x}})<J(\mathbf{y})\qquad \forall\mathbf{y}\in K\cap\mathcal{N},\quad \textit{with } \mathbf{y}\neq\bar{\mathbf{x}}. \tag{3.11}$$

Then, for h sufficiently small, the approximate problem (3.5) *has a solution* $\mathbf{x}_h$ *strongly converging toward* $\bar{\mathbf{x}}$ *into* $H^2(0, L; \mathbb{R}^3)$ *as h goes to zero.*

Proof. Let $\bar{\mathbf{x}}$ be an isolated solution of (2.1); there then exists $\delta > 0$ such that

$$J(\bar{\mathbf{x}}) < J(\mathbf{y}) \quad \forall \mathbf{y} \in \bar{B}_\delta \cap K, \quad \text{with } \mathbf{y} \neq \bar{\mathbf{x}}, \tag{3.12}$$

where, in (3.12), $\bar{B}_\delta$ denotes the closed ball of $H^2(0, L; \mathbb{R}^3)$ with center $\bar{\mathbf{x}}$ and radius δ. We consider now the finite-dimensional problem

$$\text{Minimize } J(\cdot) \text{ over } \bar{B}_\delta \cap K_h. \tag{3.13}$$

Using compactness arguments, it can easily be proved that (3.13) has at least a solution $\mathbf{x}_h$.

Let now Π_h be the interpolation operator defined by

$$\begin{cases} \Pi_h \mathbf{v} \in V_h & \forall \mathbf{v} \in C^1([0, L]; \mathbb{R}^3), \\ \Pi_h \mathbf{v}(s_i) = \mathbf{v}(s_i) & \forall i = 0, \cdots, N, \\ (\Pi_h \mathbf{v})'(s_i) = \mathbf{v}'(s_i) & \forall i = 0, \cdots, N. \end{cases}$$

If K_h is defined from (3.3), we clearly have

$$\Pi_h \mathbf{y} \in K_h \quad \forall \mathbf{y} \in K.$$

On the other hand, we have from standard results on finite-element approximation (see, for example, Strang, and Fix [1973], Oden and Reddy [1976], Ciarlet [1978], and Raviart and Thomas [1983]) that

$$\lim_{h \to 0} \|\mathbf{y} - \Pi_h \mathbf{y}\|_{2,2} = 0 \quad \forall \mathbf{y} \in H^2(0, L; \mathbb{R}^3),$$

which implies, in particular, that

$$\Pi_h \bar{\mathbf{x}} \in \bar{B}_\delta \cap K_h \quad \forall h \text{ sufficiently small}, \tag{3.14}$$

and that

$$\lim_{h \to 0} J(\Pi_h \bar{\mathbf{x}}) = J(\bar{\mathbf{x}}). \tag{3.15}$$

Let us consider now the behavior of $\{\mathbf{x}_h\}$ as h goes to zero. Since this family is bounded in $H^2(0, L; \mathbb{R}^3)$, there exists a subsequence, still denoted by $\{\mathbf{x}_h\}$, and an element $\mathbf{x}^*$ of $H^2(0, L; \mathbb{R}^3)$ such that $\{\mathbf{x}_h\}$ converges toward $\mathbf{x}^*$ weakly in $H^2(0, L; \mathbb{R}^3)$ as h goes to zero. Since $\bar{B}_\delta$ is a closed ball of $H^2(0, L; \mathbb{R}^3)$, from Lemma 3.2, this implies that $\mathbf{x}^*$ belongs to $K \cap \bar{B}_\delta$.

On the other hand, from (3.14), we have

$$J(\mathbf{x}_h) \leq J(\Pi_h \bar{\mathbf{x}}) \quad \forall h \text{ sufficiently small}.$$

Going to the limit as h goes to zero, this implies

$$J(\mathbf{x}^*) \leq \liminf J(\mathbf{x}_h) \leq \limsup J(\mathbf{x}_h) \leq J(\bar{\mathbf{x}}),$$

which, combined with (3.12) and the fact that $\mathbf{x}^* \in K \cap \bar{B}_\delta$, yields

$$\mathbf{x}^* = \bar{\mathbf{x}}.$$

Therefore, the whole sequence $\{\mathbf{x}_h\}$ converges weakly toward $\bar{\mathbf{x}}$ with

$$\lim_{h \to 0} J(\mathbf{x}_h) = J(\bar{\mathbf{x}}). \tag{3.16}$$

This weak convergence, together with (2.2) and (3.16), implies that

$$\lim_{h \to 0} \int_0^L EI|\mathbf{x}_h''|^2 \, ds = \int_0^L EI|\bar{\mathbf{x}}''|^2 \, ds,$$

which, in turn, guarantees the strong convergence of $\{\mathbf{x}_h\}$ in $H^2(0, L; \mathbb{R}^3)$. From this strong convergence property, we have, for h sufficiently small, that $\mathbf{x}_h$ belongs to the interior of $\bar{B}_\delta$ and that it therefore is a local minimizer of J on K_h. This completes the proof of the theorem. □

4. Augmented Lagrangian solution of the static problem.

4.1. Generalities. The numerical solution of problems closely related to (2.1) has been considered by several authors; let us mention among many others Hibbit, Becker, and Taylor [1979], Simo [1985], and Geradin [1984]. Problem (2.1) is actually nontrivial from the computational point of view, as can be observed by introducing the Lagrangian function associated with the energy function J and with the nonlinear inextensibility constraint

$$\mathcal{L}(\mathbf{y}, \eta) = J(\mathbf{y}) + \int_0^L \eta(|\mathbf{y}'|^2 - 1)/2 \, ds.$$

Suppose that a Lagrange multiplier function λ exists associated with a local minimizer $\bar{\mathbf{x}} \in K$. Because $\mathcal{L}$ is stationary, we have that $\{\bar{\mathbf{x}}, \lambda\}$ must satisfy

$$\begin{cases} (EI\bar{\mathbf{x}}'')'' - (\lambda \bar{\mathbf{x}}')' = \mathbf{f} \quad \text{on } [0, L] \quad \text{plus boundary conditions,} \\ |\bar{\mathbf{x}}'|^2 - 1 = 0 \quad \text{on } [0, L]. \end{cases} \tag{4.1}$$

It appears from (4.1) that λ can be seen as a generalized eigenvalue with $\bar{\mathbf{x}}$ as the corresponding generalized eigenvector associated with a fourth-order differential operator.

A possible solution strategy would be to solve a discrete version of (4.1) by variable metric methods such as Newton's method or the method developed by Powell [1979], which generalizes the celebrated Davidon, Fletcher, and Powell method. Nevertheless, due to the large number of nonlinear constraints, we believe these methods would be delicate and expensive to handle in this case.

A different approach would be to try to minimize J on the manifold K_h directly as was done by Lichnewski [1979] and Gabay [1979] using steepest descent and conjugate-gradient methods. However, although they are quite elegant in principle and very effective for many applications, since they carry out the minimization along the geodesic curves of the manifold, such methods are rather difficult to use on sets like K_h.

The methods proposed in the following section are different from those above. They do share, however, some common features with them, since they are based on the augmented Lagrangian techniques of Chapter 3, § 4, and they maintain the idea of direct minimization on a manifold associated pointwise with the inextensibility constraint.

4.2. Augmented Lagrangian formulation. Choosing $\mathbf{p}=\bar{\mathbf{x}}'$ as local variable, which appears to be a very natural choice for treating the inextensibility constraint, the minimization problem (2.1) is clearly of the form

$$(P) \qquad \mathcal{F}(B(\bar{\mathbf{x}}-\mathbf{x}_1))+\mathcal{G}(\bar{\mathbf{x}}-\mathbf{x}_1)\le \mathcal{F}(B\mathbf{y})+\mathcal{G}(\mathbf{y}) \qquad \forall \mathbf{y}\in V, \quad \bar{\mathbf{x}}-\mathbf{x}_1\in V.$$

If the boundary condition (2.4) holds, then we have the notation

$$V=\{\mathbf{y}\in H^2(0, L; \mathbb{R}^3), \mathbf{y}(0)=\mathbf{y}(L)=\mathbf{0}\}. \tag{4.2}$$

Otherwise, if the boundary condition (2.5) holds, we have

$$V=\{\mathbf{y}\in H^2(0, L; \mathbb{R}^3), \mathbf{y}(0)=\mathbf{y}'(0)=\mathbf{y}(L)=\mathbf{y}'(L)=\mathbf{0}\}. \tag{4.3}$$

In both cases, the functions $\mathcal{F}$ and $\mathcal{G}$ and the operator B are defined by

$$B\mathbf{y}=\mathbf{y}'+\mathbf{x}_1', \tag{4.4}$$

$$\mathcal{G}(\mathbf{y})=\int_0^L \frac{EI}{2}|\mathbf{y}''+\mathbf{x}_1''|^2\,ds-\int_0^L \mathbf{f}\cdot\mathbf{y}\,ds, \tag{4.5}$$

$$\mathcal{F}(\mathbf{q})=\begin{cases}0 & \text{if } |\mathbf{q}'|=1 \text{ a.e.,}\\ +\infty & \text{if not.}\end{cases} \tag{4.6}$$

In the foregoing equations, $\mathbf{x}_1$ denotes a given element of K.

Problems like (P) have been extensively studied in Chapter 3, § 4. In particular, in view of its numerical solution, (P) was associated with the following saddle-point problem.

$$\begin{cases}\text{Find } (\bar{\mathbf{x}}-\mathbf{x}_1, \mathbf{p}; \boldsymbol{\lambda})\in(V\times H)\times H \text{ such that}\\ \mathcal{L}_r(\bar{\mathbf{x}}-\mathbf{u}_1, \mathbf{p}; \boldsymbol{\mu})\le\mathcal{L}_r(\bar{\mathbf{x}}-\mathbf{x}_1, \mathbf{p}; \boldsymbol{\lambda})\le\mathcal{L}_r(\mathbf{y}, \mathbf{q}; \boldsymbol{\lambda})\\ \forall(\mathbf{y}, \mathbf{q}; \boldsymbol{\mu})\in(V\times H)\times H,\end{cases} \tag{4.7}$$

corresponding to the augmented Lagrangian $\mathcal{L}_r$ defined (with $r>0$) by

$$\mathcal{L}_r(\mathbf{y}, \mathbf{q}; \boldsymbol{\mu})=\mathcal{F}(\mathbf{q})+\mathcal{G}(\mathbf{y})+\frac{r}{2}|B\mathbf{y}-\mathbf{q}|^2+(\boldsymbol{\mu}, B\mathbf{y}-\mathbf{q}), \tag{4.8}$$

and to the Hilbert space $H=L^2(0, L; \mathbb{R}^3)$ endowed with its usual scalar product $(\cdot,\cdot)$ and the associated norm $|\cdot|$.

As in Chapter 3, it is then a simple exercise to prove that any solution $(\bar{\mathbf{x}}-\mathbf{x}_1, \mathbf{p}; \boldsymbol{\lambda})$ of (4.7) corresponds to a solution $\bar{\mathbf{x}}$ of our original minimization problem (2.1), provided that (4.7) is considered as a local saddle-point problem only.

4.3. Basic iterative method. The major interest of the augmented Lagrangian formulation (4.7) is that it can be solved numerically by one of the algorithms of Chapter 3 (ALG1–ALG4). For this problem, the algorithm that we have used in practice is ALG2, that is, the following algorithm.

ALGORITHM (4.9)–(4.12).

$$\boldsymbol{\lambda}^0 \in H \text{ and } \bar{\mathbf{x}}^{-1} \in V + \mathbf{x}_1 \text{ given;} \tag{4.9}$$

then, for $n \geq 0$, $\boldsymbol{\lambda}^n$ and $\bar{\mathbf{x}}^{n-1}$ being known, determine $\mathbf{p}^n$, $\bar{\mathbf{x}}^n$, and $\boldsymbol{\lambda}^{n+1}$ successively by

$$\mathscr{L}_r(\bar{\mathbf{x}}^{n-1} - \mathbf{x}_1, \mathbf{p}^n; \boldsymbol{\lambda}^n) \leq \mathscr{L}_r(\bar{\mathbf{x}}^{n-1} - \mathbf{x}_1, \mathbf{q}; \boldsymbol{\lambda}^n), \qquad \forall \mathbf{q} \in H, \quad \mathbf{p}^n \in H, \tag{4.10}$$

$$\mathscr{L}_r(\bar{\mathbf{x}}^n - \mathbf{x}_1, \mathbf{p}^n; \boldsymbol{\lambda}^n) \leq \mathscr{L}_r(\mathbf{y}, \mathbf{p}^n; \boldsymbol{\lambda}^n) \qquad \forall \mathbf{y} \in V, \quad \bar{\mathbf{x}}^n - \mathbf{x}_1 \in V, \tag{4.11}$$

$$\boldsymbol{\lambda}^{n+1} = \boldsymbol{\lambda}^n + \rho_n(\bar{\mathbf{x}}^n - \mathbf{p}^n). \tag{4.12}$$

Although this iterative method (which we have used computationally with $\rho_n = r$) has just been described for the continuous problem (2.1), whose formalism is much simpler than its discrete variants, in fact, we have used it on the discrete variants of (2.1) discussed in § 3. Solution strategies for solving the subproblems (4.10)–(4.11) in such a discrete framework will be described in § 4.4.

4.4. Solution of the global problem (4.11). By construction, before discretization, this problem can be expressed as follows.

$$\begin{cases} \text{Find } \bar{\mathbf{x}}^n - \mathbf{x}_1 \in V \text{ such that} \\ \displaystyle\int_0^L \{EI(\bar{\mathbf{x}}^n)'' \cdot \mathbf{y}'' + r(\bar{\mathbf{x}}^n)' \cdot \mathbf{y}'\}\, ds = \int_0^L \{\mathbf{f} \cdot \mathbf{y} + (r\mathbf{p}^n - \lambda^n) \cdot \mathbf{y}'\}\, ds \ \forall \mathbf{y} \in V. \end{cases} \tag{4.13}$$

$\in V$.

It therefore corresponds to a system of independent fourth-order, two-point boundary value problems. Its finite-element approximation can be obtained by replacing V in (4.13) by the finite-dimensional space $V \cap V_h$, with V_h given by (3.2).

If the boundary conditions are given by (2.4) (resp., (2.5)), then this discrete version of (4.13), once it is expanded on the finite-element basis generated by the degrees of freedom $\{\mathbf{v}_h(s_i)\}_{i=0}^N$ and $\{\mathbf{v}'_h(s_i)\}_{i=0}^N$, reduces to three independent linear systems of order $(2N-2)$ (respectively, $(2N-4)$) with the same matrix, which is sparse (with bandwidth 7), symmetric, positive-definite, and independent of n if r is fixed. In this case, we do a Cholesky factorization once and for all and, thus, each step (4.13) amounts to the solution of six sparse, well-posed triangular systems.

4.5. Solution of the local problem (4.10). Let us now discuss the solution of (4.10). With K_h constructed from (3.4), and $(\bar{\mathbf{x}}^n)'$ approximated by construction by piecewise polynomials of degree less than or equal to two, we only need to define in our discrete problems the variables $\mathbf{p}$ and $\boldsymbol{\lambda}$ at the three Gauss–Legendre points $s_{i+\alpha}$, $s_{i+1/2}$, and $s_{i+1-\alpha}$ of each segment $[s_i, s_{i+1}]$. Then, (4.10) becomes

$$
(4.14)\quad \begin{cases} \forall s \in \bigcup_{i=0}^{N-1} \{s_{i+\alpha}, s_{i+1/2}, s_{i+1-\alpha}\}, \text{ find } \mathbf{p}^n(s) \in \mathbb{R}^3 \text{ such that} \\ |\mathbf{p}^n(s)| = 1, \text{ and} \\ \dfrac{r}{2}|(\bar{\mathbf{x}}^{n-1})'(s) - \mathbf{p}^n(s)|^2 - \boldsymbol{\lambda}^n(s)\cdot\mathbf{p}^n(s) \le \dfrac{r}{2}|(\bar{\mathbf{x}}^{n-1})'(s) - \mathbf{q}|^2 - \boldsymbol{\lambda}^n(s)\cdot\mathbf{q} \\ \forall \mathbf{q} \in \mathbb{R}^3 \quad \text{with } |\mathbf{q}| = 1, \end{cases}
$$

and its solution is given by

$$
(4.15)\qquad \mathbf{p}^n(s) = (r(\bar{\mathbf{x}}^{n-1})'(s) + \boldsymbol{\lambda}^n(s))/|r(\bar{\mathbf{x}}^{n-1})'(s) + \boldsymbol{\lambda}^n(s)|.
$$

4.6. On the implementation of (4.9)–(4.12). For the multipler $\boldsymbol{\lambda}^0$, we have always chosen $\boldsymbol{\lambda}^0 = \mathbf{0}$. However, the initialization $\bar{\mathbf{x}}^{-1}$ of the configuration is more delicate. Indeed, the minimization problem (2.1) being nonconvex, it may have more than one solution, and, depending on the initial value $\bar{\mathbf{x}}^{-1}$, Algorithm (4.9)–(4.12) will converge to one or to the other solution. Our usual strategy has consisted of taking for $\bar{\mathbf{x}}^{-1}$ the arc of circle of length L going from $\mathbf{x}_A$ to $\mathbf{x}_B$.

The choice of the parameter r is even more fundamental. Indeed, if r is taken below a problem-dependent threshold r_c, then Algorithm (4.9)–(4.12) does not converge. If r is too large, then the algorithm converges very slowly. The optimal value for r appears numerically to be just above r_c. A good strategy for the choice of r, then, is the following.

$$
(4.16)\quad \begin{cases} \text{For } r \text{ increasing from } |\mathbf{f}|L/10 \text{ until satisfied,} \\ \quad \text{(i) compute } \varepsilon_n = \dfrac{1}{\sqrt{L}}\left(\displaystyle\int_0^L |(\bar{\mathbf{x}}^n)' - \mathbf{p}^n|^2\,ds\right)^{1/2} \text{ for } n = 30; \\ \quad \text{(ii) if } \varepsilon_{30} < 10^{-2}, \text{ accept this value of } r, \text{ and} \\ \quad \text{(iii) if not, increase } r \text{ and reiterate the process.} \end{cases}
$$

Once $\boldsymbol{\lambda}^0$, $\bar{\mathbf{x}}^{-1}$, and r have been chosen as indicated, we implement Algorithm (4.9)–(4.12) until we verify that

$$
\frac{\int_0^L \{|\bar{\mathbf{x}}^n - \bar{\mathbf{x}}^{n-1}|^2 + |(\bar{\mathbf{x}}^n - \bar{\mathbf{x}}^{n-1})'|^2\}\,ds}{\int_0^L \{|\bar{\mathbf{x}}^n|^2 + |(\bar{\mathbf{x}}^n)'|^2\}\,ds} \le 10^{-10},
$$

and

$$\frac{1}{\sqrt{L}}\left(\int_0^L |(\bar{\mathbf{x}}^n)'-\mathbf{p}^n|^2\,ds\right)^{1/2}\leq 10^{-3},$$

which ensures that the computed configuration no longer varies with n and that the inextensibility constraint is properly satisfied.

5. Applications and extensions. The following is taken from Bourgat, Le Tallec, and Mani [1988], where additional details can be found.

5.1. Cantilever in flexion. Comparison with an analytical solution. The problem of the large bending of a cantilever by a terminal load was solved analytically by Bisshop and Drucker [1945]. Considering a rod of length L and of flexural stiffness EI that was clamped at one end and loaded at the other end by a vertical concentrated load P, they computed the horizontal and vertical displacements Δx and Δz of the loaded end (Fig. 5.1).

We tested the numerical method of § 4 on this problem with $L=10$ m, $EI=10^3$ N/m^2, and $P=10, 20, \cdots, 100$ N. Figure 5.2 represents the curves $\Delta z/L$ and $(L-\Delta x)/L$ as functions of P obtained both analytically and by our numerical tests. Observe that the two solutions coincide and that they are largely different from the solution predicted by the linear theory.

Above, convergence of Algorithm (4.9)–(4.12) was obtained in 50 iterations with r chosen in the interval $[2\times10^4, 5\times10^4]$.

5.2. Nonconservative loading. Streams. We now consider a flexible rod, for example, an underwater pipeline, that is subjected to hydrodynamical forces generated by a horizontal stream. We suppose that the stream velocity is time-independent and possibly a function of the depth, and that the corresponding forces are given by the so-called Morison formulas

$$\mathbf{f}_w=\tfrac{1}{2}\rho_w\pi C_f D|\mathbf{v}_t|\mathbf{v}_t+\tfrac{1}{2}\rho_w C_d D|\mathbf{v}_n|\mathbf{v}_n. \tag{5.1}$$

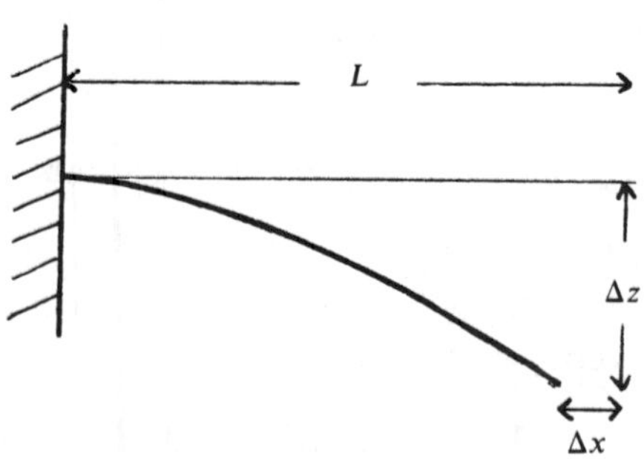

FIG. 5.1. *Cantilever in flexion.*

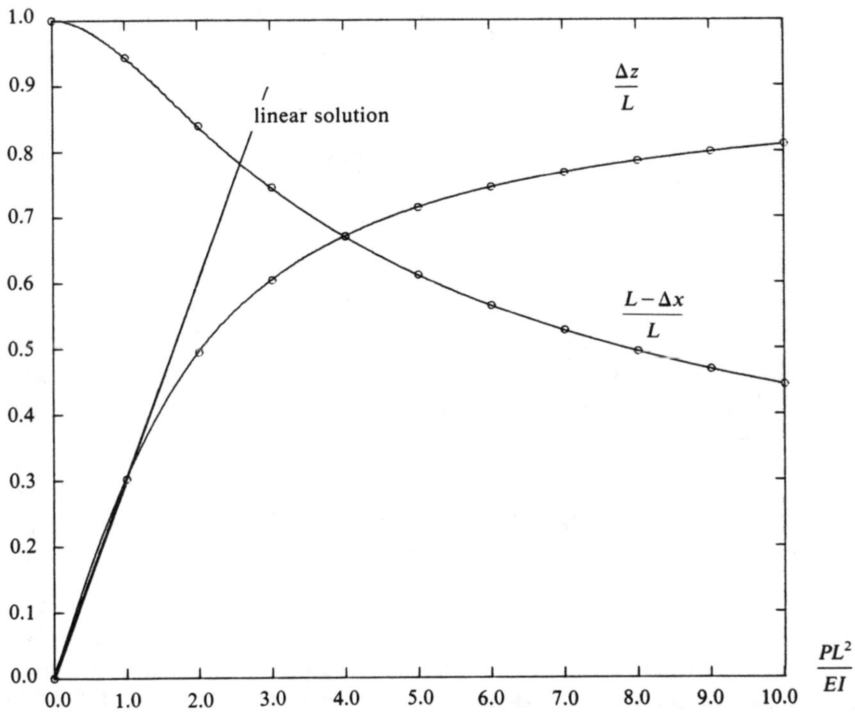

FIG. 5.2. *Results for the cantilever. Key*: —— = *analytical solution*; ○ = *computed solution.*

In (5.1), ρ_w is the volumic density of the water, D is the diameter of the rod whose cross-section is supposed to be circular, C_f and C_d are friction coefficients, and $\mathbf{v}_t$ and $\mathbf{v}_n$ are, respectively, the tangential and the normal velocity of the stream, that is,

$$\mathbf{v}_t = (\mathbf{v}_t \cdot \bar{\mathbf{x}}')\bar{\mathbf{x}}', \qquad \mathbf{v}_n = \mathbf{v} - \mathbf{v}_t. \tag{5.2}$$

Those hydrodynamical forces depend on the unknown configuration $\bar{\mathbf{x}}'$ and are no longer conservative. Therefore, exactly as in Chapter 2, § 5, the formulation (2.1) of the static problem has to be replaced by its variational counterpart, which is as follows.

$$\begin{cases} \text{Find } \bar{\mathbf{x}} \in K \text{ such that} \\ \displaystyle\int_0^L EI\bar{\mathbf{x}}'' \cdot \mathbf{y}''\, ds = \int_0^L (\mathbf{f}_w(\bar{\mathbf{x}}) + \rho\mathbf{g}) \cdot \mathbf{y}\, ds \quad \forall \mathbf{y} \in dK(\bar{\mathbf{x}}), \end{cases} \tag{5.3}$$

where $dK(\bar{\mathbf{x}})$ denotes the tangent space at $\bar{\mathbf{x}}$ to K in $H^2(0, L; \mathbb{R}^3)$. Nevertheless, Algorithm (4.9)-(4.12) (that is, (4.9), (4.14), (4.13), and (4.12)) can still be applied to the solution of (5.3) provided that, in (4.13), we compute $\mathbf{f}$ through the explicit formula

$$\mathbf{f}=\mathbf{f}_w(\bar{\mathbf{x}}^{n-1})+\rho\mathbf{g} \tag{5.4}$$

involving the solution $\bar{\mathbf{x}}^{n-1}$ computed at the previous iteration of the algorithm.

Figure 5.3 represents the computed configurations of a flexible rod subjected both to its own weight and to a horizontal stream with a velocity of, respectively, $+\mathbf{e}_1$, $\mathbf{0}$, $-\mathbf{e}_1$ (measured in m/s). The physical data were $L=32$ m, $EI=4500$ N/m^2, $\rho|\mathbf{g}|=800$ N/m, $D=0.277$ m, $\rho_w=1026$ kg/m^3, $C_f=0.03$, and $C_d=1.2$. The ends of the rod were vertically clamped in $\mathbf{x}_A=(0,0,10)$ and $\mathbf{x}_B=(3.6,0,25)$.

The behavior of the discrete version of Algorithm (4.9)-(4.12), with $\mathbf{f}$ computed by (5.4) and V_h constructed with 50 subintervals of equal size, is summarized in Table 5.1.

5.3. The case with contact. We now consider a flexible rod that can enter in contact with a soil surface located on the curve

$$x_3=\psi(x_1,x_2). \tag{5.5}$$

If we suppose this contact to be frictionless, it can then be handled mathematically by simply changing the definition of K in (2.1) to the new definition

$$\begin{cases} K=\{\mathbf{y}\,|\,\mathbf{y}\in H^2(0,L;\mathbb{R}^3), |\mathbf{y}'(s)|=1 \text{ on } [0,L], \\ \quad y_3(s)\geq\psi(y_1(s),y_2(s)) \text{ on } [0,L], \text{ plus adequate} \\ \quad \text{boundary conditions}\}. \end{cases} \tag{5.6}$$

In other words, in the presence of the contact constraint (5.5), our static problem corresponds to the minimization of the energy function J given by (2.2) over the set K of kinematically admissible configurations given in (5.6).

To solve this new problem, we can again apply the Uzawa algorithm (4.9)-(4.12), but the contact constraint must be added to the global problem (4.11), which now becomes

$$\mathcal{L}_r(\bar{\mathbf{x}}^n-\mathbf{x}_1,\mathbf{p}^n;\boldsymbol{\lambda}^n)\leq\mathcal{L}_r(\mathbf{y},\mathbf{p}^n;\boldsymbol{\lambda}^n) \qquad \forall\mathbf{y}\in K_c,\quad \bar{\mathbf{x}}^n-\mathbf{x}_1\in K_c, \tag{5.7}$$

under the definition

$$\begin{cases} K_c=\{\mathbf{y}\in H^2(0,L;\mathbb{R}^3), (\mathbf{y}+\mathbf{x}_1)_3(s)\geq\psi((\mathbf{y}+\mathbf{x}_1)_1(s),(\mathbf{y}+\mathbf{x}_1)_2(s)) \\ \qquad \text{on } [0,L], \text{ plus adequate boundary conditions}\}. \end{cases} \tag{5.8}$$

Once $H^2(0,L;\mathbb{R}^3)$ is replaced by its finite-element approximation V_h introduced in (3.2), the gobal problem (5.7) reduces to the following constrained

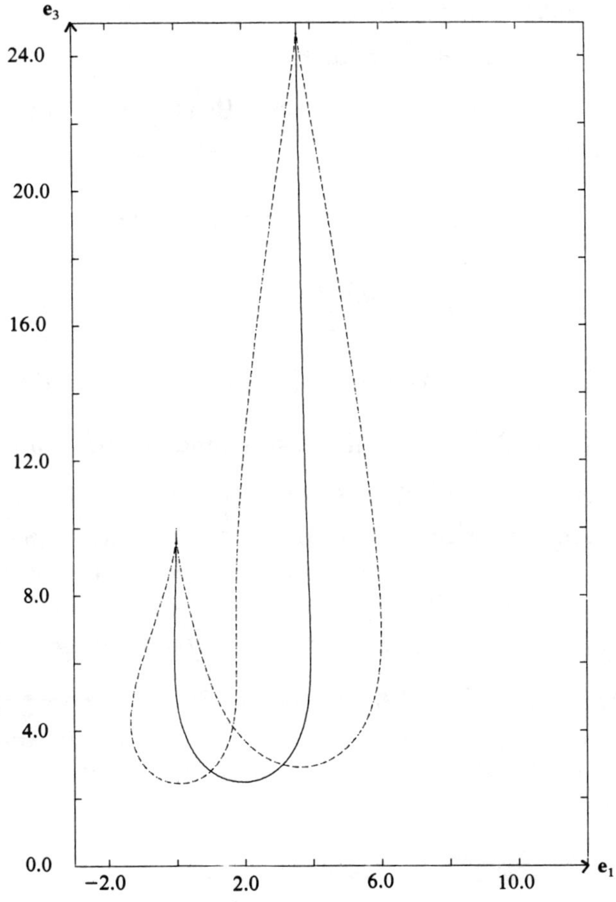

FIG. 5.3. *The computed configurations of a flexible rod subjected to its own weight and to a horizontal stream of velocity.* 1 m/s, −1 m/s, *and* 0 m/s.

TABLE 5.1
Summary of results obtained with Algorithm (4.9)–(4.12).

Stream velocity	0	1 m/s	−1 m/s
Number of iterations	68	72	154
Value of r used	10^4	5×10^4	10^5
CPU time on a Bull DPS68	44 sec	46 sec	95 sec

quadratic minimization problem.

$$(5.9)\qquad \begin{cases} \text{Find } \mathbf{u} \in K_{ch} \text{ such that} \\ \frac{1}{2}(A\mathbf{u}, \mathbf{u}) - (\mathbf{b}, \mathbf{u}) \le \frac{1}{2}(A\mathbf{v}, \mathbf{v}) - (\mathbf{b}, \mathbf{v}) \quad \forall \mathbf{v} \in K_{ch}, \end{cases}$$

under the notation

$$(5.10)\qquad (A\mathbf{v}, \mathbf{v}) = \int_0^L \{EI|\mathbf{y}_h''(\mathbf{v}) + \mathbf{x}_1''|^2 + r|\mathbf{y}_h'(\mathbf{v}) + \mathbf{x}_1'|^2\}\, ds,$$

$$(5.11)\qquad (\mathbf{b}, \mathbf{v}) = \int_0^L \{\mathbf{f} \cdot \mathbf{y}_h(\mathbf{v}) + (r\mathbf{p}^n - \boldsymbol{\lambda}^n) \cdot \mathbf{y}_h'(\mathbf{v})\}\, ds,$$

$$(5.12)\qquad \mathbf{y}_h(\mathbf{v}) = \sum_{i=0}^{N} \sum_{k=1}^{6} v_{ik} \boldsymbol{\psi}_{ik},$$

$$(5.13)\qquad \begin{cases} K_{ch} = \{\mathbf{v} \in \mathbb{R}^{6(N+1)}, \mathbf{y}_h(\mathbf{v}) \text{ satisfies the boundary conditions of } K, \\ v_{i3} \ge \psi(v_{i1} + (\mathbf{x}_1)_1(s_i), v_{i2} + (\mathbf{x}_1)_2(s_i)) - (\mathbf{x}_1)_3(s_i)\ \forall i = 0, N\}. \end{cases}$$

In (5.12), $\boldsymbol{\psi}_{ik}$ denotes the shape function of V_h associated with the degree of freedom $(\mathbf{y}_h(s_i))_k$ if $1 \le k \le 3$, or $(\mathbf{y}_h'(s_i))_{k-3}$ if $4 \le k \le 6$; in other words, $\boldsymbol{\psi}_{ik}$ is the function of V_h defined by

$$\boldsymbol{\psi}_{ik}(s_j) = \delta_{ij}\mathbf{e}_k, \qquad \boldsymbol{\psi}_{ik}'(s_j) = 0 \qquad \forall 0 \le j \le N \quad \text{if } 1 \le k \le 3,$$

$$\boldsymbol{\psi}_{ik}(s_j) = 0, \qquad \boldsymbol{\psi}_{ik}'(s_j) = \delta_{ij}\mathbf{e}_{k-3} \qquad \forall 0 \le j \le N \quad \text{if } 4 \le k \le 6.$$

Since the constraint entering the definition of K_{ch} operates separately on each i, the minimization problem (5.9) is best solved by a pointwise over-relaxation algorithm of the following type.

ALGORITHM (5.14)-(5.18).

$$(5.14)\qquad u_{ik}^0 = ((\bar{\mathbf{x}}^{n-1} - \mathbf{x}_1)(s_i))_k \quad \text{if } 1 \le k \le 3,$$

$$(5.15)\qquad u_{ik}^0 = ((\bar{\mathbf{x}}^{n-1} - \mathbf{x}_1)'(s_i))_{k-3} \quad \text{if } 4 \le k \le 6,$$

then, for $m \ge 0$, $\mathbf{u}^m$ being known, determine $\mathbf{u}^{m+1}$ by

$$(5.16)\qquad u_{ik}^{m+1/2} = \left(b_{ik} - \sum_{\substack{j \le i \\ l < k}} A_{ikjl} u_{jl}^{m+1} - \sum_{\substack{j \ge i \\ l > k}} A_{ikjl} u_{jl}^m \right) \Big/ A_{ikik},$$

$$(5.17)\qquad u_{ik}^{m+1} = u_{ik}^m + \omega(u_{ik}^{m+1/2} - u_{ik}^m) \quad \text{if } k \ne 3,$$

$$(5.18)\qquad \begin{aligned} u_{i3}^{m+1} = \sup\,[\psi[u_{i1}^{m+1} + (\mathbf{x}_1)_1(s_i), u_{i2}^{m+1} + (\mathbf{x}_1)_2(s_i)] - (\mathbf{x}_1)_3(s_i); \\ u_{i3}^m + \omega(u_{i3}^{m+1/2} - u_{i3}^m)]. \end{aligned}$$

In Algorithm (5.14)-(5.18), the parameter ω of relaxation has to be carefully adjusted in order to limit the number of iterations on m.

The final algorithm, then, for the solution of the static problem with contact is again (4.9)-(4.12), with (4.11) replaced by (5.9) and solved by (5.14)-(5.18).

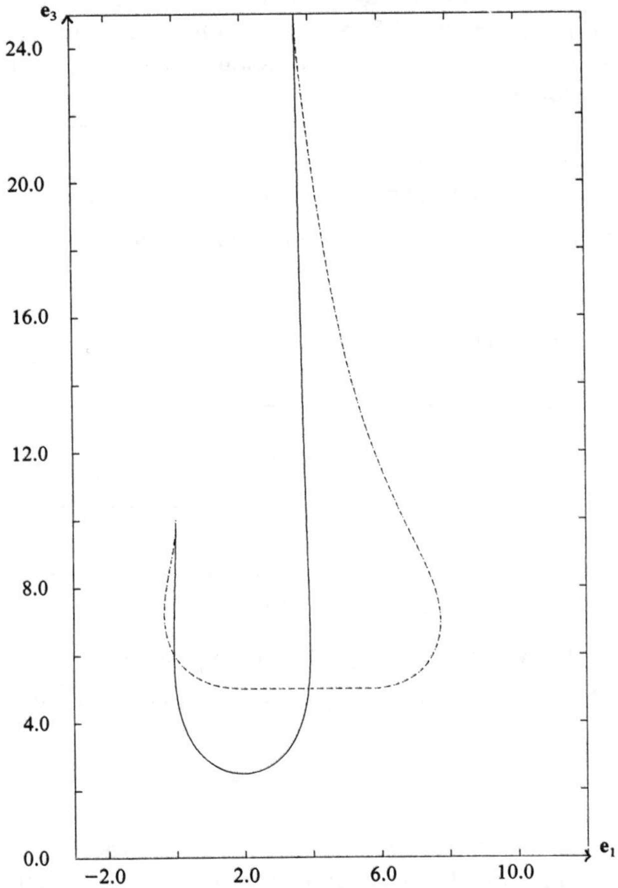

FIG. 5.4. *The computed configurations of a flexible rod in contact with soil. Key*: - - - = *case with contact*; —— = *case with no contact.*

We tested this algorithm on the physical problem of § 5.2, and the results are shown in Fig. 5.4. The number of Uzawa iterations was 300, with $r = 5 \times 10^5$ and $\omega = 1.92$. The CPU time went from 44 sec for the case without soil to 360 sec for the case with soil contact. This large increase is related to the corresponding increase of the number of Uzawa iterations and to the fact that, for each global problem, a relaxation algorithm is more expensive than a Cholesky direct inversion of the matrix **A** (speed up through the use of a multigrid method is an interesting possibility that we have not yet explored).

5.4. The case with torsion.

5.4.1. Formulation of the static problem with torsion. If we do not neglect torsional effects but keep the other assumptions of § 2.1 (inextensibility, undeformability of the cross-section, and a small strains–large displacements elastic regime), the problem of finding the stable equilibrium positions of a transversely isotropic rod under dead loading reduces to the following minimization problem.

(5.19) Minimize the total potential energy $J(\cdot)$ over the set K of kinematically admissible configurations,

where J and K are given by

$$J(\mathbf{y}, \mathbf{g}_\alpha) = \frac{1}{2}\int_0^L \{(E-G)I|\mathbf{y}''|^2 + GI(|\mathbf{g}_1'|^2 + |\mathbf{g}_2'|^2) - \mathbf{f}\cdot\mathbf{y}\}\, ds, \tag{5.20}$$

$$\begin{aligned} K = \{&(\mathbf{y}, \mathbf{g}_\alpha) \in H^2(0, L; \mathbb{R}^3) \times (H^1(0, L; \mathbb{R}^3))^2, \\ &\{\mathbf{y}'(s), \mathbf{g}_1(s), \mathbf{g}_2(s)\} \text{ is a set of direct orthonormal vectors} \\ &\text{for any } s \text{ in } [0, L], \text{ plus adequate boundary conditions} \\ &\text{on } \mathbf{y}, \mathbf{y}', \mathbf{g}_1, \text{ and } \mathbf{g}_2\}. \end{aligned} \tag{5.21}$$

In this problem EI (>0) is the *flexural stiffness* of the rod, GI ($EI > GI \geq 0$) is its *torsional stiffness*, and $\mathbf{f}$ is the *lineic density of external loads.* In the solution $(\bar{\mathbf{x}}, \mathbf{d}_\alpha)$ of (5.19), $\bar{\mathbf{x}}$ is the actual configuration of the center line of the rod, and $\mathbf{d}_\alpha$ are the current configurations of material vectors engraved on the cross-section of the rod, which are, initially, mutually orthogonal.

Formulation (5.19) is consistent with the general framework of Antman and Kenney [1981]. It has been analyzed in Bourgat, Le Tallec, and Mani [1988], and can also be found, with minor changes, in Simo [1985].

5.4.2. Augmented Lagrangian solution of (5.19). The augmented Lagrangian solution of Problem (5.19) is identical to the solution of the torsion-free problem with the replacement of $\mathbf{y}$ by $\{\mathbf{y}, \mathbf{g}_\alpha\}$ and the inextensibility constraint by the constraint

$$\{\mathbf{y}', \mathbf{g}_1, \mathbf{g}_2\} \quad \text{is a set of direct orthonormal vectors.} \tag{5.22}$$

Therefore, introducing the augmented Lagrangian $\mathscr{L}_r$ defined by

$$(5.23)\quad \mathscr{L}_r(\{\mathbf{y}', \mathbf{g}_\alpha\}, \{\mathbf{q}_i\}_{i=1}^3; \{\boldsymbol{\mu}_i\}_{i=1}^3) \begin{cases} = J(\mathbf{y}, \mathbf{g}_\alpha) + \dfrac{r}{2}\displaystyle\int_0^L |\mathbf{y}' - \mathbf{q}_3|^2\, ds \\ \quad + \dfrac{r}{2}\displaystyle\int_0^L \{|\mathbf{g}_1 - \mathbf{q}_1|^2 + |\mathbf{g}_2 - \mathbf{q}_2|^2\}\, ds \\ \quad + \displaystyle\int_0^L \{\boldsymbol{\mu}_3 \cdot (\mathbf{y}' - \mathbf{q}_3) + \boldsymbol{\mu}_1 \cdot (\mathbf{g}_1 - \mathbf{q}_1) \\ \quad + \boldsymbol{\mu}_2 \cdot (\mathbf{g}_2 - \mathbf{q}_2)\}\, ds \\ \qquad\qquad \text{if } \{\mathbf{q}_1, \mathbf{q}_2, \mathbf{q}_3\} \text{ satisfies (5.22)}, \\ = +\infty \quad \text{if not}, \end{cases}$$

Problem (5.19) can again be solved by ALG2 of Chapter 3, that is, the following algorithm.

ALGORITHM (5.24)-(5.27).

$$(5.24)\qquad \{\boldsymbol{\lambda}_i^0\}_{i=1}^3 \text{ and } \{\bar{\mathbf{x}}^{-1}, \mathbf{d}_\alpha^{-1}\} \text{ given};$$

then, for $n \geq 0$, $\{\boldsymbol{\lambda}_i^n\}_{i=1}^3$ and $\{\bar{\mathbf{x}}^{n-1}, \mathbf{d}_\alpha^{n-1}\}$ being known, determine $\{\mathbf{p}_i^n\}_{i=1}^3$, $\{\bar{\mathbf{x}}^n, \mathbf{d}_\alpha^n\}$, and $\{\boldsymbol{\lambda}_i^{n+1}\}_{i=1}^3$ successively by

$$(5.25)\quad \begin{cases} \mathscr{L}_r(\{\bar{\mathbf{x}}^{n-1}, \mathbf{d}_\alpha^{n-1}\}, \{\mathbf{p}_i^n\}_{i=1}^3; \{\boldsymbol{\lambda}_i^n\}_{i=1}^3) \leq \mathscr{L}_r(\{\bar{\mathbf{x}}^{n-1}, \mathbf{d}_\alpha^{n-1}\}, \{\mathbf{q}_i\}_{i=1}^3; \{\boldsymbol{\lambda}_i^n\}_{i=1}^3) \\ \forall \{\mathbf{q}_i\} \in (L^2(0, L; \mathbb{R}^3))^3, \quad \{\mathbf{p}_i^n\} \in (L^2(0, L; \mathbb{R}^3))^3, \end{cases}$$

$$(5.26)\quad \begin{cases} \mathscr{L}_r(\{\bar{\mathbf{x}}^n, \mathbf{d}_\alpha^n\}, \{\mathbf{p}_i^n\}_{i=1}^3; \{\boldsymbol{\lambda}_i^n\}_{i=1}^3) \leq \mathscr{L}_r(\{\mathbf{y}, \mathbf{g}_\alpha\}, \{\mathbf{p}_i^n\}_{i=1}^3; \{\boldsymbol{\lambda}_i^n\}_{i=1}^3) \\ \forall \{\mathbf{y}, \mathbf{g}_\alpha\} \in H^2(0, L; \mathbb{R}^3) \times (H^1(0, L; \mathbb{R}^3))^2 \quad \text{plus boundary conditions}, \\ \quad \text{and with} \\ \{\bar{\mathbf{x}}^n, \mathbf{d}_\alpha^n\} \in H^2(0, L; \mathbb{R}^3) \times (H^1(0, L; \mathbb{R}^3))^2 \quad \text{plus boundary conditions}, \end{cases}$$

$$(5.27)\qquad \{\boldsymbol{\lambda}_i^{n+1}\}_{i=1}^3 = \{\boldsymbol{\lambda}_i^n\}_{i=1}^3 + \rho_n\{\mathbf{d}_1^n - \mathbf{p}_1^n, \mathbf{d}_2^n - \mathbf{p}_2^n, (\bar{\mathbf{x}}^n)' - \mathbf{p}_3^n\}.$$

After finite-element discretization, using a Hermite cubic approximation for $\mathbf{y}$ and a Lagrangian quadratic approximation for $\mathbf{g}_\alpha$, the global problem (5.26) reduces to nine independent linear systems associated with sparse, symmetric, positive-definite matrices that do not vary with n if r is fixed.

As for the local problem (5.25), once it is written at the three Gauss–Legendre points $s_{i+\alpha}$, $s_{i+1/2}$, and $s_{i+1-\alpha}$ of each segment $[s_i, s_{i+1}]$, it can be expressed as

$$(5.28)\qquad -\mathbf{B} \cdot \mathbf{Q}^P(s) \leq -\mathbf{B} \cdot \mathbf{Q} \qquad \forall \mathbf{Q} \in \mathbf{0}_3^+, \quad \mathbf{Q}^P(s) \in \mathbf{0}_3^+,$$

under the notation

$$B_{\alpha j} = [r\mathbf{d}_\alpha^{n-1}(s) + \boldsymbol{\lambda}_\alpha^n(s)] \cdot \mathbf{e}_j, \qquad 1 \leq \alpha \leq 2,$$

$$B_{3j} = [r(\bar{\mathbf{x}}^{n-1})'(s) + \boldsymbol{\lambda}_3^n(s)] \cdot \mathbf{e}_j,$$

$$\mathbf{B} = \mathbf{Q}^g\mathbf{D}\mathbf{Q}^d, \quad \text{where } \mathbf{D} \text{ diagonal} \quad \text{and} \quad (\mathbf{Q}^g, \mathbf{Q}^d) \in (\mathbf{0}_3^+)^2,$$

$$Q_{ij}^P(s) = \mathbf{p}_i^n(s) \cdot \mathbf{e}_j,$$

$$\mathbf{0}_3^+ = \text{set of direct orthogonal matrices of } \mathbb{R}^{3\times 3},$$

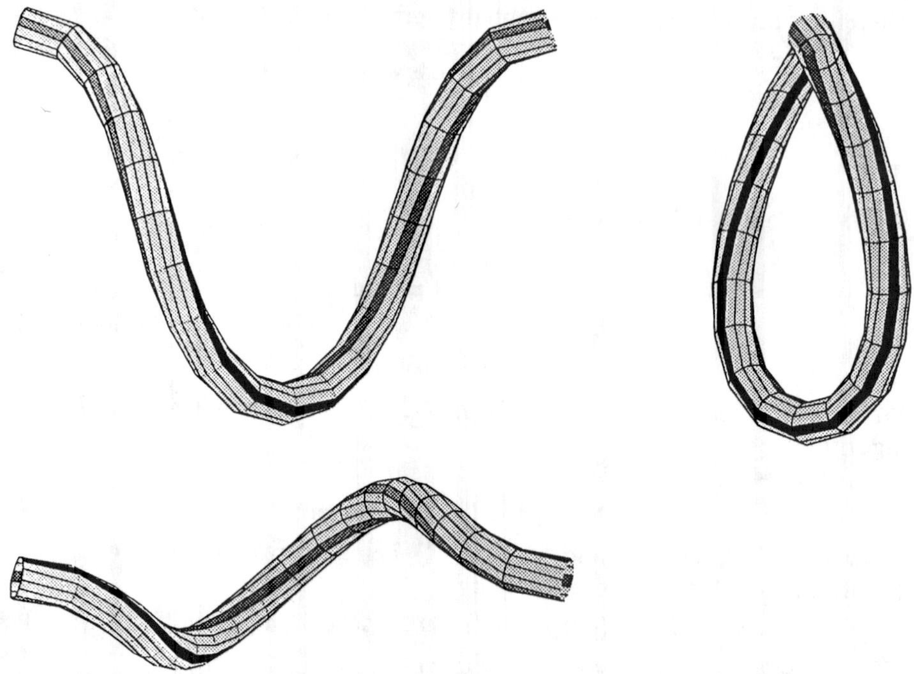

FIG. 5.5. *The computed configurations of a flexible rod subjected to torsion: Projection of the solution in three spaces for* $\beta = 10$ m.

and its solution is given by

$$\mathbf{p}_i^n(s) = \sum_{j=1}^{3} (\mathbf{Q}^g \mathbf{Q}^d)_{ij} \mathbf{e}_j. \tag{5.29}$$

5.4.3. Numerical results. We now consider a flexible rod whose initial configuration is horizontal. We first rotate one end of the rod at an angle

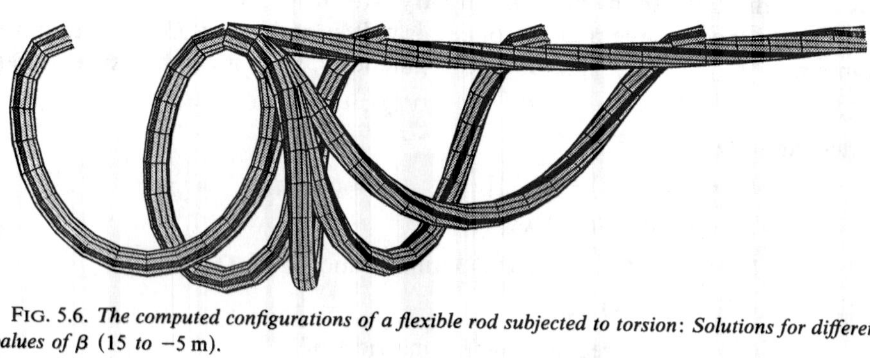

FIG. 5.6. *The computed configurations of a flexible rod subjected to torsion: Solutions for different values of* β (15 *to* −5 m).

$\theta_L = 2\pi$. At this stage, the torsion inside the rod is uniform, with a value of $2\pi/L$, and there is no flexion. We then move the extremities of the rod toward each other while keeping the values $\mathbf{d}_\alpha(0)$ and $\mathbf{d}_\alpha(L)$ fixed. In other words, we impose the boundary conditions

$$(5.30) \qquad \begin{cases} \mathbf{x}_A = 0, \quad \mathbf{x}'_A = \mathbf{e}_1, \quad \mathbf{x}_B = \beta \mathbf{e}_1, \quad \mathbf{x}'_B = \mathbf{e}_1, \\ \mathbf{d}_1(0) = \mathbf{d}_1(L) = -\mathbf{e}_2, \quad \mathbf{d}_2(0) = \mathbf{d}_2(L) = -\mathbf{e}_3. \end{cases}$$

Our problem consists of finding the final configuration of the rod subjected to the above boundary conditions, and with the physical data $L = 20$ m, $EI = 1000$ N/m^2, $GI = 900$ N/m^2, and $\rho|\mathbf{g}| = 50$ N/m. For this purpose, we used Algorithm (5.24)–(5.27), with 20 segments $[s_i, s_{i+1}]$ of equal length, and, as an initial configuration $\bar{\mathbf{x}}^{-1}$, either the configuration computed with the previous value of β, or, for the first computation, the uniformly twisted configuration. The computations were done with $r = \rho_n = 4000$ and required between 200 and 400 Uzawa iterations at an average CPU time of 6 min 30 sec per computation on a DPS 8-70.

The computed configurations for β ranging from 15 m to -5 m are shown in Fig. 5.5 and Fig. 5.6, and the corresponding values of the torsion $\theta(s) = \mathbf{d}_1(s) \cdot \mathbf{d}_2(s)$ are indicated in Table 5.2.

We observe that the value of the torsion decreased with β. Thus, without changing the boundary conditions on $\mathbf{d}_\alpha$, we were able to go from sheer torsion to sheer flexion through a sequence in which torsion and flexion were both present. This continuous evolution, with strong coupling between flexion and torsion, agrees perfectly with experimental data and therefore validates both the model and the algorithm.

TABLE 5.2
Values of the torsion $\theta(s)$ for β ranging from 20 *to* -5 m.

	s				
	0	5	10	15	20
β					
20 m	0.3126	0.3129	0.3129	0.3129	0.3126
15 m	0.2793	0.2798	0.2781	0.2798	0.2793
10 m	0.2132	0.2139	0.2088	0.2134	0.2133
5 m	0.1024	0.0927	0.0662	0.0882	0.1025
2 m	0.02555	0.0166	−0.0036	0.0126	0.0255
0 m	0.0030	0.0014	−0.0018	0.0007	0.0030
−2 m	0.0002	0.0001	−0.0002	0.0001	0.0002
−5 m	0.0000	0.0000	0.0000	0.0000	0.0000

6. Solution of the dynamical problem.

6.1. Synopsis. In order to simulate numerically the dynamical behavior of the flexible rods considered in § 1, we extend in this section the methods discussed in §§ 3, 4, and 5. For simplicity, we suppose that the boundary conditions are still given by either (2.4) or (2.5), with $\mathbf{x}_A$, $\mathbf{x}_B$, $\mathbf{x}'_A$, and $\mathbf{x}'_B$ possibly dependent on time t. After presenting the mathematical formulation of the dynamical problem in § 6.2, we shall describe in § 6.3 an implicit time-discretization scheme of the Houbolt type that reduces it to a sequence of static problems very close in nature to Problem (2.1). Since the Houbolt scheme is a multistep one, a starting procedure is needed, and this will be described in § 6.4. Finally, after a brief description of the solution procedure for the resulting static problems (§ 6.5), the results of numerical experiments will be presented in § 6.6.

6.2. Formulation of the time-dependent problem. From the virtual work theorem (Chap. 1, § 3), the dynamical behavior of an inextensible, thin elastic rod is modeled by the following initial value problem.

Find $\bar{\mathbf{x}}:[0, t_1]\to K_t$ such that

$$(6.1)\qquad \begin{cases} \displaystyle\int_0^L \{\rho\ddot{\bar{\mathbf{x}}}\cdot\mathbf{y}+EI\bar{\mathbf{x}}''\cdot\mathbf{y}''-\mathbf{f}\cdot\mathbf{y}\}\,ds=0 \qquad \forall\mathbf{y}\in dK_t(\bar{\mathbf{x}}),\quad t\in(0, t_1),\\ \bar{\mathbf{x}}(\cdot, 0)=\mathbf{x}_0,\qquad \dot{\bar{\mathbf{x}}}(\cdot, 0)=\mathbf{x}_1.\end{cases}$$

As in § 2, $\bar{\mathbf{x}}(s, t)$ denotes the present position of the generic point $M(s)$ of the rod, ρ and EI are, respectively, the lineic mass density and flexural stiffness of the rod, $\mathbf{f}$ is the lineic density of external forces, and K_t is the set of kinematically admissible configurations at time t, i.e.,

$$(6.2)\qquad \begin{cases} K_t=\{\mathbf{y}\,|\,\mathbf{y}\in H^2(0, L; \mathbb{R}^3), |\mathbf{y}'(s)|=1 \text{ on } [0, L];\\ \qquad \text{either } \mathbf{y}(0)=\mathbf{x}_A(t) \text{ and } \mathbf{y}(L)=\mathbf{x}_B(t) \text{ or}\\ \qquad \mathbf{y}(0)=\mathbf{x}_A(t), \mathbf{y}'(0)=\mathbf{x}'_A(t), \mathbf{y}(L)=\mathbf{x}_B(t) \text{ and } \mathbf{y}'(L)=\mathbf{x}'_B(t)\}.\end{cases}$$

Then, the tangent set $dK_t(\bar{\mathbf{x}})$ to K_t at $\bar{\mathbf{x}}$ is given by

$$(6.3)\qquad \begin{cases} dK_t(\bar{\mathbf{x}})=\{\mathbf{y}\,|\,\mathbf{y}\in H^2(0, L; \mathbb{R}^3), \bar{\mathbf{x}}'(s, t)\cdot\mathbf{y}'(s)=0 \text{ on } [0, L];\\ \qquad \text{either } \mathbf{y}(0)=\mathbf{y}(L)=\mathbf{0} \text{ or}\\ \qquad \mathbf{y}(0)=\mathbf{y}'(0)=\mathbf{y}(L)=\mathbf{y}'(L)=\mathbf{0}\}.\end{cases}$$

Finally, as usual, for a given function $\psi(s, t)$, we denote

$$\dot{\psi}=\frac{\partial\psi}{\partial t},\qquad \ddot{\psi}=\frac{\partial^2\psi}{\partial t^2},\qquad \psi'=\frac{\partial\psi}{\partial s},\qquad \psi''=\frac{\partial^2\psi}{\partial s^2}.$$

Remark 6.1. From the definition of K_t, the initial data $\mathbf{x}_0$ and $\mathbf{x}_1$ must satisfy the following compatibility conditions.

$$\begin{cases} \mathbf{x}_0 \in K_0, \quad \mathbf{x}_0' \cdot \mathbf{x}_1' = 0 \quad \text{on } [0, L], \\ \mathbf{x}_1(0) = \dot{\mathbf{x}}_A(0) \quad \text{and} \quad \mathbf{x}_1(L) = \dot{\mathbf{x}}_B(0) \quad \text{or} \\ \mathbf{x}_1(0) = \dot{\mathbf{x}}_A(0), \quad \mathbf{x}_1'(0) = \dot{\mathbf{x}}_A'(0), \quad \mathbf{x}_1(L) = \dot{\mathbf{x}}_B(0), \quad \text{and} \quad \mathbf{x}_1'(L) = \dot{\mathbf{x}}_B'(0). \end{cases}$$

Remark 6.2. The external loading exerted on a flexible rod subjected to hydrodynamical forces is traditionally modeled by the Morison formula

$$\mathbf{f} = \left(\rho - \frac{\pi D^2}{4}\rho_w\right)\mathbf{g} + \frac{1}{2}\rho_w \pi C_f D|\mathbf{v}_t|\mathbf{v}_t$$
$$+ \frac{1}{2}\rho_w C_d D|\mathbf{v}_n|\mathbf{v}_n + \rho_w \pi \frac{D^2}{4}(\bar{\mathbf{x}}' \times (1.9\ddot{\mathbf{H}} - 0.9\ddot{\bar{\mathbf{x}}})) \times \bar{\mathbf{x}}'.$$

Above, ρ_w is the volumic density of the water, D is the diameter of the rod whose section is supposed to be circular, C_f and C_d are friction coefficients, $\ddot{\mathbf{H}}$ is the surface wave acceleration, and $\mathbf{v}_t$ and $\mathbf{v}_n$ are, respectively, the tangential and the normal relative velocity of the water with respect to the rod.

6.3. Time-integration scheme. The numerical integration of dynamical structural problems has motivated many studies (see Belytschko and Hughes [1983], Bathe and Wilson [1976], and Hughes [1987], and the references therein). The solution of the dynamical problem (6.1) in this case is complicated by the presence of the inextensibility condition. To solve (6.1), we therefore select a Houbolt time-integration scheme, because it is well suited to match the difficulty associated with the inextensibility condition and because the well-known numerical dissipation associated with it will be dominated by the physical dissipation due to the hydrodynamical forces. This scheme reduces (6.1) to the following sequence of static problems.

$$\bar{\mathbf{x}}^j \in K_j \text{ is given for } j = 0, 1, 2; \tag{6.4}$$

then, for $n \geq 2$, assuming that $\bar{\mathbf{x}}^j \in K_j$ are known for $j = n-2$, $n-1$, n, we obtain $\bar{\mathbf{x}}^{n+1} \in K_{n+1}$ as the solution of the following problem.

$$\begin{cases} \text{Find } \bar{\mathbf{x}}^{n+1} \in K_{n+1} \text{ such that} \\ \displaystyle\int_0^L \left\{\frac{\rho}{\Delta t^2}(2\bar{\mathbf{x}}^{n+1} - 5\bar{\mathbf{x}}^n + 4\bar{\mathbf{x}}^{n-1} - \bar{\mathbf{x}}^{n-2}) \cdot \mathbf{y} \right. \\ \qquad \left. + EI(\bar{\mathbf{x}}^{n+1})'' \cdot \mathbf{y}'' - \mathbf{f} \cdot \mathbf{y}\right\} ds = 0 \quad \forall \mathbf{y} \in dK_{n+1}(\bar{\mathbf{x}}^{n+1}). \end{cases} \tag{6.5}$$

Here, Δt is a time step, $\bar{\mathbf{x}}^i$ is an approximation of $\bar{\mathbf{x}}(\cdot, i\Delta t)$, K_i is the set of admissible configurations at time $i\Delta t$, and $dK_i(\bar{\mathbf{x}}^i)$ is the tangent set to K_i at $\bar{\mathbf{x}}^i$.

6.4. Starting procedure. To compute the values $\bar{\mathbf{x}}^1$ and $\bar{\mathbf{x}}^2$ used in the initialization step (6.4), a starting procedure is needed. For that purpose, we approximate (6.1) at times Δt and $2\Delta t$ by the Crank–Nicholson scheme below.

$$(6.6)\qquad \begin{cases} \displaystyle\int_0^L \left\{\frac{\rho}{\Delta t^2}(\bar{\mathbf{x}}^2+\bar{\mathbf{x}}^0-2\bar{\mathbf{x}}^1)\cdot\mathbf{y}+EI(\hat{\mathbf{x}}^1)''\cdot\mathbf{y}''-\mathbf{f}\cdot\mathbf{y}\right\} ds=0,\\ \forall\mathbf{y}\in dK_1(\hat{\mathbf{x}}^1),\quad \text{where } \hat{\mathbf{x}}^1=(\bar{\mathbf{x}}^2+2\bar{\mathbf{x}}^1+\bar{\mathbf{x}}^0)/4\in K_1,\end{cases}$$

$$(6.7)\qquad \begin{cases} \displaystyle\int_0^L \left\{\frac{\rho}{\Delta t^2}(\bar{\mathbf{x}}^3+\bar{\mathbf{x}}^1-2\bar{\mathbf{x}}^2)\cdot\mathbf{y}+EI(\hat{\mathbf{x}}^2)''\cdot\mathbf{y}''-\mathbf{f}\cdot\mathbf{y}\right\} ds=0\\ \forall\mathbf{y}\in dK_2(\hat{\mathbf{x}}^2),\quad \text{where } \hat{\mathbf{x}}^2=(\bar{\mathbf{x}}^3+2\bar{\mathbf{x}}^2+\bar{\mathbf{x}}^1)/4\in K_2.\end{cases}$$

We have too many unknowns in (6.6)–(6.7), which we eliminate using a Taylor expansion at $t=0$ and $t=\Delta t$. This yields

$$(6.8)\qquad \bar{\mathbf{x}}^2+\bar{\mathbf{x}}^0-2\bar{\mathbf{x}}^1=\tfrac{4}{3}(\hat{\mathbf{x}}^1-\bar{\mathbf{x}}^0-\Delta t\dot{\bar{\mathbf{x}}}^0),$$

$$(6.9)\qquad \bar{\mathbf{x}}^3+\bar{\mathbf{x}}^1-2\bar{\mathbf{x}}^2=\tfrac{4}{3}(3\hat{\mathbf{x}}^2-8\hat{\mathbf{x}}^1+5\bar{\mathbf{x}}^0+2\Delta t\dot{\bar{\mathbf{x}}}^0).$$

Substituting (6.8)–(6.9) and the initial data back into (6.6)–(6.7), we obtain the following starting procedure.

$$(6.10)\qquad \begin{cases} \text{Find } \hat{\mathbf{x}}^1\in K_1 \text{ such that } \forall\mathbf{y}\in dK_1(\hat{\mathbf{x}}^1) \text{ we have}\\ \displaystyle\int_0^L \left\{\frac{4\rho}{3\Delta t^2}(\hat{\mathbf{x}}^1-\mathbf{x}_0-\Delta t\mathbf{x}_1)\cdot\mathbf{y}+EI(\hat{\mathbf{x}}^1)''\cdot\mathbf{y}''-\mathbf{f}\cdot\mathbf{y}\right\} ds=0,\end{cases}$$

$$(6.11)\qquad \begin{cases} \text{Find } \hat{\mathbf{x}}^2\in K_2 \text{ such that } \forall\mathbf{y}\in dK_2(\hat{\mathbf{x}}^2) \text{ we have}\\ \displaystyle\int_0^L \left\{\frac{4\rho}{3\Delta t^2}(3\hat{\mathbf{x}}^2-8\hat{\mathbf{x}}^1+5\mathbf{x}_0+2\Delta t\mathbf{x}_1)\cdot\mathbf{y}+EI(\hat{\mathbf{x}}^2)''\cdot\mathbf{y}''-\mathbf{f}\cdot\mathbf{y}\right\} ds=0.\end{cases}$$

$$(6.12)\qquad \text{Set } \{\bar{\mathbf{x}}^0,\bar{\mathbf{x}}^1,\bar{\mathbf{x}}^2\}=\{\mathbf{x}_0,\hat{\mathbf{x}}^1,\hat{\mathbf{x}}^2\} \text{ in (6.4).}$$

6.5. Solution of the static problem (6.5). If, in the case where $\mathbf{f}$ is solution-dependent, we estimate it from the values of the solution obtained at the previous time steps, problem (6.5) reduces to the following minimization problem.

$$(6.13)\qquad \begin{cases} \text{Minimize the function } J_{n+1} \text{ over the set } K_{n+1} \text{ of}\\ \text{kinematically admissible configurations at time } t_{n+1},\end{cases}$$

with J_{n+1} given by

$$J_{n+1}(\mathbf{y})=\int_0^L \left\{\frac{EI}{2}|\mathbf{y}''|^2+\frac{\rho}{\Delta t^2}|\mathbf{y}|^2-\mathbf{f}\cdot\mathbf{y}\right\} ds+\frac{\rho}{\Delta t^2}\int_0^L (-5\bar{\mathbf{x}}^n+4\bar{\mathbf{x}}^{n-1}-\bar{\mathbf{x}}^{n-2})\cdot\mathbf{y}\, ds.$$

Problem (6.13) is very close to the static problem (2.1) and can be solved by iterative methods identical (only the Lagrangian functions are slightly different) to those described in § 4; it is therefore not necessary to describe them again.

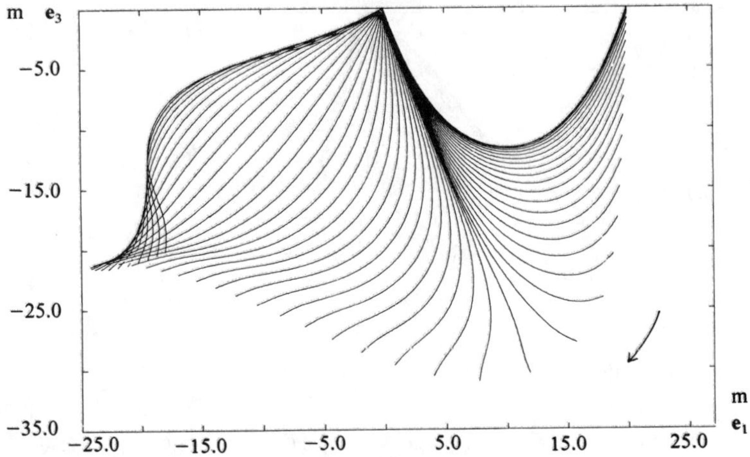

FIG. 6.1. *The free oscillations of a pipeline when one end was becoming free starting from a position of equilibrium: The position of the pipe during the first* 5 *sec.*

6.6. Numerical experiments. Consider a pipeline defined by the following parameters: $L = 32.6$ m, $EI = 700$ N/m^2, $\rho = 7.67$ kg/m, $D = 0.05$ m.

All of the calculations were done with the Houbolt scheme (6.4)-(6.5), taking $\mathbf{x}_1 = \mathbf{0}$ and $\Delta t = 5 \times 10^{-2}$ sec. Problem (6.5) was solved by Algorithm (4.9)-(4.12) with $\rho_n = r \simeq 10^5$. These experiments were run at INRIA by Dr. J. F. Bourgat on a DPS 8/Multics.

6.6.1. A first numerical experiment. Starting from an equilibrium position corresponding to $\mathbf{x}_A = \mathbf{0}$ and $\mathbf{x}_B = 20\mathbf{e}_1$, the free oscillations of the pipeline

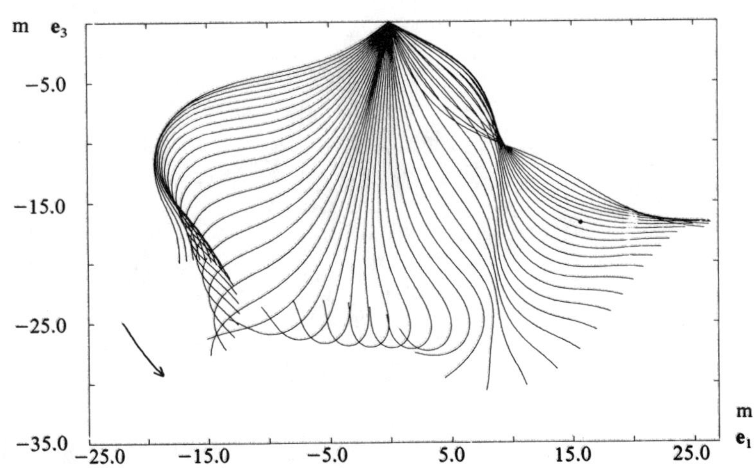

FIG. 6.2. *The position of the pipe between* 5 *and* 10 *sec.*

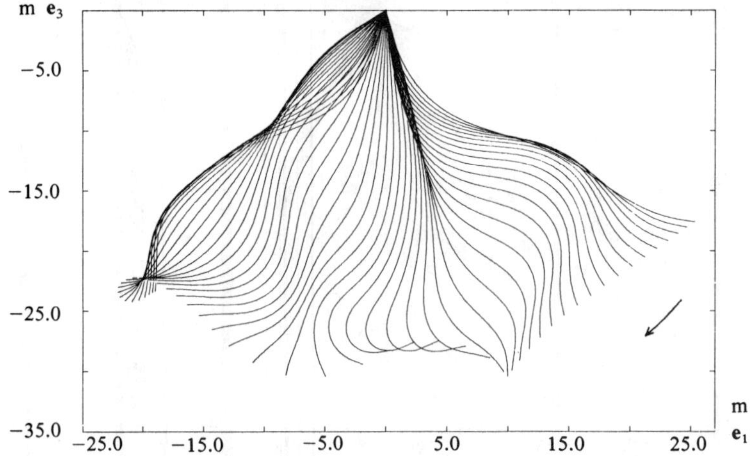

FIG. 6.3. *The position of the pipe between* 10 *and* 15 *sec.*

were studied in the case in which extremity B was becoming free at time $t=0$. We show in Figs. 6.1–6.4 the oscillations of the pipeline during the time intervals, [0–5 sec], [5–10 sec], [10–15 sec], and [15–20 sec], respectively (the different positions are shown every 0.15 sec).

6.6.2. A second family of numerical experiments. Consider the motion of the pipeline under water. At $t=0$, we supposed that A and B were as in the experiment in § 6.6.1, and, again, that B was becoming free. Because of the water, the equilibrium positions were not exactly the same as they were in the previous experiment. Friction forces, due to water, were

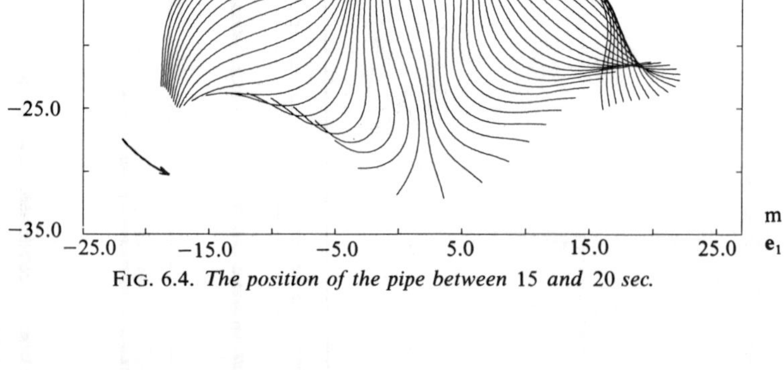

FIG. 6.4. *The position of the pipe between* 15 *and* 20 *sec.*

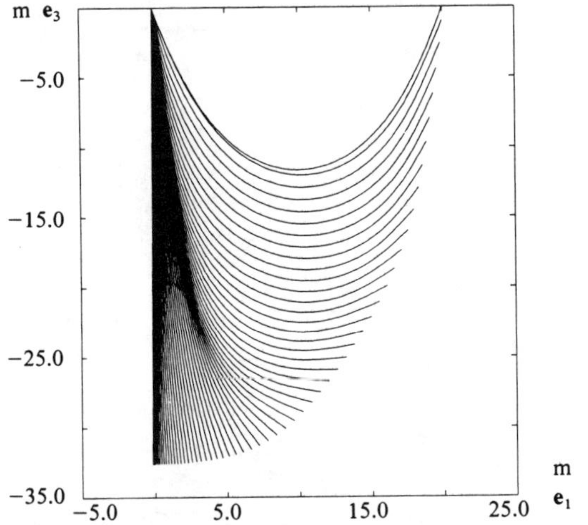

FIG. 6.5. *The motion of a pipeline under water at $t = 0$ when one end was becoming free at $t = 0$ and no stream was present.*

included in the mathematical model of the motion, and, as can be seen in Figs. 6.5 and 6.6, they seriously damped the motion of the pipeline, since a new equilibrium situation was reached in a finite time, practically speaking. Figure 6.5 corresponds to a no-stream situation, so that the new equilibrium position is vertical; Fig. 6.6 corresponds to a horizontal underwater stream of velocity −1 m/s leading to an oblique new equilibrium position.

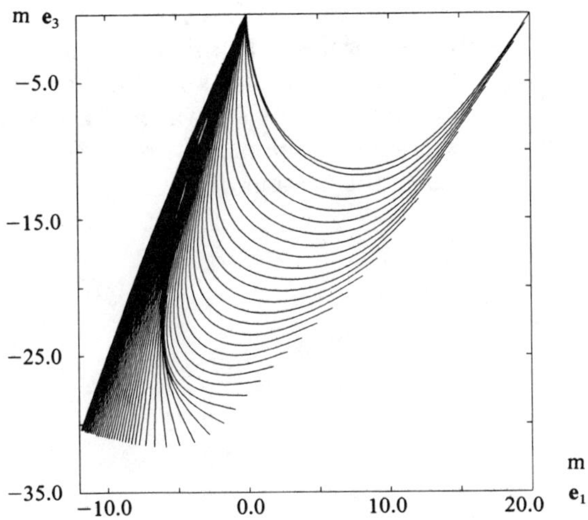

FIG. 6.6. *The motion of a pipeline under water at $t = 0$ when one end was becoming free, subjected to a horizontal stream of velocity* −1 m/s.

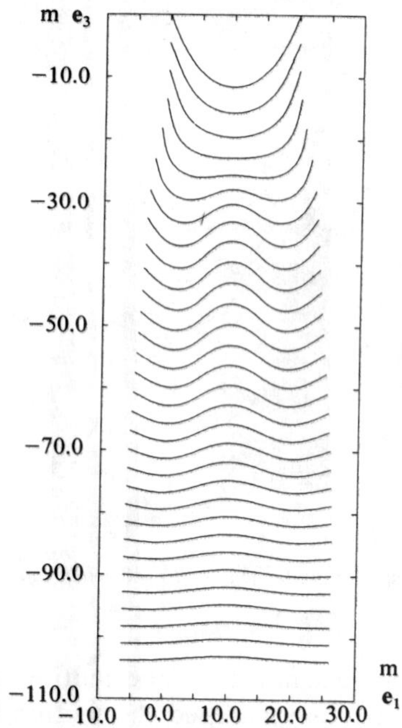

FIG. 6.7. *The motion of a pipeline under water at $t = 0$ when both ends were becoming free.*

The various positions are shown every 0.5 sec. The drag coefficients were $C_d = 1.2$ and $C_f = 0.03$ and the apparent weight was taken to be equal to 57.1 N/m.

6.6.3. A third numerical experiment. In this experiment, we still consider an underwater motion. At $t = 0$, $\mathbf{x}_A = \mathbf{0}$ and $\mathbf{x}_B = 20\mathbf{e}_1$, and we supposed that both extremities were becoming free. As shown in Fig. 6.7 (which shows the location of the pipeline every 2 sec) the motion reduces quickly to a vertical translation motion directed to the bottom of the sea.

References

R. A. ADAMS [1975], *Sobolev Spaces*, Academic Press, New York.

M. A. AJIZ AND A. JENNINGS [1984], *A robust incomplete Cholesky conjugate gradient algorithm*, Internat. J. Numer. Methods Engrg., 20, pp. 949–966.

S. S. ANTMAN AND C. S. KENNEY [1981], *Large buckled states of nonlinearly elastic rods under torsion, thrust and gravity*, Arch. Rational Mech. Anal., 76, pp. 289–338.

S. S. ANTMAN AND G. ROSENFELD [1978], *Global behavior of buckled states of nonlinearly elastic rods*, SIAM Rev., 20, pp. 513–566.

K. J. ARROW, L. HURWICZ, AND H. UZAWA [1958], *Studies in Nonlinear Programming*, Stanford University Press, Stanford, CA.

O. AXELSSON [1976], *A class of iterative methods for finite element equations*, Comput. Methods Appl. Mech. Engrg., 19, pp. 123–138.

I. BABUSKA [1973], *The finite element method with Lagrange multipliers*, Numer. Math., 20, pp. 179–192.

J. M. BALL [1977], *Convexity conditions and existence theorems in nonlinear elasticity*, Arch. Rational Mech. Anal., 63, pp. 337–403.

——— [1982], *Discontinuous equilibrium solutions and cavitation in nonlinear elasticity*, Philos. Trans. Roy. Soc. London Ser. A, 306, pp. 557–611.

K. J. BATHE AND E. L. WILSON [1976], *Numerical Methods in Finite Element Analysis*, Prentice-Hall, Englewood Cliffs, NJ.

D. BEGIS [1973], *Etude numérique du comportement d'un fluide de Bingham*, Rapport INRIA-Laboria 42.

——— [1979], *Etude numérique de l'écoulement d'un fluide de Bingham par une méthode de Lagrangien augmenté*, Rapport INRIA-Laboria 355.

T. BELYTSCHKO AND P. G. HODGE [1970], *Plane stress limit analysis by finite elements*, J. Engrg. Mech. Division, 96, EM6, pp. 931–944.

T. BELYTSCHKO AND T. J. R. HUGHES [1983], *Computational methods for transient analysis*, in Computational Methods in Mechanics, 1, North-Holland, Amsterdam, New York.

A. BERMUDEZ AND C. MORENO [1981], *Duality methods for solving variational inequalities*, Comput. Math. Appl., 7, pp. 43–58.

D. P. BERTSEKAS [1982], *Constrained Optimization and Lagrange Multiplier Methods*, Academic Press, New York.

K. E. BISSHOP AND D. C. DRUCKER [1945], *Large deflections of cantilever beams*, Quart. Appl. Math., 3, pp. 272–275.

D. BLANCHARD AND P. LE TALLEC [1986], *Numerical analysis of the equations of small strains quasistatic elastoviscoplasticity*, Numer. Math., 50, pp. 147–169.

J. F. BOURGAT [1984], Personal communication.

J. F. BOURGAT, J. M. DUMAY, AND R. GLOWINSKI [1980], *Large displacement calculation of flexible pipes by finite element and nonlinear programming methods*, SIAM J. Sci. Statist. Comput., 1, pp. 34–81.

J. F. BOURGAT, P. LE TALLEC, AND S. MANI [1988], *Modélisation et calcul des grands déplacements de tuyaux élastiques en flexion-torsion*, J. Méc. Théor. Appl., 1, pp. 1–30.

H. BREZIS, J. M. CORON, AND E. LIEB [1986], *Harmonic maps with defects*, Comm. Math. Phys., 107, pp. 649–705.

F. BREZZI [1974], *On the existence, uniqueness, and approximation of saddle-point problems arising from Lagrange multipliers*, RAIRO Anal. Numér., 8, R-2, pp. 129–151.

——— [1979], *Nonstandard finite elements for fourth order elliptic problems*, in Energy Methods in Finite Element Analysis, Wiley, Chichester.

J. CAHOUET AND J. P. CHABARD [1986], *Multidomains and multisolvers finite element approach for the Stokes problem*, in Innovative Numerical Methods in Engineering, Springer-Verlag, Berlin, New York.

J. CAHOUET AND A. HAUGUEL [1986], *Finite element methods for incompressible Navier Stokes and shallow water equations*, in Computational Fluids Dynamic Lecture Series of the von Kármán Institute, March 1986, von Kármán Institute, Rhodes Sainte Genese, Belgium.

R. CASCIARO AND L. CASCINI [1982], *A mixed formulation and mixed finite elements for limit analysis*, Internat. J. Numer. Methods Engrg., 18, pp. 211–243.

J. CEA [1971], *Optimisation: Théorie et Algorithmes*, Dunod, Paris.

——— [1978], *Optimization: Theory and Algorithms*, in Lectures on Mathematics and Physics, 53, Tata Institute of Fundamental Research, Bombay, Springer-Verlag, Berlin, New York.

J. CEA AND R. GLOWINSKI [1973], *Sur des méthodes d'optimisation par relaxation*, Rev. Française Automat. Informat. Recherche Opérationnelle, Sér. Rouge 7, R3, pp. 5–31.

P. CHADWICK AND E. W. HADDON [1972], *Inflation-extension and eversion of a tube of incompressible isotropic elastic material*, J. Inst. Math. Appl., 10, pp. 258–278.

T. F. CHAN AND R. GLOWINSKI [1978], *Finite element approximation and iterative solution of a class of mildly nonlinear elliptic equations*, Report STAN-CS-78-674, Stanford University, Stanford, CA.

S. CHANDRASEKHAR [1977], *Liquid Crystals*, Cambridge University Press, London.

F. CHATELIN [1983], *Spectral Approximation of Linear Operators*, Academic Press, New York.

P. G. CIARLET [1978], *The Finite Element Method for Elliptic Problems*, North-Holland, Amsterdam, New York.

——— [1982], *Introduction à l'analyse numérique matricielle et à l'optimisation*, Masson, Paris.

——— [1988], *Mathematical Elasticity. Vol. 1: Three-dimensional Elasticity*, North-Holland, Amsterdam, New York.

P. G. CIARLET AND G. GEYMONAT [1982], *Sur les lois de comportement en élasticité nonlinéaire compressible*, C.R. Acad. Sci. Paris, Sér. A.

R. COHEN, R. HARDT, D. KINDERLEHRER, S. Y. LIN, AND M. LUSKIN [1987], *Minimum energy configurations for liquid crystals: computational results*, in Theory and Applications of Liquid Crystals, pp. 99–121, Springer-Verlag, New York, Berlin.

P. CONCUS AND G. H. GOLUB [1976], *A generalized conjugate gradient method for nonsymmetric systems of linear equations*, in Computing Methods in Applied Sciences and Engineering, R. Glowinski and J. L. Lions, eds., Lecture Notes in Economics and Mathematical Systems, 134, pp. 56–65, Springer-Verlag, Berlin, New York.

C. A. COULOMB [1773], *Essai sur une application des règles de Maxima et Minimia à quelques problèmes de statique relatifs à l'architecture*, Mémoire destiné à l'Académie Royale des Sciences.

K. CULMANN [1866], *Die graphische Statik*, Zurich.

A. DAMLAMIAN [1974], *Nonlinear evolution equations with variable norms*, Ph.D Thesis, Harvard University, Boston.

J. DANIEL [1970], *The Approximate Minimization of Functionals*, Prentice-Hall, Englewood Cliffs, NJ.

P. G. DE GENNES [1974], *The Physics of Liquid Crystals*, Oxford University Press, London.

M. DE JOSSELIN DE JONG [1977], *Improvement of the lowerbound solution for the vertical cutoff in a cohesive, frictionless soil*, Géotechnique, 28, pp. 197–201.

L. DE LOURA [1986], *A numerical method for the Hartree equation of the helium atom*, Calcolo, 23, pp. 185–207.

J. DOUGLAS AND H. H. RACHFORD [1956], *On the numerical solution of the heat conduction problem in 2d and 3d space variables*, Trans. Amer. Math. Soc., 82, pp. 421–439.

G. DUVAUT AND J. L. LIONS [1972], *Les Inéquations en Mécanique et en Physique*, Dunod, Paris.

N. DYN AND W. E. FERGUSON [1983], *The numerical solution of equality constrained quadratic programming problems*, Math. Comp., 41, pp. 165–170.

I. EKELAND AND R. TEMAM [1974], *Analyse Convexe et Problèmes Variationnels*, Dunod, Paris.

——— [1976], *Convex Analysis and Variational Problems*, North-Holland, Amsterdam, New York.

J. L. ERICKSEN [1976], *Equilibrium theory of liquid crystals*, in Advances in Liquid Crystals, G. H. Brown, ed., 2, pp. 233–298, Academic Press, New York.

G. FICHERA [1972], *Existence Theorems in Elasticity*, Handbuch der Physik, Bd.VIa/2, pp. 347–389, Springer-Verlag, New York, Berlin.

M. FORTIN [1975], *Utilisation de la méthode des éléments finis en Mécanique des Fluides*, Calcolo, 4, pp. 406–441.

M. FORTIN AND R. GLOWINSKI, eds. [1982], *Méthodes de Lagrangien Augmenté. Application à la résolution numérique de problèmes aux limites*, Dunod-Bordas, Paris.

——— [1983], *Augmented Lagrangian Methods*, North-Holland, Amsterdam, New York.

M. FREMOND AND A. FRIAA [1982], *Les méthodes statique et cinématique en calcul à la rupture et en analyse limite*, J. Mécanique, 1, pp. 881–905.

A. FRIAA [1979], *La loi de Norton–Hoff en plasticité et viscoplasticité*, Thèse, Université de Paris VI.

D. GABAY [1979], *Méthodes numériques pour l'optimisation non linéaire*, Thèse d'Etat, Université Pierre et Marie Curie.

——— [1982], *Application de la méthode des multiplicateurs aux inéquations variationnelles*, in Méthodes de Lagrangien Augmenté, M. Fortin and R. Glowinski, eds., Dunod-Bordas, Paris.

——— [1983], *Application of the method of multipliers to variational inequalities*, in Augmented Lagrangian Methods, M. Fortin and R. Glowinski, eds., North-Holland, Amsterdam, New York.

G. GALILEO [1638], *Discorzi e demostrazioni mathematiche interno a due nueve scienze*, Dialogo secundo, Leyden.

F. A. GAYDON AND A. W. MACCRUM [1954], *A theoretical investigation of the yield point loading of a square plate with a central circular hole*, J. Mech. Phys. Solids, 2, pp. 156–169.

M. GERADIN [1984], *Finite element approach to kinematics and dynamics analysis of mechanisms using Euler parameters*, in Numerical Methods for Nonlinear Problems, Taylor et al., eds., Pineridge Press, Swansea, United Kingdom.

P. GERMAIN [1973], *Mécanique des Milieux Continus*, Masson, Paris.

G. GEYMONAT AND P. SUQUET [1986], *Functional spaces for Norton–Hoff materials*, Math. Methods Appl. Sci., 8, pp. 206–222.

P. E. GILL AND W. MURRAY, eds. [1974], *Numerical Methods for Constrained Optimization*, Academic Press, New York.

P. E. GILL, W. MURRAY, AND M. H. WRIGHT [1981], *Practical Optimization*, Academic Press, New York.

V. GIRAULT AND P. A. RAVIART [1986], *Finite Element Approximation of the Navier Stokes Equations*, Springer-Verlag, Berlin, New York.

R. GLOWINSKI [1984], *Numerical Methods for Nonlinear Variational Problems*, Springer-Verlag, Berlin, New York.

——— [1986], *Splitting methods for the numerical solution of the incompressible Navier-Stokes equations*, in Vistas in Applied Mathematics, Optimization Software, New York, pp. 57-95.

R. GLOWINSKI, J. GOUSSEBAÏLE, AND G. LABADIE [1990], to appear.

R. GLOWINSKI AND P. LE TALLEC [1982], *Numerical solution of problems in incompressible finite elasticity*, SIAM J. Appl. Math., 42, pp. 400-425.

R. GLOWINSKI, J. L. LIONS, AND R. TREMOLIÈRES [1976], *Analyse Numérique des Inéquations Variationnelles*, Dunod-Bordas, Paris.

——— [1981], *Numerical Analysis of Variational Inequalities*, North-Holland, Amsterdam, New York.

R. GLOWINSKI, D. MARINI, AND M. VIDRASCU [1984], *Finite element approximations and iterative solution of a fourth order elliptic variational inequality*, IMA J. Numer. Anal., 4, pp. 127-167.

R. GLOWINSKI AND A. MARROCO [1974], *Sur l'approximation par éléments finis d'ordre un et la résolution par pénalité-dualité d'une classe de problèmes de Dirichlet non-linéaires*, C.R. Acad. Sci. Paris, 287A, pp. 327-333.

——— [1975] *Sur l'approximation par éléments finis d'ordre un et la résolution par pénalité-dualité d'une classe de problèmes de Dirichlet non-linéaires*, Rev. Française Automat. Informat. Recherche Operationnelle Anal. Numér. 9, R-2, pp. 41-76.

D. GOGNY AND P. L. LIONS [1986], *Hartree-Fock theory in nuclear physics*, Rev. Française Automat. Informat. Recherche Operationnelle Modél. Math. Anal. Numér., vol. 20, pp. 571-637.

C. H. GOLUB AND G. MEURANT [1983], *Résolution Numérique des Grands Systèmes Linéaires*, Eyrolles.

T. GUENNOUNI [1982], *Matériau de Norton-Hoff pour divers critères de la plasticité de mécanique des sols*, Thèse, Ecole Nationale des Ponts et Chaussées, Paris.

T. GUENNOUNI AND P. LE TALLEC [1982], *Calcul à la rupture: Régularisation de Norton-Hoff et Lagrangien augmenté*, J. Mécanique, 2, pp. 75-99.

M. D. GUNZBURGER, C. H. LIU, AND R. A. NICOLAIDES [1983], *A finite element method for diffusion dominated unsteady viscous flows*, Comput. Methods Appl. Mech. Engrg., 39, pp. 55-67.

M. GURTIN [1981], *Topics in Finite Elasticity*, CBMS/NSF Conf. Series 35, SIAM, Philadelphia, PA.

D. Q. HAINES AND W. D. WILSON [1979], *Strain energy density of rubber-like materials*, J. Mech. Phys. Solids, 27, pp. 345-360.

R. HART AND D. KINDERLEHRER [1986], *Mathematical questions of liquid crystal theory*, in Theory and Application of Liquid Crystals, J. L. Ericksen, and D. Kinderlehrer, eds., IMA Volumes in Mathematics and Applications, 5.

R. HART, D. KINDERLEHRER, AND F. H. LIN [1986], *Existence and partial regularity of static liquid crystal configurations*, Comm. Math. Phys., 105, pp. 547-570.

D. R. HARTREE [1928], *The wave mechanics of an atom with a non-coulomb central field, Part I. Theory and methods*, Proc. Camb. Phil. Soc., 24, pp. 89-312.

——— [1957], *The Calculation of Atomic Structures*, John Wiley, New York, London.

M. HESTENES [1969], *Multiplier and gradient methods*, J. Optim. Theory Appl., 4, pp. 303-320.

——— [1980], *Conjugate Direction Methods in Optimization*, Springer-Verlag, Berlin, New York.

H. D. HIBBIT, E. B. BECKER, AND L. M. TAYLOR [1979], *Nonlinear analysis of some slender pipelines*, Comput. Methods Appl. Mech. Engrg., 17/18, pp. 203-225.

P. G. HODGE [1959], *Plastic Analysis of Structures*, McGraw-Hill, New York, Toronto, London.

T. J. R. HUGHES [1987], *The Finite Element Method*, Prentice-Hall, Englewood Cliffs, NJ.

R. D. JAMES [1985], *Displacive phase transformation in solids*, Report of Brown University, Division of Engineering, Providence, RI.

J. K. KNOWLES AND E. STERNBERG [1980], *Discontinuous gradient near the tip of a crack in finite antiplane shear*, J. Elasticity, 10, pp. 81–110.

W. T. KOITER [1960], *General theorems for elastic plastic solids*, in Progress in Solid Mechanics, North-Holland, Amsterdam, New York.

O. KUSAKABE, T. KIMURA, AND H. YAMAGUCHI [1981], *Bearing capacity of slopes under strips loads on the top surfaces*, Soils and Foundations, 21, pp. 29–40.

O. LADYSHENKAYA [1969], *The Mathematical Theory of Viscous Incompressible Flows*, Gordon and Breach, New York.

H. LE DRET [1985], *Constitutive laws and existence questions in incompressible nonlinear elasticity*, J. Elasticity, 15, pp. 369–387.

P. LE TALLEC [1981], *Compatibility condition and existence results in discrete finite incompressible elasticity*, Comput. Methods Appl. Mech. Engrg., 27, pp. 239–259.

——— [1982], *Existence and approximation results for nonlinear mixed problems: Application to incompressible finite elasticity*, Numer. Math., 38, pp. 365–382.

P. LE TALLEC AND J. T. ODEN [1981], *Existence and characterization of hydrostatic pressure in finite deformations of incompressible elastic bodies*, J. Elasticity, 11, pp. 341–357.

P. LE TALLEC AND M. RAVACHOL [1987], *Approximation par éléments finis d'écoulements progressivement incompressibles*, C.R. Acad. Sci. Paris, 305, Série I, pp. 697–700.

P. LE TALLEC AND M. VIDRASCU [1984], *Une méthode numérique pour les problèmes d'équilibre de corps hyperélastiques compressibles en grandes déformations*, Numer. Math., 43, pp. 199–224.

A. LICHNEWSKI [1979], *Une méthode de gradient conjugué sur des variétés. Applications à certains problèmes de valeurs propres nonlinéaires*, Numer. Funct. Anal. Optim., 1, pp. 515–560.

J. L. LIONS [1969], *Quelques méthodes de résolution de problèmes aux limites non linéaires*, Dunod-Gauthier-Villars, Paris.

P. L. LIONS AND B. MERCIER [1979], *Splitting algorithms for the sum of two nonlinear operators*, SIAM J. Numer. Anal., 16, pp. 964–979.

D. G. LUENBERGER [1970], *The conjugate residual method for constrained minimisation problems*, SIAM J. Numer. Anal., 7, pp. 390–398.

G. I. MARCHOUK [1975], *Methods of Numerical Mathematics*, Springer-Verlag, New York, Berlin.

G. I. MARCHOUK AND Y. A. KUNETZOV [1974], *Méthodes itératives et fonctionnelles quadratiques*, in Sur les Méthodes Numériques en Sciences Physiques et Economiques, J. L. Lions and G. I. Marchouk, eds., pp. 1–132, Dunod, Paris.

J. E. MARSDEN AND T. J. R. HUGHES [1983], *Mathematical Foundations of Elasticity*, Prentice-Hall, Englewood Cliffs, NJ.

J. A. MEIJERINK AND H. A. VAN DER VORST [1977], *An iterative solution method for linear systems of the coefficient matrix is a symmetric M-matrix*, Math. Comp., 31, pp. 148–162.

B. MERCIER [1977], *Une méthode pour résoudre le problème de charges limites*, J. Mécanique, 16, pp. 567–591.

R. MIYOSHI [1973], *A finite element method for the solution of fourth-order partial differential equations*, Kumamoto J. Sci. (Math.), 9, pp. 87–116.

J. NECAS [1967], *Les méthodes directes en théorie des équations elliptiques*, Masson, Paris.

J. T. ODEN [1972], *Finite Elements of Nonlinear Continua*, McGraw-Hill, New York, Toronto, London.

J. T. ODEN AND J. N. REDDY [1976], *An Introduction to the Mathematical Theory of Finite Elements*, Wiley Interscience, New York.

R. W. OGDEN [1972], *Large deformation isotropic elasticity. On the correlation of theory and experiment for incompressible rubber-like solids*, Proc. Roy. Soc. London, A326, pp. 567–583.

——— [1976], *Volume changes associated with the deformation of rubber-like solids*, J. Mech. Phys. Solids, 24, pp. 323–338.

Z. OPIAL [1967], *Weak convergence of the successive approximation for nonexpansive mappings in Banach spaces*, Bull. Amer. Math. Soc., 73, pp. 591–597.

C. C. PAIGE AND M. A. SAUNDERS [1975], *Solution of sparse indefinite systems of linear equations*, SIAM J. Numer. Anal., 12, pp. 617–629.

B. N. PARLETT [1980], *The Symmetric Eigenvalue Problem*, Prentice-Hall, Englewood Cliffs, NJ.

J. PASTOR [1978], *Analyse limite: Détermination numérique des solutions statiques complètes. Application au talus vertical*, J. Méc. Appl., 2, pp. 167–196.

D. H. PEACEMAN AND H. H. RACHFORD [1955], *The numerical solution of parabolic and elliptic differential equations*, J. Soc. Ind. Appl. Math., 3, pp. 28–41.

E. POLAK [1971], *Computational Methods in Optimisation*, Academic Press, New York.

B. T. POLYAK [1979], *On the Bertsekas minimization method of composite functions*, in International Symposium on Systems Optimization and Analysis, A. Bensoussan and J. L. Lions, eds., Lecture Notes in Control and Information Sciences, 14, pp. 179–186, Springer-Verlag, Berlin, New York.

M. J. D. POWELL [1969], *A method for nonlinear constraints in minimization problems*, in Optimisation, Chap. 19, R. Fletcher, ed., Academic Press, New York.

——— [1979], *Variable metric methods for constrained optimization*, in Computing Methods in Applied Science and Engineering, R. Glowinski and J. L. Lions, eds., Lecture Notes in Mathematics, 704, pp. 62–72, Springer-Verlag, Berlin, New York.

P. A. RAVIART AND J. M. THOMAS [1983], *Introduction à l'Analyse Numérique des Equations aux Dérivées Partielles*, Masson, Paris.

V. RUAS [1981], *A class of asymmetric simplicial finite element method for solving finite incompressible elasticity problems*, Comput. Methods Appl. Mech. Engrg., 27, pp. 319–343.

J. SALENCON [1983], *Calcul à la rupture et analyse limite*, Presses de l'Ecole Nationale des Ponts et Chaussées, Paris.

A. SEGAL [1979], *On the numerical solution of Stokes equations using the finite element method*, Comput. Methods Appl. Mech. Engrg., 19, pp. 165–185.

J. SIMO [1985], *A finite strain beam formulation: the three-dimensional dynamic problem*, Comput. Methods Appl. Mech. Engrg., 49, pp. 55–70.

G. STRANG AND G. J. FIX [1973], *An Analysis of the Finite Element Method*, Prentice-Hall, Englewood Cliffs, NJ.

G. STRANG AND R. TEMAM [1980], *Duality and relaxation in the variational problems of plasticity*, J. Mécanique, 19, pp. 493–527.

P. SUQUET [1981], *Sur les équations de la plasticité: Existence et régularité de la solution*, J. Mécanique, 20, pp. 1–39.

——— [1982] *Plasticité et homogénéisation*, Thèse, Université de Paris VI.

L. TARTAR [1978], *Topics in nonlinear analysis*, Public. Math. d'Orsay, Université Paris Sud, Département de Mathématiques.

R. TEMAM [1977], *Theory and Numerical Analysis of the Navier Stokes Equations*, North-Holland, Amsterdam, New York.

——— [1986], *A generalized Norton–Hoff model and the Prandtl–Reuss law of plasticity*, Arch. Rational Mech. Anal., 95, pp. 137–183.

C. TRUESDELL AND W. NOLL [1965], *The Nonlinear Field Theories of Mechanics*, Handbuch der Physik, III/3, Springer-Verlag, Berlin, New York.

J. VON NEUMANN [1937], *Some matrix inequalities and metrization of matrix space*, Tomsk Univ. Rev., 1, pp. 286–300, reprinted in Collected Works, IV, Pergamon Press, Elmsford, NY, Oxford, 1962.

K. YOSHIDA [1968], *Functional Analysis*, Springer-Verlag, Berlin, New York.

O. ZIENKIEWICZ [1977], *The Finite Element Method*, McGraw-Hill, New York, Toronto, London.

Index